AF443735

RELIABLE METHODS FOR COMPUTER SIMULATION

Error Control and A Posteriori Estimates

P. NEITTAANMÄKI

Department of Mathematical Information Technology
University of Jyväskylä
Jyväskylä
Finland

and

S. REPIN

V.A. Steklov Institute of Mathematics
St. Petersburg
Russia

ELSEVIER

AMSTERDAM – BOSTON – LONDON – NEW YORK – OXFORD – PARIS
SAN DIEGO – SAN FRANCISCO – SINGAPORE – SYDNEY – TOKYO

ELSEVIER B.V.
Sara Burgerhartstraat 25
P.O. Box 211, 1000 AE Amsterdam
The Netherlands

ELSEVIER Inc.
525 B Street, Suite 1900
San Diego, CA 92101-4495
USA

ELSEVIER Ltd
The Boulevard, Langford Lane
Kidlington, Oxford OX5 1GB
UK

ELSEVIER Ltd
84 Theobalds Road
London WC1X 8RR
UK

1st edition 2004

Library of Congress Cataloging in Publication Data
A catalog record is available from the Library of Congress.

British Library Cataloguing in Publication Data
A catalogue record is available from the British Library.

ISBN: 0-444-51376-0
ISSN: 0168-2024

04739111

Contents

Preface

Recent decades have seen a very rapid success in developing numerical methods based on explicit control over approximation errors. It may be said that nowadays a new direction is forming in numerical analysis, the main goal of which is to develop methods of *reliable computations*. In general, a reliable numerical method must solve two basic problems:

(a) generate a sequence of approximations that converges to a solution,

(b) verify the accuracy of these approximations.

A computer code for such a method must consist of two respective blocks: *solver* and *checker*.

An intensive investigation of the problem (a) (developing correct and efficient numerical methods) started in the middle of the 20th century. At present, there is a vast amount of literature devoted to this subject. For this reason, in Chapter 2, we recall only some principal results. Readers can find a detailed exposition of these questions in the literature cited. This chapter also includes mathematical knowledge used in subsequent parts.

In this book, we are chiefly concerned with the problem (b) and try to present the main approaches developed for a posteriori error estimation in various problems.

The material of the book is organized as follows. Chapter 3 is devoted to an analysis of two-sided a posteriori error estimates for iteration methods. Here, the derivation of estimates is based upon the Banach fixed point theorem. In Chapter 4, we outline the methods of the a posteriori error estimation used for finite element approximations.

Subsequent parts of the book are concentrated on a posteriori error estimates of the functional type (also called Duality Error Majorants) for differential equations. They provide computable bounds of errors for all types of conforming approximations. Originally, these estimates were derived by methods of duality theory in the modern calculus of variations. For convenience of the readers, we expose all necessary mathematical background in Chapter 5. A special but important case of linear elliptic problems is considered in Chapter 6. Here, we present a respective theory and also expose a series of numerical results intended to demonstrate the performance of Duality Error Majorants. Chapter 7 is concerned with nonlinear variational problems. It gives a general approach to a posteriori error estimation for variational problems with uniformly convex functionals. In Chapter 8, we apply this theory to an important class of variational problems arising in the theory of variational inequalities.

The authors try to retain a rigorous mathematical style, however proofs are constructive whenever possible and additional mathematical knowledge is presented when necessary. The book contains a number of new mathematical results and lists a posteriori error estimation methods that have been developed in the very recent time.

Some parts of the text have been used in lectures at the State Technical University of St.-Petersburg, the University of Jyväskylä, and the University of Houston. We hope that the book will be useful for advanced specialists, as well as for students and PhD students specialized in applied mathematics and scientific computing.

Pekka Neittaanmäki   Sergey Repin

Jyväskylä–Saint-Petersburg, October 2003

Chapter 1

Introduction

1.1 Sources of errors affecting the reliability of numerical solutions

In mathematical modeling, a physical object (or process) is analyzed by a certain mathematical model. Let U be a physical value that characterizes this process and u be a respective value obtained from the mathematical model. Then the quantity

$$\varepsilon_1 = |U - u|$$

is an *error of the mathematical model*. Here, $|\cdot|$ is understood in a broad sense. In a particular problem, it may denote the absolute value of the difference or a norm in a suitable functional space. For example, deformations of solid bodies are often described by linear elasticity theory. In this case, U is a vector-valued function of displacements that arise in a body under the action of given forces and u is a solution of the system of differential equations used by this theory. Then, ε_1 represents the error of the above mathematical model.

In a great majority of mathematical models arising in physics, mechanics, biology and other sciences one cannot directly obtain "exact" (analytic) solutions. The reason for this is that adequate models of complicated processes usually lead to systems of partial differential equations or to coupled systems that could also include algebraic relations, integral equations, and additional conditions. An "exact" solution of such a mathematical problem is usually understood in a rather abstract sense – as an element of a certain functional space. Properties of such a solution may be investigated by purely mathematical methods, however the quantitative analysis inevitably leads to the necessity of approximating u by a sequence of "simple" (e.g., piece wise polynomial) functions. Thus, a sequence of "approximate" problems associated with the original one arises. Let u_h denote a solution of such an approximate problem defined on a mesh of character size h. Then, u_h encompasses the *approximation error*

$$\varepsilon_2 = |u - u_h|.$$

Recalling the linear elasticity model, we see that ε_2 corresponds to errors that arise if differential equations are approximated by systems of algebraic equations (e.g., in the finite element or Ritz–Galerkin methods).

However, finite–dimensional problems are also solved approximately, so that instead of u_h we obtain u_h^ε. The quantity

$$\varepsilon_3 = |u_h - u_h^\varepsilon|$$

shows an *error of the numerical algorithm* performed with a concrete computer. This error includes

- roundoff errors,

- errors arising in iteration processes forcibly stopped by some stopping criteria,

- errors caused by possible defects in computer codes.

It is worth noting that the detection of the errors of the latter type may be very difficult and often causes serious troubles in developing computer codes. Therefore, having a method able to verify the numerical results one can significantly increase the reliability of a computer program.

All what we have said above is schematically presented in Fig. 1.1.1. We see that in place of the desired U we obtain u_h^ε. The difference between them meets the principal inequality

$$\boxed{\;\|U - u_h^\varepsilon\| \leq \quad \varepsilon_1 + \varepsilon_2 + \varepsilon_3.\;} \qquad (1.1.1)$$

The inequality (1.1.1) can be used if the accuracy of the mathematical model has been earlier confirmed and it is clear that the error ε_1 is negligible with respect to ε_2 and ε_3. This situation may be referred to as "computations on the basis of a reliable (certified) model". In this case, the function u_h^ε (obtained as a result of computer simulation) and realistic estimates of ε_2 and ε_3 provide bounds for the function U.

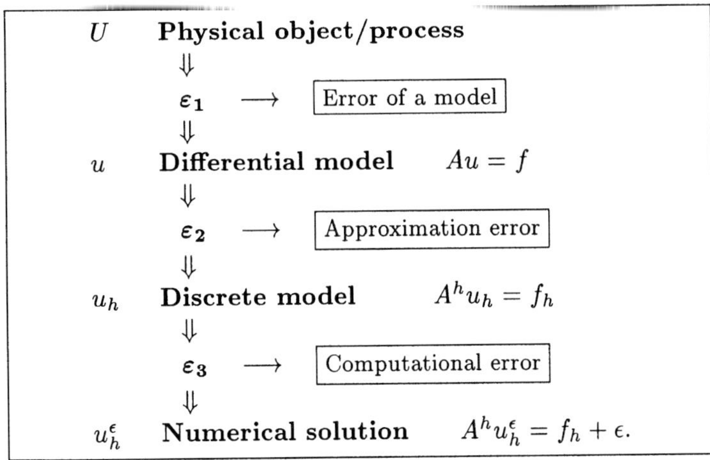

Figure 1.1.1: Errors arising in the mathematical modeling.

Another principal inequality is

$$\|\varepsilon_1\| \leq \|U - u_h^\varepsilon\| + \varepsilon_2 + \varepsilon_3. \tag{1.1.2}$$

This inequality can be used to verify the accuracy of a mathematical model. Indeed, the accuracy ε_1 is bounded from above by the difference in the results of physical and mathematical experiments (which are known) and by the errors ε_2 and ε_3.

Thus, two major problems of mathematical modeling, namely,

- reliable computer simulation,

- numerical verification of mathematical models,

require efficient methods able to provide explicitly computable and realistic estimates of the errors ε_2 and ε_3. Below, we focus our attention on this problem.

1.2　The main approaches to error estimation

In the context of a completely reliable approach, it is necessary not only to obtain an approximate solution but also to estimate explicitly the error $\varepsilon_2 + \varepsilon_3$. However, in practical computations this second part of the work is often ignored. Sometimes this happens, forcibly because the problem analyzed is very complicated and computable error estimates are simply unknown. In other cases such negligence may reflect the belief that approximations obtained with the help of fine meshes and powerful computers have values of ε_1 and ε_2 too small to be taken into account. In general, such a concept does not lead to reliable numerical results (see, e.g., [58] Chapter 11, where it is discussed that approximate solutions of complicated boundary–value problems are often "mesh–dependent" so that restructuring of a mesh may lead to a different result). Strictly speaking, approximate solutions presented without an error control can be viewed only as preliminary ones.

Nowadays, there are two main approaches to error estimation associated with *a priori* and *a posteriori* error estimates.

1.2.1   A priori error estimates

These estimates are intended to estimate the behavior of errors *a priori*, i.e., before computations. For example, consider a boundary–value problem in an abstract form $Au = f$. Assume that it is approximated by a discrete problem $A_h u_h = f_h$ with the help of a sampling $\mathcal{F}_h$. Let $\mathcal{D}$ denote the set of given data associated with our boundary-value problem, i.e., $\mathcal{D}$ includes properties of the domain, coefficients, and the boundary conditions. Assume that it is found a functional M dependent on $\mathcal{D}$, $\mathcal{F}_h$, h, and the exact solution u such that

$$\|u - u_h\| \leq M(\mathcal{D}, \mathcal{F}_h, h, u), \tag{1.2.1}$$

where $\| \cdot \|$ is a suitable (e.g., "energy") norm. If u_h tends to u as $h \to 0$, then a consistent *a priori error majorant M* must meet the condition

$$M(\mathcal{D}, \mathcal{F}_h, h, u) \longrightarrow 0 \quad \text{as } h \to 0. \tag{1.2.2}$$

If, in addition,

$$M(\mathcal{D}, \mathcal{F}_h, h, u) \leq C(\mathcal{D}, u) h^{\beta}, \tag{1.2.3}$$

where $C(\mathcal{D}, u)$ is a positive constant and $\beta > 0$, then we say that a *rate convergence estimate* is obtained.

It is worth noting that the right-hand side of (1.2.1) includes an unknown exact solution u. For this reason, a priori error estimates have mainly a theoretical meaning: they justify the asymptotic behavior of an error. This justification is, by no means, of profound importance for all numerical methods. It shows that the approximations used are *correct in principle*.

The estimate (1.2.3) establishes a qualified convergence estimate. As a rule, it can be derived only if the exact solution u possesses additional regularity properties and belongs to a certain subset of the "natural" energy space.

At present, the theory of a priori convergence estimates is well-studied (see, e.g., [53, 204]).

1.2.2 A posteriori error estimates

A posteriori error estimates use computed solutions instead of the exact ones. Thus, an *a posteriori* counterpart of the estimate (1.2.1) is

$$\|u - u_h\| \leq \mathrm{M}\,(\mathcal{D}, \mathcal{F}_h, h, u_h),\qquad(1.2.4)$$

In this book, we consider a posteriori error estimates for iteration and finite element methods and analyze a new class of estimates (a posteriori error estimates of the functional type) that can be applied for various variational problems.

A posteriori error estimates for iteration methods follow from the classical Banach fixed point theorem. They are considered in Chapter 3.

Historically, a posteriori error estimates for finite element approximations arose as a needful tool for mesh–adaptive methods, in which a new (refined) mesh is constructed by analyzing an approximate solution computed on a coarse mesh. These estimates were mainly used as error indicators intended to detect subdomains with excessively high errors. Two widely used types of such estimators are based either on evaluating of the negative norm of the residual or on analyzing the difference between the computed solution and its post–processed image usually obtained by various gradient averaging methods. The first method brings its origin in the papers by I. Babuška and W. C. Rheinboldt [15, 16]. Later it was investigated and extended by many authors (see, e.g., the books [3, 18, 213]). Residual type error estimate have the following common form:

$$\|u - u_h\| \leq \mathrm{M}\,(u_h, c_1, c_2, ...c_N, \mathcal{D}),$$

where $\| \cdot \|$ denotes a natural (energy) norm of a boundary-value problem, u_h is a Galerkin approximation and c_i, $i = 1, 2, ...N$ are the so-called interpolation constants that depend on properties of a special type interpolation operator acting from V to the finite–dimensional space V_h (see [54]).

Another method was used in the works of O. C. Zienkiewicz and J. Z. Zhu [226, 227] who noticed that the difference between the computed gradients and their averaged values often successfully presents the distribution of modeling errors. Mathematically, this method is based on the so-called *superconvergence* phenomenon (see, e.g., [155, 229, 216]) that arises if the exact solution of a problem has an extra regularity.

Estimates of the above mentioned types are widely used in computational practice. However, these estimates are valid only for exact solutions (Galerkin approximations) of respective finite–dimensional problems.

In Chapters 6-8, we introduce a posteriori error estimates of the functional type. They are derived on the functional level by the methods of functional analysis and calculus of variations and provide upper and lower estimates of the deviation from exact solution for any conforming approximation. A posteriori error majorant M of this type satisfies the following conditions:

- for any approximation v from an admissible (energy) set V, it yields an upper bound of the error, i.e,

$$\|v - u\| \leq \mathrm{M}\,(\mathcal{D}, v); \tag{1.2.5}$$

- the functional M vanishes if and only if v coincides with u, i.e.,

$$\mathrm{M}\,(\mathcal{D}, v) = 0 \quad \Longleftrightarrow \quad v = u; \tag{1.2.6}$$

- the estimate (1.2.5) is *consistent*, which means that

$$\mathrm{M}\,(\mathcal{D}, v_k) \to 0 \quad \text{for any sequence } v_k \xrightarrow[k \to \infty]{} u. \tag{1.2.7}$$

Functional estimates contain only *global constants*, which usually depend on the *constants in the embedding inequalities* for functional spaces arising in the problem considered. We present mathematical justifications of these estimates and discuss their computational properties.

Chapter 2

Mathematical background

2.1 Basic notation

2.1.1 Vectors and matrices

First, we introduce the notation used throughout the book. By $\mathbb{R}^n$ we denote the n-dimensional Euclidean space whose elements are vectors with n components (i.e., $x = (x_1, x_2, ..., x_n)$). Domains in $\mathbb{R}^n$ are denoted by the letters Ω or ω. Normally, these domains are connected, open, bounded, and their boundaries $\partial\Omega$ (or $\partial\omega$) are assumed to be Lipschitz continuous. The latter condition means that for every $x \in \partial\Omega$ there exists a neighborhood in which $\partial\Omega$ is represented as a Lipschitz continuous function in some local Cartesian cordinates.

The closure of sets is denoted by the bar, so that $\bar{\Omega}$ is the closure of Ω. The Lebesgue measure of a set is denoted by the symbol $|\cdot|$. The symbol $\subset\subset$ means strict inclusion ($\omega \subset\subset \Omega$ if ω is contained in Ω together with the closure).

By $\mathbb{M}^{n\times m}$ we denote the space of real $n \times m$ matrices and $\mathbb{M}_s^{n\times n}$ is a subspace of this space containing $n \times n$ symmetric matrices (tensors) with real entries. Any matrix A can be viewed as a linear operator mapping $\mathbb{R}^n$ to itself, so that any $x \in \mathbb{R}^n$ is transformed to a vector $z = Ax$ with components $z_i = \sum_{j=1}^{n} A_{ij}x_j$. For elements of $\mathbb{R}^n$ and $\mathbb{M}^{n\times n}$, the following operations are used:

$$a \cdot b = \sum_{i=1}^{n} a_i b_i \in \mathbb{R}, \quad a, b \in \mathbb{R}^n \qquad \text{(scalar product of vectors)},$$

$$a \otimes b = \sum_{i,j=1}^{n} a_i b_j \in \mathbb{M}^{n \times n} \qquad \text{(tensor product of vectors)},$$

$$\sigma : \varepsilon = \sum_{i,j=1}^{n} \sigma_{ij} \varepsilon_{ij} \in \mathbb{R}, \quad \sigma, \varepsilon \in \mathbb{M}^{n \times n} \qquad \text{(scalar product of tensors)}.$$

The norms of vectors and tensors are defined with the help of the products introduced above:

$$|a| := \sqrt{a \cdot a}, \qquad |\sigma| := \sqrt{\sigma : \sigma},$$

where the symbol $:=$ means "equals by definition". The unit matrix is denoted by $\mathbb{I}$. If $\tau \in \mathbb{M}^{n \times n}$, then $\tau^D = \tau - \frac{1}{n} \mathbb{I}$ is the deviator of τ.

2.1.2   Spaces of functions

For the sake of convenience, we briefly recall some basic facts from the theory of functions and functional analysis (see, e.g., [121, 190]) that are necessary for the understanding of subsequent results.

A linear space V is called a *normed* space if for any element v there is a real number $\|v\|$ satisfying the following axioms:

- $\|v\| = 0$ if and only if v is the zero element of V (i.e., $v = 0_V$),

- for any pair of elements u and v, the "triangle inequality" $\|u + v\| \leq \|u\| + \|v\|$ holds,

- for any v and real number α, the relation $\|\alpha v\| = |\alpha| \|v\|$ holds.

Normed spaces are a special case of *metric* spaces. A metric space is defined as a pair (X, d), where X is a set and d is a real-valued function defined for any two elements of X and subject to the conditions

- $d(x, y) = 0$ if and only if $x = y$,

- $d(x, y) \leq d(x, z) + d(y, z)$,

- $d(x, y) = d(y, x)$.

A normed space is converted to a metric one if we set $d(x, y) = \|x - y\|$.

Let $\{u_k\}$ be a sequence of elements of V. If $\|u_k - u\|$ tends to zero as $k \to +\infty$, then we say that this sequence is *strongly convergent* to u.

A subspace $V' \subset V$ is said to be *dense* in V if any element of V can be represented as a strong limit of a sequence of elements in V'. A space V is called *separable* if it contains a countable dense subset V'. A sequence $\{u_k\}$ is called *fundamental* or Cauchy sequence if $\|u_k - u_m\| \to 0$ as $k, m \to +\infty$. A space V is called *complete* if any fundamental sequence converges to a certain element of this space. A complete normed space is called a *Banach space.*

Hilbert spaces form an important subclass of Banach spaces. In a Hilbert space, any pair of elements u and v can be associated with a number (u, v) called the *scalar product* of these two elements. It must satisfy the following axioms:

- $(u, v) = (v, u)$,

- $(u, v + w) = (u, v) + (u, w)$,

- $(\alpha u, v) = \alpha(u, v)$.

The quantity (u, u) is nonnegative and vanishes if and only if $u = 0_V$, where 0_V denotes the zero element of V. Obviously, such a scalar product generates the norm $\|u\| = \sqrt{(u, u)}$. The Cauchy–Buniakovsky–Schwartz inequality for vectors is extended to the Hilbert spaces, i.e.,

$$|(u, v)| \leq \|u\|\|v\|.$$

The space of functions integrable in Ω with power p is denoted by $L^p(\Omega)$. This space is a Banach space with respect to the norm

$$\|v\|_{p,\Omega} := \left(\int_\Omega |v|^p \right)^{1/p}.$$

If $p = 2$, we use the simplified notation $\|v\|_\Omega$ (or $\|v\|$) for the L^2-norm, which is associated with the scalar product

$$(u, v) := \int_\Omega uv\,dx.$$

Analogous spaces of vector- and tensor-valued functions are denoted by $L^p(\Omega, \mathbb{R}^n)$ and $L^p(\Omega, \mathbb{M}^{n \times n})$, respectively. For their norms we keep the same notation, i.e.,

$$\|\sigma\|_{p,\Omega} := \left(\int_\Omega |\sigma|^p \right)^{1/p}, \quad \forall \sigma \in L^p(\Omega, \mathbb{M}^{n \times n}).$$

The space of measurable essentially bounded functions is denoted by $L^\infty(\Omega)$. It is equipped with the norm

$$\|u\|_{\infty,\Omega} = \operatorname*{ess\,sup}_{x \in \Omega} |u(x)|.$$

By $\overset{\circ}{C}{}^\infty(\Omega)$ we denote the space of all infinitely differentiable functions with compact supports in Ω. The spaces of k-times differentiable scalar- and vector-valued functions are denoted by $C^k(\Omega)$ and $C^k(\Omega, \mathbb{R}^n)$, respectively; $\overset{\circ}{C}{}^k(\Omega)$ is the subspace of $C^k(\Omega)$ that contains functions vanishing at the boundary; $P^k(\Omega)$ denotes the set of polynomial functions defined in $\Omega \subset \mathbb{R}^n$, i.e., $u \in P^k(\Omega)$ if

$$u = \sum_{|\alpha| \le m} a_\alpha x^\alpha, \quad m \le k,$$

where $\alpha := (\alpha_1, \ldots \alpha_n)$ is the multi-index, $|\alpha| = \alpha_1 + \alpha_2 + \ldots + \alpha_n$, $a_\alpha = a_{\alpha_1, \ldots \alpha_n}$, and $x^\alpha = x^{\alpha_1} x^{\alpha_2} \ldots x^{\alpha_n}$.

For a space V, we define the space V^* of all linear continuous functionals on V, which is called *topologically dual* to V. The value of $v^* \in V^*$ on $v \in V$ is denoted by $\langle v^*, v \rangle$. If V is a Banach space, then V^* can also be normed by setting

$$\|v^*\|_* := \sup_{v \in V} \frac{\langle v^*, v \rangle}{\|v\|}.$$

Here and later on, we assume that the supemum (or infimum) of a quotient is taken with respect to all elements of V except 0_V. A functional space is called *reflexive* if it coincides with the bidual space V^{**} (i.e., if there exists a one-to-one mapping of V to V^{**} and back that preserves the metric). All the Hilbert spaces are reflexive. The same is true for spaces L^p with $1 < p < +\infty$.

The theorem of F. Reisz asserts that for Hilbert spaces, any functional $v^* \in V^*$ can be written in the form of a scalar product introduced in such a space, i.e.,

$$(u, v) = \langle v^*, v \rangle, \qquad \forall v \in V,$$

where u is uniquely determined.

A sequence $\{v_k\}$ *weakly converges* to $v \in V$ if

$$\langle v^*, v_k \rangle \to \langle v^*, v \rangle, \qquad \forall v^* \in V^*.$$

Weak convergence is denoted by the symbol $\rightharpoonup$.

For the sake of completeness, we also recall some known algebraic and functional relations. The first of them is the inequality

$$|pq| \leq \frac{1}{2\beta}|p|^2 + \frac{\beta}{2}|q|^2,\tag{2.1.1}$$

which is valid for any positive β. This inequality can be viewed as the simplest version of the Young–Fenchel inequality for the convex functionals considered in the Chapter 5. Similar inequalities hold for vectors or tensors.

The inequality

$$|a \cdot b| \leq \left(\sum_{i=1}^{n}|a_i|^{\alpha}\right)^{\frac{1}{\alpha}}\left(\sum_{i=1}^{n}|b_i|^{\alpha^*}\right)^{\frac{1}{\alpha^*}},\tag{2.1.2}$$

where $\frac{1}{\alpha^*} + \frac{1}{\alpha} = 1$, is the discrete Hölder inequality whose functional form is

$$(u,v) \leq \|u\|_{\alpha,\Omega}\|v\|_{\alpha^*,\Omega}.\tag{2.1.3}$$

The operators (mappings) are denoted by calligraphic letters ($\mathcal{A}$, $\mathcal{B}$, $\mathcal{C}$, etc). An operator $\mathcal{A}$ acting from a Banach space X to a Banach space Y is called *linear* and *bounded* if

$$\mathcal{A}(x_1 + x_2) = \mathcal{A}x_1 + \mathcal{A}x_2,$$
$$\mathcal{A}(\lambda x_1) = \lambda\mathcal{A}x_1, \qquad \forall x_1, x_2 \in X, \lambda \in \mathbb{R},$$
$$\exists C > 0 \text{ such that } \|\mathcal{A}x\| \leq C\|x\|, \qquad \forall x \in X.$$

The lowest constant C is called the norm of $\mathcal{A}$. The set of all linear continuous operators acting from X to Y is denoted by $\mathcal{L}(X,Y)$.

2.2   Sobolev spaces

2.2.1   Generalized derivatives and Sobolev spaces

Let Ω be an open bounded domain in $\mathbb{R}^n$ with Lipschitz continuous boundary.

Definition 2.2.1. *We say that f is locally integrable in Ω and write $f \in L^{1,loc}(\Omega)$, if $f \in L^1(\omega)$ for any $\omega \subset\subset \Omega$.*

Similarly, one can define the space $L^{p,loc}(\Omega)$ that consists of functions locally integrable with degree $p \geq 1$.

Definition 2.2.2. *Let* $f, g \in L^{1,loc}(\Omega)$ *and*

$$\int_\Omega g\varphi\, dx = -\int_\Omega f\frac{\partial\varphi}{\partial x_i}\, dx, \quad \forall\varphi \in \overset{\circ}{C}{}^1(\Omega).$$

Then g is called a generalized derivative (in the sense of Sobolev) of the function f with respect to x_i and we write

$$g = \frac{\partial f}{\partial x_i}.$$

Higher order generalized derivatives are defined in a similar way. For example, if $f, g \in L^{1,loc}(\Omega)$ and

$$\int_\Omega g\varphi\, dx = \int_\Omega f\frac{\partial^2\varphi}{\partial x_i\partial x_j}\, dx, \quad \forall\varphi \in \overset{\circ}{C}{}^2(\Omega),$$

then g is a generalized derivative of f with respect to x_i and x_j. For generalized derivatives we keep the classical notation and write $g = \partial^2 f/\partial x_i\partial x_j$ or $g = f_{,ij}$. If f is differentiable in the classical sense, then its generalized derivatives coincide with the classical ones.

To extend this definition further, we use the multi-index notation and write $D^\alpha f$ in place of $\partial^k f/\partial x_1^{\alpha_1}\partial x_2^{\alpha_2}\ldots\partial x_n^{\alpha_n}$.

Definition 2.2.3. *Let* $f, g \in L^{1,loc}(\Omega)$ *and*

$$\int_\Omega g\varphi\, dx = (-1)^{|\alpha|}\int_\Omega fD^\alpha\varphi\, dx, \quad \forall\varphi \in \overset{\circ}{C}{}^k(\Omega).$$

Then, g is called a generalized derivative of f of degree $|\alpha|$, and we write $g = D^\alpha f$.

The spaces of functions that have of integrable generalized derivatives up to a certain order are called Sobolev spaces.

Definition 2.2.4. *A function f lies in the Sobolev space $W^{1,p}(\Omega)$ if $f \in L^p$ and all the generalized derivatives of f of the first order are integrable with power p, i.e.,*

$$f_{,i} = \frac{\partial f}{\partial x_i} \in L^p(\Omega).$$

The norm in $W^{1,p}$ is defined as follows:

$$\|f\|_{1,p,\Omega} := \left(\int_\Omega (|f|^p + \sum_{i=1}^n |f_{,i}|^p)dx\right)^{1/p}.$$

The other Sobolev spaces are defined quite similarly: $f \in W^{l,p}(\Omega)$ if all generalized derivatives up to the order l are integrable with power p and the quantity

$$\|f\|_{l,p,\Omega} := \left(\int_{\Omega} \sum_{|\alpha| \le l} |D^{\alpha} f|^p \, dx \right)^{1/p}$$

is finite. For the Sobolev spaces $W^{l,2}(\Omega))$ we also use simplified notation $H^l(\Omega)$. The Sobolev spaces of vector- and tensor-valued functions are introduced by obvious extensions of the above definitions. We denote them by $W^{l,p}(\Omega, \mathbb{R}^n)$ and $W^{l,p}(\Omega, \mathbb{M}^{n \times n})$, respectively.

It is important to establish relationships between the Sobolev spaces and other frequently used functional classes such as $L^p(\Omega)$ and $C^k(\Omega)$. Respective results are known as *embedding theorems* (see [201]). They establish that if $p, q \ge 1$, $l > 0$ and $l + \frac{n}{q} \ge \frac{n}{p}$, then $W^{l,p}(\Omega)$ is continuously embedded in $L^q(\Omega)$. Moreover, if $l + \frac{n}{q} > \frac{n}{p}$, then the embedding operator is compact.

If $l - k > \frac{n}{p}$, then $W^{l,p}(\Omega)$ is compactly embedded in $C^k(\overline{\Omega})$.

The functions in Sobolev spaces have counterparts on $\partial\Omega$ (and on other manifolds of lower dimensions) that are associated with spaces of *traces*. Thus, there exist some bounded operators mapping the functions defined in Ω to functions defined on the boundary. For example, the operator $\gamma : H^1(\Omega) \to L^2(\partial\Omega)$ is called the *trace* operator if it satisfies the following conditions:

$$\gamma v = v \,|_{\partial\Omega}, \qquad \forall v \in C^1(\Omega),$$
$$\|\gamma v\|_{2,\partial\Omega} \le c\|v\|_{1,2,\Omega},$$

where c is a positive constant independent of v. From these relations, we observe that such a trace is a natural generalization of the trace defined for a continuous function. It was established that the image of γ forms a subset of $L^2(\partial\Omega)$, which is the space $H^{1/2}(\partial\Omega)$. The functions from other Sobolev spaces also are known to have traces in Sobolev spaces with fractional indices. Henceforth, we understand the boundary values of functions in the sense of traces, so that the phrase "$u = \phi$ on $\partial\Omega$" means that the trace γu of a function u defined in Ω coincides with a given function ϕ defined on $\partial\Omega$. If for two functions u and v defined in Ω we say that $u = v$ on $\partial\Omega$, then we mean that $\gamma(u - v) = 0$ on $\partial\Omega$. The spaces of functions that have zero traces on the boundary are marked by the symbol $\circ$ (e.g., $\overset{\circ}{W}{}^{l,p}(\Omega)$ and $\overset{\circ}{H}{}^{1}(\Omega)$).

Remark 2.2.1. For a function $f \in W^{l,p}(\Omega)$ we can define the quantity

$$|f|_{l,p,\Omega} = \left(\int_{\Omega} \sum_{|\alpha|=l} |D^{\alpha}f)|^{p} \right)^{1/p},$$

which is a seminorm in $W^{l,p}(\Omega)$. For functions of the space $\overset{\circ}{W}{}^{l,p}(\Omega)$ it is a norm equivalent to the standard norm $\|f\|_{1,p,\Omega}$. In the simplest case this fact follows from the Friederichs inequality

$$\|w\|_{\Omega} \leq C_{\Omega}\|\nabla w\|_{\Omega}, \quad \forall w \in \overset{\circ}{H}{}^{1}(\Omega), \tag{2.2.1}$$

where C_{Ω} is a positive constant independent of w.

Remark 2.2.2. In continuum mechanics, of importance is the following assertion known as the Korn's inequality. Let Ω be an open, bounded domain with Lipschitz continuous boundary. Then

$$\int_{\Omega} \left(|v|^{2} + |\varepsilon(v)|^{2} \right) dx \geq \mu_{\Omega} \|v\|_{1,2,\Omega}^{2}, \quad \forall v \in H^{1}(\Omega, \mathbb{R}^{n}), \tag{2.2.2}$$

where μ_{Ω} is a positive constant independent of v and $\varepsilon(v)$ denotes the symmetric part of the tensor ∇v, i.e.,

$$\varepsilon_{ij}(v) = \frac{1}{2}\left(\frac{\partial v_{i}}{\partial x_{j}} + \frac{\partial v_{j}}{\partial x_{i}} \right).$$

It is not difficult to verify that the left-hand side of (2.2.2) is bounded from above by the H^{1}-norm of v. Thus, it represents a norm equivalent to $\|\cdot\|_{1,2,\Omega}$.

2.2.2 Sobolev spaces with negative indices

Let $\Omega \in \mathbb{R}^{n}$ be a bounded connected domain with Lipschitz continuous boundary. By $\overset{\circ}{C}{}^{\infty}(\Omega)$ we denote the space of all continuous functions with compact supports in Ω that have continuous (classical) derivatives of any order. The sequence $\{\varphi_{i}\} \in \overset{\circ}{C}{}^{\infty}(\Omega)$ is said to be convergent to zero if

1. there exists a set $\Omega_{1} \subset \Omega$ such that $\operatorname{supp}\varphi_{i} \subset \Omega_{1}$ for all $i \in \mathbb{N}$,

2. all derivatives of φ_{i} tend to zero uniformly as $i \to \infty$.

Definition 2.2.5. *Linear functionals defined on the functions of the space* $\overset{\circ}{C}{}^{\infty}(\Omega)$ *are called distributions.*

Traditionally, the space of distributions is denoted by $\mathcal{D}'(\Omega)$. The value of a distribution g on a function $\varphi \in \overset{\circ}{C}{}^\infty(\Omega)$ is denoted by $\langle g, \varphi \rangle$. We say that the distributions g_1 and g_2 are equal in Ω if

$$\langle g_1, \varphi \rangle = \langle g_2, \varphi \rangle, \quad \forall \varphi \in \overset{\circ}{C}{}^\infty(\Omega).$$

We say that a distribution g is the sum of distributions g_1 and g_2 if

$$\langle g, \varphi \rangle = \langle (g_1 + g_2), \varphi \rangle, \quad \forall \varphi \in \overset{\circ}{C}{}^\infty(\Omega).$$

If a distribution can be identified with a locally integrable function, then it is called *regular*. In this case, the action of g is given by the Lebesgue integral

$$\langle g, \varphi \rangle = \int_\Omega g\varphi \, dx.$$

Other distributions are called *singular*. As an example of such a distribution, we mention the well-known δ-function, which is a linear functional defined by the relation $\langle \delta, \varphi \rangle = \varphi(0)$.

Distributions possess an important property: they have derivatives of any order if differentiation is understood in a special (generalized) sense.

Definition 2.2.6. *Let g be a distribution. Its generalized derivative $D^\alpha g$ is a linear functional defined for any $\varphi \in \overset{\circ}{C}{}^\infty(\Omega)$ by the following rule:*

$$\langle D^\alpha g, \varphi \rangle := (-1)^{|\alpha|} \langle g, D^\alpha \varphi \rangle. \tag{2.2.3}$$

Consider a simple example. Let $g \in \mathcal{D}'(\Omega)$, then the quantity $-\langle g, \frac{\partial \varphi}{\partial x_i} \rangle$ is another linear functional defined on $\mathcal{D}(\Omega)$. This functional is viewed as a generalized partial derivative of g taken over the i-th independent variable. It is not difficult to see that the set of distributions is closed with respect to the differentiation operation introduced above.

Any function from the space $L^q(\Omega)$ ($q \geq 1$) defines a certain distribution and, therefore, has generalized derivatives of any order. The sets of distributions, which are derivatives of q-integrable functions, are called *Sobolev spaces with negative indices* or, simply, *negative Sobolev spaces*. One can define this class as follows.

Definition 2.2.7. *The space $W^{-l,q}(\Omega)$ is the space of distributions $g \in \mathcal{D}'(\Omega)$ such that*

$$g = \sum_{|\alpha| \leq l} D^\alpha g_\alpha,$$

where $g_\alpha \in L^q(\Omega)$.

In particular, the space $W^{-1,q}(\Omega)$ contains distributions that can be viewed as generalized derivatives of L^q-functions. For $f \in L^q(\Omega)$, the functional

$$\left\langle \frac{\partial f}{\partial x_i}, \varphi \right\rangle := - \int_\Omega f \frac{\partial \varphi}{\partial x_i}\, dx$$

is linear and continuous not only for functions from $\overset{\circ}{C}{}^{\infty}(\Omega)$ but, also, for those from $\overset{\circ}{W}{}^{1,p}(\Omega)$, where $1/p + 1/q = 1$. This fact follows from the density of smooth functions in $\overset{\circ}{W}{}^{1,p}(\Omega)$ and from known theorems on the continuation of linear functionals. Hence, first generalized derivatives of f lies in the space dual to $\overset{\circ}{W}{}^{1,p}(\Omega)$. The latter is denoted by $W^{-1,p}(\Omega)$. Since $\overset{\circ}{W}{}^{1,2}(\Omega)$ has the other notation $\overset{\circ}{H}{}^{1}(\Omega)$, the respective dual space is denoted by $H^{-1}(\Omega)$.

Let $\xi \in H^{-1}(\Omega)$. By definition, there exist functions $g_i \in L^2(\Omega)$, $i = 0, 1, 2, ..., n$, such that

$$\langle \xi, \varphi \rangle = \int_\Omega \left(g_0 \varphi - \sum_{i=1}^{n} g_i \varphi_{,i} \right) dx, \quad \forall \varphi \in \overset{\circ}{H}{}^{1}(\Omega). \tag{2.2.4}$$

It is easy to see that

$$\nu = \sup_{\varphi \in \overset{\circ}{H}{}^{1}(\Omega)} \frac{|\langle \xi, \varphi \rangle|}{\|\varphi\|_{1,2,\Omega}} \leq \left(\|g_0\|_\Omega^2 + \sum_{i=1}^{n} \|g_i\|_\Omega^2 \right)^{1/2}.$$

Thus, for any $\xi \in H^{-1}(\Omega)$ the quantity ν is nonnegative and finite. It defines the norm $\|\xi\|_{(-1),\Omega}$. Henceforth, we use the term "negative norms" for the norms in Sobolev spaces with negative indexes (this, of cause, does not mean that their values are negative).

For $\xi \in H^{-1}(\Omega)$ and $\varphi \in \overset{\circ}{H}{}^{1}(\Omega)$, the quantity $\langle \xi, \varphi \rangle$ may be viewed as a generalization of the concept of integral, and thus we formally write $\int_\Omega \xi \varphi\, dx$ instead of $\langle \xi, \varphi \rangle$. Then, directly from the definition of the negative norm, we find that

$$\int_\Omega \xi \varphi\, dx \leq \|\xi\|_{(-1),\Omega} \|\varphi\|_{1,2,\Omega}. \tag{2.2.5}$$

In addition to $\|\xi\|_{(-1),\Omega}$, we introduce another negative norm

$$|\xi| := \sup_{\varphi \in \overset{\circ}{H}{}^{1}(\Omega)} \frac{|\langle \xi, \varphi \rangle|}{\|\nabla \varphi\|_\Omega}. \tag{2.2.6}$$

By the Friederichs inequality we obtain

$$(C_\Omega^2 + 1)^{-1/2}\|\varphi\|_{1,2,\Omega} \leq \|\nabla\varphi\|_\Omega \leq \|\varphi\|_{1,2,\Omega}, \quad \forall\varphi \in \overset{\circ}{H}^1(\Omega).$$

Therefore,

$$\|\xi\|_{(-1),\Omega} \leq |\xi| \leq (C_\Omega^2 + 1)^{1/2}\|\xi\|_{(-1),\Omega}. \tag{2.2.7}$$

By (2.2.7) we observe that these two norms are equivalent and we have an estimate analogous to (2.2.5):

$$\int_\Omega \xi\varphi\,dx \leq |\xi|\|\nabla\varphi\|_\Omega. \tag{2.2.8}$$

By similar arguments we can introduce spaces $H^{-l}(\Omega)$, which are topologically dual to $\overset{\circ}{H}^l(\Omega)$. These spaces are complete with respect to the norm

$$\|\xi\|_{(-l),\Omega} := \sup_{\varphi \in \overset{\circ}{H}^l(\Omega)} \frac{|\langle \xi, \varphi \rangle|}{\|\varphi\|_{l,2,\Omega}} = \sup_{\varphi \in \overset{\circ}{H}^l(\Omega)} \frac{\left| \sum_{|\alpha|=l} \int_\Omega (-1)^{|\alpha|} g_\alpha D^\alpha \varphi\,d\Omega \right|}{\|\varphi\|_{l,\Omega}}.$$

Finally, we note that $\overset{\circ}{H}^l \subset \overset{\circ}{H}^m$ for $l > m$ and

$$\overset{\circ}{H}^l(\Omega) \subset L^2(\Omega) \subset H^{-l}(\Omega).$$

2.3 Generalized solutions and their approximation

In this section, we briefly recall definitions of generalized solutions of partial differential equations, prove their existence for one class of linear elliptic problems, and give some facts that form a basis for approximation theory. Readers can find a systematic exposition of the existence and regularity theory for partial differential equations, e.g., in the books by D. Gilbarg and N.S. Trudinger [87], O. Ladyzhenskaya [132], O. Ladyzhenskaya and N. Uraltseva [135] J.-L. Lions and E. Magenes [138], and many other publications.

In general terms, the boundary-value problem is stated as follows: to find a function u such that

$$\mathcal{L}u + f = 0 \qquad \text{in } \Omega \tag{2.3.1}$$

$$u = u_0 \qquad \text{on } \partial\Omega, \tag{2.3.2}$$

where an operator $\mathcal{L}$ is usually formed by partial derivatives of u. For example, in this book we will often consider problems with

$$\mathcal{L}u = \frac{\partial}{\partial x_i} a_{ij} \frac{\partial u}{\partial x_j} = \operatorname{div} A \nabla u. \tag{2.3.3}$$

In this case, it is always assumed that $A = \{a_{ij}\}$ is a symmetric matrix subject to the condition

$$c_1 |\xi|^2 \le A\xi \cdot \xi \le c_2 |\xi|^2, \qquad \forall \xi \in \mathbb{R}^n. \tag{2.3.4}$$

The problem (2.3.1)–(2.3.4) may have no solution understood as a twice differentiable function satisfying the differential equation in the classical sense. The question of how one can correctly define solutions of the boundary-value problems for partial differential equations has a long history dated back to the 19-th century. It was finally solved only in the 20-th century, when it was formed the concept of generalized (weak) solutions to boundary–value problems (see, e.g., [132]. In a sense, this concept can be viewed as a generalization of the idea that forms a basis of the Bubnov–Galerkin method [39, 83]. Within the framework of this concept, u is viewed as a distribution such that

$$\langle \mathcal{L}u + f, w \rangle = 0, \qquad \forall w \in C_0^\infty. $$

However, in many cases generalized solutions are defined in narrower functional classes. In particular, for the operator (2.3.3), a generalized solution can be defined as an element of $H^1(\Omega)$ satisfying (2.3.2) and the relation

$$\int_\Omega A\nabla u \cdot \nabla w \, dx = \int_\Omega f w \, dx, \qquad \forall w \in \overset{\circ}{H}{}^1(\Omega). \tag{2.3.5}$$

We see that the operator (2.3.3) is associated with a symmetric bilinear form

$$a(u,v) = \int_\Omega A\nabla u \cdot \nabla v \, dx,$$

which is defined on the functions from $H^1(\Omega)$. Many other operators are also associated with some bilinear forms satisfying the conditions

$$a(v,v) \ge c_1 \|v\|_V^2, \qquad \forall v \in V, \tag{2.3.6}$$

$$a(v,w) \le c_2 \|v\|_V \|w\|_V, \quad \forall v, w \in V, \tag{2.3.7}$$

where $\|\cdot\|_V$ is the norm of V. Such forms are called *V-elliptic*. By the form a, we introduce another norm in V:

$$\| v \| := \sqrt{a(v,v)},$$

which is equivalent to the original one.

Assume that V is a Hilbert space with scalar product $(\cdot,\cdot)$. The following statement plays the key role.

Proposition 2.3.1. *Let $a : V \times V$ be a bilinear V-elliptic form. Then there exist a bounded operator $\mathcal{A} \in \mathcal{L}(V,V)$ such that*

$$(\mathcal{A}v, w) = a(v,w), \qquad \forall v,w \in V, \tag{2.3.8}$$

and the inverse operator $\mathcal{A}^{-1} \in \mathcal{L}(V,V)$. Moreover, the norms of these operators satisfy the estimates

$$\|\mathcal{A}\| \le c_2, \qquad \|\mathcal{A}^{-1}\| \le \frac{1}{c_1}. \tag{2.3.9}$$

A proof of this proposition (which is often called the Lax–Milgram lemma) can be found, e.g., in [53].

Consider the problem: to find $u \in V$ such that

$$a(u,v) = \langle v^*, v \rangle, \qquad \forall v \in V, \tag{2.3.10}$$

where $v^* \in V^*$. Note that this problem coincides with (2.3.5) if $\langle v^*, v \rangle$ is defined by the integral.

Proposition 2.3.2. *Problem (2.3.10) has a unique solution that satisfies the estimate*

$$\|u\| \le \frac{1}{c_1}\|v^*\|_*. \tag{2.3.11}$$

Proof. By virtue of the Reisz's theorem, there exists an element $w \in V$ such that $\langle v^*, v \rangle = (w,v)$ and $\|w\| = \|v^*\|_*$. By Proposition 2.3.1, we find that the problem (2.3.10) has the form

$$(\mathcal{A}u - w, v) = 0, \qquad \forall v \in V,$$

which implies the relation $\mathcal{A}u = w$. Since the inverse operator exists, this uniquely determines the solution $u = \mathcal{A}^{-1}w$. It is clear that

$$\|u\| \le \|\mathcal{A}^{-1}\|\,\|w\| \le \frac{1}{c_1}\,\|v^*\|_*.$$

$\square$

The next assertion states that the solution u defined above is a minimizer of the following variational problem: to find $u \in V$ such that

$$J(u) = \inf_{v \in V} J(v), \quad J(v) := \frac{1}{2} \parallel v \parallel^2 - \langle v^*, v \rangle. \qquad (2.3.12)$$

Proposition 2.3.3. *The function u is a solution of (2.3.10) if and only if it is a solution of (2.3.12).*

Proof. Note that

$$J(v) - J(u) = a(u, v - u) - \langle v^*, v \rangle + \frac{1}{2} a(v - u, v - u).$$

If u satisfies (2.3.10), then this relation means that

$$J(v) - J(u) = \frac{1}{2} \parallel v - u \parallel^2 > 0, \quad \forall v \in V, \qquad (2.3.13)$$

and, therefore, u is a minimizer of (2.3.12).

Assume that u minimizes the functional J. Then

$$J(u + \lambda v) - J(u) \geq 0, \quad \forall v \in V, \; \lambda > 0,$$

and we obtain

$$a(u, v) - \langle v^*, v \rangle \geq \lambda \frac{1}{2} \parallel v \parallel^2.$$

Since λ is an arbitrary positive number and v lies in a linear space, we conclude that the relation (2.3.10) holds. $\qquad \Box$

To solve the problem numerically, we replace the space V by a certain subspace V_k, where the index k shows the dimension of the subspace.

Definition 2.3.1. *A function $u_k \in V_k$ satisfying the relation*

$$a(u_k, v_k) = \langle v^*, v_k \rangle, \quad \forall v_k \in V_k,$$

is called a Galerkin approximation *of u.*

By (2.3.13), we see that for any $v_k \in V_k$,

$$\frac{1}{2} \parallel u - u_k \parallel^2 = J(u_k) - J(u) \leq J(v_k) - J(u) = \frac{1}{2} \parallel u - v_k \parallel^2.$$

From here, it follows the basic relation for the difference between the Galerkin approximation and the exact solution

$$\parallel u - u_k \parallel = \inf_{v_k \in V_k} \parallel u - v_k \parallel. \qquad (2.3.14)$$

It can be rewritten in a different form (which is called the *projection theorem* or Céa's lemma)

$$\|u - u_k\| \leq \frac{c_2}{c_1} \inf_{v_k \in V_k} \|u - v_k\| = \frac{c_2}{c_1}\|u - \pi_k u\|, \qquad (2.3.15)$$

where $\pi_k u$ is the projection of u onto V_k. The estimate (2.3.15) shows that the error is bounded from above by the distance between the exact solution u and its orthogonal projection on the finite-dimensional space V_k. If the approximation spaces $\{V_k\}$ are constructed in such a way that for any $v \in V$,

$$\inf_{v_k \in V_k} \|v - v_k\| \to 0, \quad \text{as } k \to +\infty, \qquad (2.3.16)$$

then from (2.3.15) and (2.3.16) it immediately follows that the Galerkin approximations u_k converge to the exact solution u.

Estimates of the convergence rate require a more profound analysis of interpolation properties of the spaces V_k. Certainly, piecewise polynomial functions form a class, which is one of the most convenient classes for constructing V_k. Since u (the exact solution of a boundary–value problem) is an element of a certain Sobolev space, the minimization problem on the right-hand side in (2.3.14) leads to the question: how accurately can the elements of a Sobolev space be interpolated by piecewise polynomial functions? This question is studied by *interpolation theory* (see, e.g., P. G. Ciarlet [53]), which, together with projection type estimates, yields *a priori error estimates* qualified in terms of a mesh size. Such estimates are obtained not only for the energy norm, but also for some other norms. For example pointwise error estimates were derived in C. I. Goldstain [86], R. Rannacher [164], A. H. Schatz [193], A. H. Schatz and L. B. Wahlbin [194, 195, 196]. Typically, such estimates are obtained under the condition that the exact solutions possess additional regularity. Without an additional regularity an estimate can be derived in a weaker norm (see, e.g., A.H. Schatz and J.Wang [197]).

Readers can find various aspects of approximation theory for differential equations in J. P. Aubin [9], D. Braess [32], F. Brezzi and M. Fortin [38], P. G. Ciarlet [53], V. Girault and P. A. Raviart [88], R. Glowinski [89], R. Glowinski and O. Pironneau [90], K. Eriksson, D. Estep, P. Hansbo and C. Johnson [68], M. Feistauer [73], C. Johnson [108], H. Kardestuncer and D. H. Norrie [117], M. Křížek and P. Neittaanmäki [126], S. Mikhlin [143], J. T. Oden and J. N. Reddy [153], D. R. J. Owen and E. Hinton [157], J. L. Segerlind [191], G. Strang and G. Fix [204], O. C. Zienkiewicz and K . Morgan [225], and in many other publications. The progress achieved in this field and some unsolved problems are discussed in the papers by I. Babuška [12] and O. C. Zienkiewicz [223].

Finally, we consider an interpolation operator that is used in some a posteriori error estimation methods when one must map the functions of the energy space to the respective finite dimensional space. A way for constructing interpolation operators mapping the H^1–functions to the space of piecewise affine continuous functions was suggested by P. Clement [54]. For plane simplexes, this construction is as follows. Let Ω be a polygonal domain, which is the union $\mathcal{F}_h$ of simplexes T_i, and v be a given function in $\overset{\circ}{H}{}^1(\Omega)$. By E_{ij} we denote the common edge of the simplexes T_i and T_j and $\operatorname{diam} T_i$, $|T_i|$, and $|E_{ij}|$ stand for the length of the maximal edge of T_i, the square of T_i, and the length of E_{ij}, respectively. If s is an inner node of the triangulation $\mathcal{F}_h$, then ω_s denotes the set of all simplexes having this node.

For any s, we find a polynomial $p_s(x) \in P^1(\omega_s)$ such that

$$\int_{\omega_s} (v - p_s)q\, dx = 0 \quad \forall q \in P^1(\omega_s). \tag{2.3.17}$$

It is easy to see that p_s is the L^2-projection of v. Now, the interpolation operator π_h is defined by setting

$$\pi_h v(x_s) = p(x_s), \quad \forall x_s \in \operatorname{int} \Omega, \tag{2.3.18}$$

$$\pi_h v(x_s) = 0, \quad \forall x_s \in \partial\Omega. \tag{2.3.19}$$

Such an interpolation operator is linear and continuous. It maps $\overset{\circ}{H}{}^1(\Omega)$ to the space of piecewise affine continuous functions. Moreover (see [54]), it is subject to the relations

$$\|v - \pi_h v\|_{2,T_i} \le c_{1i}\operatorname{diam}(T_i)\|v\|_{1,2,\omega_N(T_i)}, \tag{2.3.20}$$

$$\|v - \pi_h v\|_{2,E_{ij}} \le c_{2ij}|E_{ij}|^{1/2}\|v\|_{1,2,\omega_E(T_i)}, \tag{2.3.21}$$

where $\omega_N(T_i)$ is the union of all simplexes having at least one common node with T_i and $\omega_E(T_i)$ is the union of all simplexes having a common edge with T_i. In Chapter 4, these estimates will be used for deriving a posteriori error estimates for finite element approximations.

Chapter 3

A posteriori estimates for iteration methods

3.1 The Banach fixed point theorem

Let X be a complete metric space with a metric function d and $\mathcal{T} : X \to X$ be a continuous operator.

Definition 3.1.1. *A point $x_\odot$ is called a fixed point of $\mathcal{T}$ if*

$$x_\odot = \mathcal{T} x_\odot. \tag{3.1.1}$$

Many mathematical problems can be stated as fixed point problems for some operators. Typically, approximations of a fixed point are constructed by the iteration sequence

$$x_i = \mathcal{T} x_{i-1} \qquad i = 1, 2, \dots. \tag{3.1.2}$$

In this connection, the following two problems must be solved:

   a) find the conditions that guarantee convergence of x_i to $x_\odot$,

   b) find computable estimates of the error $e_i = d(x_i, x_\odot)$.

These problems posess solutions provided, that $\mathcal{T}$ is subject to the following additional condition.

Definition 3.1.2. *An operator $\mathcal{T} : X \to X$ is called q-contractive on a set $S \subset X$ if there exists a positive real number q such that the inequality*

$$d(\mathcal{T} x, \mathcal{T} y) \leq q\, d(x, y) \tag{3.1.3}$$

holds for any elements x and y of the set S.

Theorem 3.1.1 (Banach). *Let $\mathcal{T}$ be a q-contractive mapping of a closed nonempty set $S \subset X$ to itself with some $q < 1$. Then, $\mathcal{T}$ has a unique fixed point that belongs to S and the sequence x_i obtained by (3.1.2) converges to this point.*

Proof. It is easy to see that

$$d(x_{i+1}, x_i) = d(\mathcal{T}x_i, \mathcal{T}x_{i-1}) \leq qd(x_i, x_{i-1}) \leq \ldots \leq q^i d(x_1, x_0).$$

Therefore, for any $m > 1$ we have

$$d(x_{i+m}, x_i) \leq$$
$$\leq d(x_{i+m}, x_{i+m-1}) + d(x_{i+m-1}, x_{i+m-2}) + \ldots + d(x_{i+1}, x_i) \leq$$
$$\leq q^i(q^{m-1} + q^{m-2} + \ldots + 1)d(x_1, x_0). \quad (3.1.4)$$

Since

$$\sum_{k=0}^{m-1} q^k \leq \frac{1}{1-q},$$

(3.1.4) implies the estimate

$$d(x_{i+m}, x_i) \leq \frac{q^i}{1-q}d(x_1, x_0). \quad (3.1.5)$$

The right-hand side of (3.1.5) tends to zero as $i \to \infty$ and, consequently, $\{x_i\}$ is a Cauchy sequence, which has a limit in X. Denote this limit by y. Then,

$$d(\mathcal{T}x_i, \mathcal{T}y) \leq qd(x_i, y) \to 0$$

and by (3.1.2) we conclude that $\mathcal{T}y = y$. Hence, it follows that any limit of such a sequence is a fixed point.

It remains to prove that a fixed point is unique. Assume that there are two different fixed points $x_\odot^1$ and $x_\odot^2$, i.e.

$$\mathcal{T}x_\odot^k = x_\odot^k, \qquad k = 1, 2.$$

Therefore,

$$d(x_\odot^1, x_\odot^2) = d(\mathcal{T}x_\odot^1, \mathcal{T}x_\odot^2) \leq qd(x_\odot^1, x_\odot^2).$$

But $q < 1$, and thus such an inequality cannot be true. This contradiction shows that a fixed point is unique and $y = x_\odot$. $\qquad\qquad\square$

It is easy to see that

$$e_j = d(x_j, x_\odot) = d(\mathcal{T}x_{j-1}, \mathcal{T}x_\odot) \le qe_{j-1} \le q^j e_0.$$

Thus, the errors of a convergent sequence tend to zero with ratio q. However, the quantity e_0 is unknown and, in view of this, such an estimate has mainly a theoretical meaning. The proposition below furnishes upper and lower estimates of e_j, which are easy to compute provided, that the number q (or a good estimate of it) is known.

Proposition 3.1.1. *Let $\{x_j\}_{j=0}^{\infty}$ be a sequence obtained by the iteration process (3.1.2) with a mapping $\mathcal{T}$ satisfying the condition (3.1.3). Then, for any x_j, $j > 1$, the following estimate holds:*

$$M_\ominus^j \le e_j \le M_\oplus^j, \tag{3.1.6}$$

where $e_j = d(x_j, x_\odot)$ and

$$M_\ominus^j := \frac{1}{1+q}d(x_{j+1}, x_j), \qquad M_\oplus^j := \frac{q}{1-q}d(x_j, x_{j-1}).$$

Proof. The upper estimate in (3.1.6) follows from (3.1.5). Indeed, put $i = 1$ in this relation. We have

$$d(x_{1+m}, x_1) \le \frac{q}{1-q}d(x_1, x_0).$$

Since $x_{1+m} \to x_\odot$ as $m \to +\infty$, we pass to the limit with respect to m and obtain

$$d(x_\odot, x_1) \le \frac{q}{1-q}d(x_1, x_0).$$

Take the point x_{j-1} as the starting point of the sequence (i.e., set $x_0 = x_{j-1}$). Then, the above relation implies

$$d(x_\odot, x_j) \le \frac{q}{1-q}d(x_j, x_{j-1}).$$

The lower estimate of the error follows from the relation

$$d(x_j, x_{j-1}) \le d(x_j, x_\odot) + d(x_{j-1}, x_\odot) \le (1+q)d(x_{j-1}, x_\odot),$$

which shows that

$$d(x_{j-1}, x_\odot) \ge \frac{1}{1+q}d(x_j, x_{j-1}).$$

$$\square$$

Remark 3.1.1. Since

$$\frac{M_{\oplus}^{j}}{M_{\ominus}^{j}} = \frac{q(1+q)}{1-q}\frac{d(x_j, x_{j-1})}{d(x_{j+1}, x_j)} \geq \frac{1+q}{1-q},$$

we see that that the efficiency of the upper and lower bounds given by (3.1.6) deteriorates as $q \to 1$.

Remark 3.1.2. If X is a normed space, then

$$d(x_{j+1}, x_j) = \|R(x_j)\|,$$

where $R(x_j) := \mathcal{T}x_j - x_j$ is the residual of the basic equation (3.1.1). Thus, the upper and lower estimates of errors are expressed in terms of the *residuals of the respective iteration equation* computed for two neighbor steps:

$$\frac{1}{1+q}\|R(x_j)\| \leq e_j \leq \frac{q}{1-q}\|R(x_{j-1})\|.$$

$\square$

Sometimes it is easier to prove that a sequence of approximate solutions is obtained by a contractive mapping if we consider the operator

$$\hat{\mathcal{T}} = T^n := \underbrace{TT...T}_{n \text{ times}}$$

instead of T. The assertion below shows that in this case we arrive at similar results.

Proposition 3.1.2. *Let $\mathcal{T} : S \to S$ be a continuous mapping such that $\hat{\mathcal{T}}$ is a q-contractive mapping with $q \in (0, 1)$. Then, the equations*

$$x = \mathcal{T}x \qquad \text{and} \qquad \hat{x} = \hat{\mathcal{T}}\hat{x}$$

have one and the same fixed point, which is unique and can be found by the iteration procedure.

Proof. By Theorem 3.1.1, we observe that the operator $\hat{\mathcal{T}}$ has a unique fixed point $\hat{x}_{\ominus}$. To show that $\hat{x}_{\ominus}$ is a fixed point of $\mathcal{T}$, we note that

$$\mathcal{T}\hat{x}_{\ominus} = \mathcal{T}\hat{\mathcal{T}}\hat{x}_{\ominus} = \mathcal{T}\hat{\mathcal{T}}^2\hat{x}_{\ominus} = ...$$
$$= \mathcal{T}\hat{\mathcal{T}}^i\hat{x}_{\ominus} = \mathcal{T}^{(1+in)}\hat{x}_{\ominus} = \mathcal{T}^{in}\mathcal{T}\hat{x}_{\ominus}. \quad (3.1.7)$$

Denote $x_0 = \mathcal{T}\hat{x}_{\ominus}$. By (3.1.7) we conclude that for any i

$$\mathcal{T}\hat{x}_{\ominus} = \hat{\mathcal{T}}^i x_0. \quad (3.1.8)$$

Passing to the limit on the right-hand side in (3.1.8), we arrive at the relation $\mathcal{T}\hat{x}_\odot = \hat{x}_\odot$, which means that $\hat{x}_\odot$ is a fixed point of the operator $\mathcal{T}$.

Assume that $x_\odot$ is a fixed point of $\mathcal{T}$. Then,

$$x_\odot = \mathcal{T}^2 x_\odot = .. = \mathcal{T}^n x_\odot = \hat{\mathcal{T}} x_\odot$$

and $x_\odot$ is a fixed point of $\mathcal{T}$. Since $\hat{x}_\odot$ exists and is unique, we conclude that $\mathcal{T}$ has a unique fixed point, which coincides with $\hat{x}_\odot$. $\qquad\square$

Remark 3.1.3. We can derive more accurate bounds of errors if we use more terms of the sequence $\{x_j\}$. Indeed,

$$\begin{aligned} d(x_j, x_\odot) \ &\leq d(x_j, x_{j+1}) + d(x_{j+1}, x_\odot) \leq \\ &\leq d(x_j, x_{j+1}) + \frac{q}{1-q} d(x_j, x_{j+1}), \end{aligned}$$

and we obtain another upper bound

$$d(x_j, x_\odot) \leq \frac{1}{1-q} d(x_j, x_{j+1}). \tag{3.1.9}$$

Since

$$\frac{1}{1-q} d(x_j, x_{j+1}) \leq \frac{q}{1-q} d(x_{j-1}, x_j),$$

we observe that, in general, this bound is sharper than $M_\oplus^j$. Obviously, (3.1.9) can also be applied to subsequences of $\{x_j\}$ (e.g., to $\{x_{ls}\}$, $s = 0, 1, 2...$). In this case, we obtain various upper bounds of $d(x_j, x_\odot)$ computed on the basis of some terms of the sequence $\{x_j\}$ that have indices greater than j. They are as follows:

$$d(x_j, x_\odot) \leq M_\oplus^{j,l} := \frac{1}{1-q^l} d(x_j, x_{j+l}).$$

Note that the right-hand side of this estimate tends to $d(x_j, x_\odot)$ as $l \to +\infty$. Thus, for a sufficiently large l the bound will be accurate even if q is close to 1.

The lower estimates can be improved by similar arguments. We have the estimate

$$d(x_j, x_\odot) \geq M_\ominus^{j,l} := \frac{1}{1+q^l} d(x_j, x_{j+l})$$

whose right-hand side also tends to the exact value of the error as $l \to +\infty$.

Let L be a given number that indicates the number of successive elements used in the evaluation of error bounds for x_j. Compute the quantities

$$\bar{M}_{\ominus}^{j,L} := \sup_{l=1,2,\dots L} \left\{ \frac{1}{1+q^l}\, d(x_j, x_{j+l}) \right\}, \qquad (3.1.10)$$

$$\bar{M}_{\oplus}^{j,L} := \inf_{l=1,2,\dots L} \left\{ \frac{1}{1-q^l}\, d(x_j, x_{j+l}) \right\}. \qquad (3.1.11)$$

They yield upper and lower estimates, which are the sharper, the greater the number L.

Another sequence of upper bounds follows from the relation

$$d(x_j, x_{\odot}) \le d(x_j, x_{j+1}) + d(x_{j+1}, x_{\odot}) \le$$

$$\le \mathcal{M}_{\oplus}^{j,2}(x_j, x_{j+1}, x_{j+2}) := d(x_j, x_{j+1}) + \frac{1}{1-q} d(x_{j+1}, x_{j+2}). \quad (3.1.12)$$

Note that

$$\mathcal{M}_{\oplus}^{j,2} \le d(x_j, x_{j+1}) + \frac{q}{1-q} d(x_j, x_{j+1}) = M_{\oplus}^{j,1},$$

so that the majorant that includes x_j, x_{j+1}, and x_{j+2} provides a more exact upper bound. Similarly, we can obtain lower bounds of the error computed by x_j, x_{j+1}, and x_{j+2}.

3.2   Iteration methods for bounded linear operators

3.2.1   Two-sided estimates

Consider a bounded linear operator $\mathcal{L} : X \to X$, where X is a Banach space. Given $b \in X$, the iteration process is defined by the relation

$$x_j = \mathcal{L}\, x_{j-1} + b. \qquad (3.2.1)$$

Let $x_{\odot}$ be a fixed point of (3.2.1) and

$$\|\mathcal{L}\| = q < 1. \qquad (3.2.2)$$

By applying Theorem 3.1.1, it is easy to show that $\{x_j\} \to x_{\odot}$. Indeed, let $\bar{x}_j = x_j - x_{\odot}$. Then

$$\bar{x}_j = \mathcal{L}x_{j-1} + b - x_{\odot} = \mathcal{L}(x_{j-1} - x_{\odot}) = \mathcal{L}\bar{x}_{j-1}. \qquad (3.2.3)$$

Since $0_X = \mathcal{L}0_X$, we note that 0_X is a unique fixed point of $\mathcal{L}$. Therefore,

$$\|x_j - x_\odot\|_X = \|\bar{x}_j - 0_X\|_X \leq$$

$$\leq \frac{q^j}{1-q}\|\bar{x}_1 - \bar{x}_0\|_X = \frac{q^j}{1-q}\|R(x_0)\|_X \quad (3.2.4)$$

and

$$\|x_j - x_\odot\|_X \leq \frac{q}{1-q}\|R(x_{j-1})\|_X, \quad (3.2.5)$$

where $R(z) = \mathcal{L}z + b - z$. Also, (3.1.6) implies a lower bound of the error

$$\|x_j - x_\odot\|_X \geq \frac{1}{1+q}\|x_{j+1} - x_j\|_X = \frac{1}{1+q}\|R(x_j)\|_X. \quad (3.2.6)$$

Hence, we arrive at the estimate

$$\frac{1-q}{q}\|x_j - x_\odot\|_X \leq \|R(x_{j-1})\|_X \leq (1+q)\|x_{j-1} - x_\odot\|_X.$$

Other estimates follow from (3.1.9)–(3.1.12).

3.2.2 Iteration schemes

Important applications of the above results are associated with systems of linear simultaneous equations and other algebraic problems. In this section, we consider them and present two-sided bounds of errors that follow from (3.2.5) and (3.2.6).

Set $X = \mathbb{R}^n$ and assume that $\mathcal{L}$ is defined by a nondegenerate matrix $A \in \mathbb{M}^{n \times n}$ decomposed into three matrixes

$$A = A_l + A_d + A_r, \quad (3.2.7)$$

where A_l, A_r, and A_d are certain lower, upper, and diagonal matrices, respectively.

Iteration methods for systems of linear simultaneous equations associated with A are often represented in the form

$$B\frac{x_i - x_{i-1}}{\tau} + A\,x_{i-1} = f. \quad (3.2.8)$$

In (3.2.8), the matrix B and the parameter τ may be taken in various ways (depending on the properties of A). We consider three frequently encountered cases:

(a) $B = A_d$,

(b) $B = A_d + A_l$,

(c) $B = A_d + \omega A_l$, $\tau = \omega$.

For $\tau = 1$, (a) and (b) lead to the methods of Jacobi and Zeidel, respectively. In (c), the parameter ω must be in the interval $(0, 2)$. If $\omega > 1$, we have the so-called "upper relaxation method", and $\omega < 1$ corresponds to the "lower relaxation method".

The method (3.2.8) is reduced to (3.2.1) if we set

$$\mathcal{L} = \mathbb{I} - \tau B^{-1} A \qquad \text{and} \qquad b = \tau B^{-1} f, \qquad (3.2.9)$$

where $\mathbb{I}$ is the unit matrix. It is known that x_i converges to $x_\odot$ that is a solution of the system

$$A x_\odot = f \qquad (3.2.10)$$

if an only if all the eigenvalues of $\mathcal{L}$ are less than one. Obviously, B and τ should be taken in such a way that they guarantee the fulfillment of this condition in the concrete example considered. Assume that this is done and, moreover, we have established the condition (3.2.2). In view of (3.2.4)-(3.2.6), the quantities

$$M_\oplus^i \;\; = \;\; q(1 - q)^{-1} \, \|R(x_{i-1})\| , \qquad (3.2.11)$$
$$M_\oplus^{0i} \;\; = \;\; q^i (1 - q)^{-1} \, \|R(x_0)\| , \qquad (3.2.12)$$
$$M_\ominus^i \;\; = \;\; (1 + q)^{-1} \, \|R(x_i)\| \qquad (3.2.13)$$

furnish upper and lower bounds of the error for the vector x_i. The validity of them is demonstrated with an example below.

Remark 3.2.1. If q is very close to 1, then the convergence of an iteration process may be very slow. As we have seen, in this case, the quality of error estimates is also degraded. A well–accepted way for accelerating the convergence consists of using a modified system obtained from the original one by means of a suitable *preconditioner* P^{-1} and solving the system $\left(P^{-1} A\right) x = P^{-1} f$ with a smaller condition number. Of cause, the best preconditioner is the unknown matrix A^{-1}. Therefore, a preconditioner is often constructed from the parts of A that are not difficult to invert (e.g., in the simplest case it is taken as the matrix inverse to the diagonal part of A). This iteration technique is well presented in the literature (see, e.g., [11]) and the references therein).

Remark 3.2.2. In some cases, one can obtain two-sided estimates for each component of a solution. Let $x_\odot$ be a solution of the system of linear simultaneous equations

$$x_\odot = \mathrm{A} x_\odot + f,$$

where $A = A^{\oplus} - A^{\ominus}$ and

$$A^{\ominus} = \{a_{ij}^{\ominus}\} \in \mathbb{M}^{n \times n}, \qquad a_{ij}^{\ominus} \geq 0,$$
$$A^{\oplus} = \{a_{ij}^{\oplus}\} \in \mathbb{M}^{n \times n}, \qquad a_{ij}^{\oplus} \geq 0.$$

Henceforth, we partially order the space of vectors and write $x \leq y$ if $x_i \leq y_i$ for $i = 1, 2, \ldots n$. Assume that the vectors $x_0^{\ominus}$ and $x_0^{\oplus}$ are such that

$$x_0^{\ominus} \leq x_{\odot} \leq x_0^{\oplus}.$$

Compute $x_1^{\ominus}$ and $x_1^{\oplus}$ by the relations

$$x_1^{\ominus} = A^{\oplus} x_0^{\ominus} - A^{\ominus} x_0^{\oplus} + f,$$
$$x_1^{\oplus} = A^{\oplus} x_0^{\oplus} - A^{\ominus} x_0^{\ominus} + f.$$

It is easy to see that

$$x_1^{\ominus} - x_{\odot} = A^{\oplus}(x_0^{\ominus} - x_{\odot}) - A^{\ominus}(x_0^{\oplus} - x_{\odot}) \leq 0,$$
$$x_1^{\oplus} - x_{\odot} = A^{\oplus}(x_0^{\oplus} - x_{\odot}) - A^{\ominus}(x_0^{\ominus} - x_{\odot}) \geq 0.$$

Hence,

$$x_1^{\ominus} \leq x_{\odot} \leq x_1^{\oplus}.$$

Quite similarly, we observe that the subsequent elements of the iteration process

$$x_{k+1}^{\ominus} = A^{\oplus} x_k^{\ominus} - A^{\ominus} x_k^{\oplus} + f$$
$$x_{k+1}^{\oplus} = A^{\oplus} x_k^{\oplus} - A^{\ominus} x_k^{\ominus} + f,$$

possess the same properties. Therefore, for the ith component we find the following two-sided bounds:

$$\max_{j=0,1,\ldots k+1} \left(x_j^{\ominus}\right)_i \leq (x_{\odot})_i \leq \min_{j=0,1,\ldots k+1} \left(x_j^{\oplus}\right)_i.$$

This method can also be applied to functional equations, provided that $A^{\oplus}$ and $(-A^{\ominus})$ are monotone operators defined on a partially ordered space (see, e.g., [55]).

3.2.3 Example

Consider the problem $Ax = f$ for a symmetric matrix A with coefficients $a_{ij} = 0.8/ij$ if $i \neq j$ and $a_{ii} = i$. The system is solved by the method

$$x_{i+1} = \left(\mathbb{I} - \tau B^{-1} A\right) x_i + \tau B^{-1} F$$

with $B = A_D$ and $x_0 = \{0, 0, ...0\}$. In this example $n = 200$, $q = 0.662$, and $\tau = 0.760$. The values of the error and the estimates are presented in Table 3.2.1 depicted in Fig. 3.2.1. These results show that the two-sided estimates (3.1.6) give a correct representation on the value of $\|x_\odot - x_i\|$, while the *a priori* majorant $M_\oplus^{0i}$ leads to an unacceptable overestimation of the error. We note that this situation is quite typical and is observed in the majority of other problems.

Table 3.2.1:

i	$M_\ominus^i$	$\|e\|$	$M_\oplus^i$	$M_\oplus^{0i}$
1	.187145E+03	.412471E+03	.245893E+04	.245893E+04
2	.452820E+02	.104019E+03	.610732E+03	.162904E+04
3	.123433E+02	.311517E+02	.147774E+03	.107924E+04
4	.405504E+01	.116679E+02	.402813E+02	.714995E+03
5	.166633E+01	.517711E+01	.132333E+02	.473684E+03
6	.767379E+00	.244532E+01	.543792E+01	.313815E+03
7	.366283E+00	.117450E+01	.250428E+01	.207902E+03
8	.176340E+00	.566166E+00	.119533E+01	.137735E+03
16	.515722E-03	.165576E-02	.349042E-02	.511127E+01
17	.248671E-03	.798371E-03	.168302E-02	.338621E+01
18	.119903E-03	.384956E-03	.811515E-03	.224336E+01
19	.578146E-04	.185617E-03	.391295E-03	.148623E+01
20	.278769E-04	.895001E-04	.188673E-03	.984624E+00

3.3   Iterational methods for integral equations

Many problems in science and engineering can be stated in terms of integral equations. We consider one of the most frequently encountered cases where the problem is as follows: to find a function $x_\odot(t) \in C[a, b]$ such that

$$x_\odot(t) = \lambda \int_a^b K(t, s)\, x_\odot(s)\, ds + f(t), \qquad (3.3.1)$$

where $\lambda \geq 0$, K (the kernel) is a continuous function for

$$(x, t) \in Q := \{a \leq s \leq b,\ a \leq t \leq b\}$$

and

$$|K(t, s)| \leq M, \qquad \forall (t, s) \in Q.$$

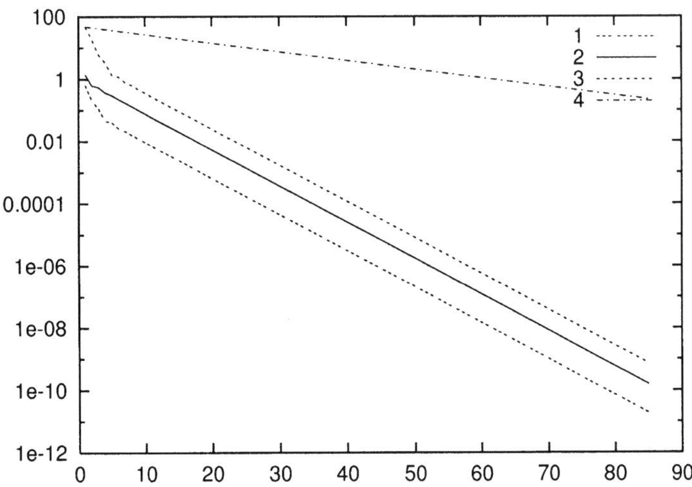

Figure 3.2.1: A priori and a posteriori estimates for an iteration process: $1 - M^i_\ominus$, $2 - \|e\|$, $3 - M^i_\oplus$, $4 - M^{0i}_\oplus$.

Also, we assume that $f \in C[a, b]$.

Let us define the operator $\mathcal{T}$ as follows:

$$\mathcal{T}x := \lambda \int_a^b K(t, x)x(s)\,ds + f(t) \qquad (3.3.2)$$

and show that $\mathcal{T}$ maps continuous functions to continuous ones. Assume that t_0 is an arbitrary point of the interval $[a, b]$ and $t_0 + \Delta t$ also lies in this interval. By $y(t)$ we denote the function $\mathcal{T}x(t)$. It is easy to see that

$$|y(t_0 + \Delta t) - y(t_0)| \leq$$

$$\leq |\lambda| \int_a^b |K(t_0 + \Delta t, s) - K(t_0, s)||x(s)|\,ds +$$

$$+ |f(t_0 + \Delta t) - f(t_0)|.$$

Since K and f are continuous on the compact sets Q and $[a, b]$, respectively, they are uniformly continuous on these sets, and for any given ε one can find a small number δ such that

$$|f(t_0 + \Delta t) - f(t_0)| < \varepsilon$$

and

$$|K(t_0 + \Delta t, s) - K(t_0, s)| < \varepsilon,$$

provided that $|\Delta t| < \delta$. Thus, we have

$$|y(t_0 + \Delta t) - y(t_0)| \leq \varepsilon(|x||b - a| \max_{s \in [a,b]} |x(s)| + 1) = C\varepsilon,$$

and, consequently, $y(t_0 + \Delta t)$ tends to $y(t_0)$ as $|\Delta t| \to 0$.

Let us now prove that $\mathcal{T} : C[a,b] \to C[a,b]$ is a contractive mapping. This fact directly follows from the definition of this operator. Indeed,

$$d(\mathcal{T}x, \mathcal{T}y) = \max_{a \leq t \leq b} |\mathcal{T}x(t) - \mathcal{T}y(t)| =$$

$$= \max_{a \leq t \leq b} \left| \lambda \int_a^b K(t,s)(x(s) - y(s))\, ds \right| \leq$$

$$\leq |\lambda| M(b - a) \max_{a \leq s \leq b} |x(s) - y(s)| = |\lambda| M(b - a) d(x,y),$$

so that $\mathcal{T}$ is a q-contractive operator with

$$q = |\lambda| M(b - a), \tag{3.3.3}$$

provided that

$$|\lambda| < \frac{1}{M(b - a)}. \tag{3.3.4}$$

Now, an approximate solution of the integral equation (3.3.1) may be found by the iteration method

$$x_{i+1}(t) = \lambda \int_a^b K(t,s) x_i(s)\, ds + f(t). \tag{3.3.5}$$

If (3.3.4) holds, then from the Banach theorem it follows that the sequence $\{x_i\}$ converges to the solution of (3.3.1). By (3.1.6), we find that the accuracy of x_i is subject to the estimate

$$\frac{1}{1+q} \int_a^b K(t,s)(x_{i+1}(s) - x_i(s))\, ds \leq$$

$$\leq \max_{a \leq t \leq b} |x_i(t) - x_{\odot}(t)| \leq \frac{q}{1-q} \int_a^b K(t,s)(x_i(s) - x_{i-1}(s))\, ds, \tag{3.3.6}$$

where q is defined by the relation (3.3.3).

Another example is presented by the Volterra equation

$$x_{\odot}(t) = \lambda \int_a^t K(t,s)\, x_{\odot}(s)\, ds + f(t). \tag{3.3.7}$$

Quite similarly, we establish that

$$d(\mathcal{T}x, \mathcal{T}y) \le |\lambda| M(t-a) d(x,y).$$

Moreover,

$$d(\mathcal{T}^n x, \mathcal{T}^n y) \le |\lambda|^n M^n \frac{(t-a)^n}{n!} d(x,y),$$

so that the operator $\mathcal{T}$ is q-contractive with a certain $q < 1$, provided that n is large enough. In view of Proposition 3.1.2, we conclude that the iteration method converges to a solution of (3.3.7) and the errors are controlled by the estimates (3.1.6).

3.4 Iteration methods for ordinary differential equations

Iteration methods are also applied to the computation of approximate solutions of stationary and evolutionary problems associated with differential equations. These methods are based upon certain transformations of differential problems into integral ones. If one can prove that the respective integral operator is subject to the conditions of the Banach theorem, then all the results of the general theory (existence, convergence, and error estimates) are applicable to the problem considered. Below we examine only one simple example taken from [121]. For a detailed investigation of this subject we refer to [55, 222] and numerous publications on the theory of integral equations.

Let u be a solution of the simplest initial boundary-value problem

$$\frac{du}{dt} = \varphi(t, u(t)), \quad u(t_0) = a, \tag{3.4.1}$$

where the solution $u(t)$ is to be found on the interval $[t_0, t_1]$. Assume that the function $\varphi(t, p)$ is continuous on the set

$$Q = \{t_0 \le t \le t_1,\ a - \Delta \le p \le a + \Delta\}$$

and

$$|\varphi(t, p_1) - \varphi(t, p_2)| \le L|p_1 - p_2|, \quad \forall (t, p) \in Q. \tag{3.4.2}$$

Then the problem (3.4.1) can be reduced to the one for the integral equation

$$u(t) = \int_{t_0}^{t} \varphi(s, u(s))\, ds + a \tag{3.4.3}$$

and it is natural to solve the latter problem by the iteration method

$$u_j(t) = \int_{t_0}^{t} \varphi(s, u_{j-1}(s))\, ds + a. \tag{3.4.4}$$

To justify this procedure, we must verify that the operator

$$\mathcal{T}u := \int_{t_0}^{t} \varphi(s, u(s))\, ds + a$$

is q-contractive with respect to the norm

$$\|u\| := \max_{t \in [t_0, t_1]} |u(t)|. \tag{3.4.5}$$

Then

$$\|\mathcal{T}z - \mathcal{T}y\| = \max_{t \in [t_0,t_1]} \left| \int_{t_0}^{t} (\varphi(s, z(s)) - \varphi(s, y(s)))\, ds \right| \leq$$

$$\leq \max_{t \in [t_0,t_1]} L \int_{t_0}^{t} |z(s) - y(s)|\, ds \leq L \int_{t_0}^{t_1} |z(s) - y(s)|\, ds \leq$$

$$\leq L(t_1 - t_0) \max_{s \in [t_0,t_1]} |z(s) - y(s)| = L(t_1 - t_0)\|z - y\|.$$

We see that if

$$t_1 < t_0 + L^{-1}, \tag{3.4.6}$$

then the operator $\mathcal{T}$ is q-contractive if

$$q := L(t_1 - t_0) < 1.$$

Therefore, if the interval $[t_0, t_1]$ is small enough, then the existence and uniqueness of a continuous solution $u(t)$ follows from the Banach theorem. Moreover, this solution can be found by the iteration procedure whose accuracy is controlled by the estimates (3.1.6).

3.5   Notes for the Chapter

The theory of contractive mappings with applications of fixed–point theorems to various problems is exposed in the books by A. N. Kolmogorov and S. V. Fomin [121], L. Collatz [55], and E. Zeidler [222].

Iteration methods for problems in linear algebra are exposed, e.g., in U. Axelsson [11] and R.S. Varga [209].

Two-sided a posteriori estimates (3.1.6) are presented in A. Ostrovski [156]. More information on a posteriori error control for various problems solved by iteration methods can be found in, e.g., H. W. Engl and O. Scherzer [67], Jin Qi-nian [107], A. S. Leonov [137], P. Meyer [142], F. Potra [160], Y. Hayashi [99], E. G. Geisler, A. A. Tal, and D. P. Garg [85], H. Schultz and O. Steinbach [198], K. Tsuruta and K. Ohmori [207], and in the literature cited therein.

Chapter 4

A posteriori estimates for finite element approximations

Finite element method is one of the main numerical tools in modern numerical analysis. From the mathematical point of view, this method has its origin in the works of W. Ritz [187] and R. Courant [57]. A detailed exposition of its mathematical basis and applications can be found in numerous books (see, e.g., P. Ciarlet [53], G. Streng and G. Fix [204]). In what follows, we assume that readers are familiar with foundations of the finite element method and focus our attention on a posteriori error estimates. This subject has been intensively investigated in the last decades and generated a vast amount of publications. The goal of this chapter is to give a concise overview of several methods developed for controlling errors in terms of energy norms and other quantities. We restrict ourselves with consideration of only one class of linear elliptic problems. On this paradigm, we try to emphasize principal mathematical ideas that form the basis of each method. References to publications that contain a detailed exposition of various approaches to a posteriori error estimation for FEM are given at the end of the chapter.

4.1 Methods based upon estimates of the negative norm of the equation residual

4.1.1 Approximation errors and residuals

We begin by recalling basic relations between residuals and errors that hold for systems of linear simultaneous equations. Let $\mathcal{A} \in \mathbb{M}^{n \times n}$, $\det \mathcal{A} \neq 0$, and f be a given n-vector. Then, there exists a vector $u \in \mathbb{R}^n$ satisfying the system

$$\mathcal{A}u + f = 0. \tag{4.1.1}$$

Assume that v is an approximation of u. We have

$$\mathcal{A}(v - u) = \mathcal{A}v + f$$

and, therefore,

$$\|v - u\| \leq \|\mathcal{A}^{-1}\| \|\mathcal{A}v + f\|. \tag{4.1.2}$$

This is the simplest residual type estimate that provides an upper bound of the error $e := v - u$ in terms of the norm of the residual

$$r := \mathcal{A}v + f.$$

Upper and lower estimates of $\|e\|$ evaluated in the norm of r can easily be obtained if the quantities

$$\lambda_{\min} = \min_{\substack{y \in \mathbb{R}^n \\ y \neq 0}} \frac{\|\mathcal{A}y\|}{\|y\|} \quad \text{and} \quad \lambda_{\max} = \max_{\substack{y \in \mathbb{R}^n \\ y \neq 0}} \frac{\|\mathcal{A}y\|}{\|y\|}$$

are known. Since $\mathcal{A}e = r$, we see that

$$\lambda_{\min} \leq \frac{\|\mathcal{A}e\|}{\|e\|} = \frac{\|r\|}{\|e\|} \leq \lambda_{\max} \tag{4.1.3}$$

and, therefore,

$$\lambda_{\max}^{-1}\|r\| \leq \|e\| \leq \lambda_{\min}^{-1}\|r\|. \tag{4.1.4}$$

Since u is a solution, we have

$$\lambda_{\min} \leq \frac{\|\mathcal{A}u\|}{\|u\|} = \frac{\|f\|}{\|u\|} \leq \lambda_{\max} \tag{4.1.5}$$

and

$$\lambda_{\max}^{-1}\|f\| \leq \|u\| \leq \lambda_{\min}^{-1}\|f\|. \tag{4.1.6}$$

The estimates (4.1.4) and (4.1.6) imply the basic error relation:

$$\frac{\lambda_{\min}}{\lambda_{\max}} \frac{\|r\|}{\|f\|} \leq \frac{\|v - u\|}{\|u\|} \leq \frac{\lambda_{\max}}{\lambda_{\min}} \frac{\|r\|}{\|f\|}. \tag{4.1.7}$$

The quantity $\frac{\lambda_{\max}}{\lambda_{\min}}$ is called the *condition number* of the matrix $\mathcal{A}$ and is denoted by $\mathrm{Cond}\,\mathcal{A}$. Now, we represent (4.1.7) in the form

$$\boxed{(\mathrm{Cond}\,\mathcal{A})^{-1} \frac{\|r\|}{\|f\|} \leq \frac{\|v - u\|}{\|u\|} \leq \mathrm{Cond}\,\mathcal{A} \, \frac{\|r\|}{\|f\|}.} \tag{4.1.8}$$

The estimate (4.1.8) reads as follows: the relative value of the error is bounded from above and below by the value of the residual normalized with respect to the norm of f. The factors depend on the condition number of the matrix $\mathcal{A}$. Some generalizations of these estimates can be found in [10].

In principle, the above consideration can also be applied to a wider set of linear problems, where $\mathcal{A} \in \mathcal{L}(X, Y)$ is a coercive linear operator acting from a Banach space X to another space Y and f is a given element of Y. However, if $\mathcal{A}$ is related to a boundary-value problem, then one should properly define the spaces X and Y and find a practically meaningful analog of the estimate (4.1.8).

4.1.2 Residual type error estimates for approximate solutions of boundary-value problems

Let $\mathcal{A} : X \to Y$ be a linear elliptic operator. Consider the boundary-value problem

$$\mathcal{A}u + f = 0 \quad \text{in } \Omega, \qquad u = u_0 \quad \text{on } \partial\Omega. \tag{4.1.9}$$

Assume that $v \in X$ is an approximation of u. Then, we should measure the error in X and the residual in Y, so that the principal form of the a estimate is

$$\|v - u\|_X \leq C\|\mathcal{A}v + f\|_Y, \tag{4.1.10}$$

where the constant C is independent of v. The key question is as follows: *Which spaces X and Y should we choose for a particular boundary-value problem?*

Consider the classical problem

$$-\Delta u = f \quad \text{in } \Omega, \qquad f \in L^2(\Omega), \tag{4.1.11}$$

$$u = 0 \quad \text{on } \partial\Omega, \tag{4.1.12}$$

whose generalized solution satisfies the relation

$$\int_\Omega \nabla u \cdot \nabla w \, dx = \int_\Omega f w \, dx \quad \forall w \in V_0 := \overset{\circ}{H}{}^1(\Omega) \tag{4.1.13}$$

and is subject to the energy estimate

$$\|\nabla u\|_{2,\Omega} \leq C_\Omega \|f\|_{2,\Omega}, \tag{4.1.14}$$

where C_Ω is a constant in the Friederichs inequality. Assume that the approximation $v \in V_0$ possesses extra regularity properties, so that $\Delta v \in L^2(\Omega)$. Then, from (4.1.13) it follows that

$$\int_\Omega \nabla(u - v) \cdot \nabla w \, dx = \int_\Omega (f + \Delta v) w \, dx, \quad \forall w \in V_0. \tag{4.1.15}$$

Setting here $w = u - v$, we obtain the estimate

$$\|\nabla(u - v)\|_{2,\Omega} \leq C_\Omega \|f + \Delta v\|_{2,\Omega}. \tag{4.1.16}$$

The right-hand side of (4.1.16) is formed by the L^2-norm of the residual of (4.1.11). However, this estimate may be useless for approximations constructed by conforming approximations of the energy space V_0. Indeed, the general convergence theory can guarantee that a sequence $\{v_k\}$ of such approximations converges to an exact solution in H^1, but no convergence of higher derivatives could be observed. Therefore, the quantity $\|\Delta v_k + f\|$ does not necessarily converges to zero, unlike the left-hand side of (4.1.16), which decreases and tends to zero. This means that the consistency (the key property of any practically meaningful estimate) is lost.

Now, the following question arises: *Which norm of the residual leads to a consistent estimate of the error in the energy norm?*

To answer it, we replace (4.1.13) by the integral identity

$$\int_\Omega \nabla u \cdot \nabla w \, dx = \langle f, w \rangle, \quad \forall \, w \in V_0, \tag{4.1.17}$$

where $\langle f, w \rangle$ denotes the value of $f \in H^{-1}$ on $w \in V_0$. If $f \in L^2(\Omega)$, then the right-hand side of (4.1.17) is expressed by the integral and we arrive at (4.1.13). However, as we have seen, the set of linear continuous functionals defined on V_0 is wider than the set of the "regular" functionals presented as Lebesgue integrals of summable functions. It also includes the functionals from H^{-1}. Since

$$\langle f, w \rangle \leq |\,f\,| \|\nabla w\|_{2,\Omega}, \tag{4.1.18}$$

we set $w = u$ in (4.1.17) and find that

$$\|\nabla u\|_{2,\Omega} \leq |f|. \tag{4.1.19}$$

This estimate gives an upper bound of the energy norm of u in terms of the norm of f in H^{-1} and suggests an answer to the question stated.

Assume that $v \in V_0$ is an approximation of u. By (4.1.13), we have

$$\int_{\Omega} \nabla(u - v) \cdot \nabla w \, dx = \int_{\Omega} (fw - \nabla v \cdot \nabla w) \, dx. \tag{4.1.20}$$

The right-hand side of (4.1.20) is a linear continuous functional defined on the space $\overset{\circ}{H}^1(\Omega)$, so that it belongs to $H^{-1}(\Omega)$. Thus, (4.1.20) has the form

$$\int_{\Omega} \nabla(u - v) \cdot \nabla w \, dx = \langle \Delta v + f, w \rangle, \tag{4.1.21}$$

where $f + \Delta v \in H^{-1}(\Omega)$. Setting $w = v - u$, we obtain the estimate

$$\|\nabla(u - v)\|_{2,\Omega} \leq |f + \Delta v|. \tag{4.1.22}$$

Note that

$$|f + \Delta v| = \sup_{\varphi \in \overset{\circ}{H}^1(\Omega)} \frac{|\langle f + \Delta v, \varphi \rangle|}{\|\nabla \varphi\|_{2,\Omega}} =$$

$$= \sup_{\varphi \in \overset{\circ}{H}^1(\Omega)} \frac{|\int_{\Omega} \nabla(u - v) \cdot \nabla \varphi|}{\|\nabla \varphi\|_{2,\Omega}} \leq \|\nabla(u - v)\|_{2,\Omega}$$

and we arrive at the relation

$$\|\nabla(u - v)\|_{2,\Omega} = |f + \Delta v|, \tag{4.1.23}$$

which shows that the quantity $|f + \Delta v_k|$ tends to zero for any sequence $\{v_k\}$ that tends to u in H^1. Thus, the estimate (4.1.23) is *consistent*. From (4.1.23) it follows the estimate

$$\|f + \Delta v\|_{(-1),\Omega} \leq \|\nabla(u - v)\|_{2,\Omega} \leq (C_{\Omega}^2 + 1)^{1/2}\|f + \Delta v\|_{(-1),\Omega},$$

which yields error bounds in terms of another negative norm.

Similar estimates can be derived for any uniformly elliptic equation of the divergent type. Indeed, let

$$\mathcal{A}u = \sum_{i,j=1}^{d} \frac{\partial}{\partial x_i}\left(a_{ij}(x)\frac{\partial u}{\partial x_j}\right),$$

where $a_{ij}(x) = a_{ji}(x)$ are measurable bounded functions satisfying the condition

$$\lambda_{\min}|\eta|^2 \le a_{ij}(x)\eta_i\eta_j \le \lambda_{\max}|\eta|^2, \quad \forall \eta \in \mathbb{R}^n, \ x \in \Omega \tag{4.1.24}$$

with positive constants $\lambda_{\min}$ and $\lambda_{\max}$ independent of x and η. Henceforth, we will also use a the compact notation $\mathcal{A}u = \operatorname{div} A\nabla u,$.

Assume that $f \in L^2(\Omega)$ and consider the boundary-value problem

$$\mathcal{A}u + f = 0, \qquad \text{in } \Omega,$$
$$u = 0 \qquad \text{on } \partial\Omega,$$

whose generalized solution $u \in V_0$ is defined by the integral identity

$$\int_\Omega A\nabla u \cdot \nabla w \, dx = \int_\Omega fw \, dx, \quad \forall w \in V_0. \tag{4.1.25}$$

Let $v \in V_0$ be an approximation of u. Then,

$$\int_\Omega A\nabla(u - v) \cdot \nabla w \, dx = \int_\Omega (fw - A\nabla v \cdot \nabla w) \, dx, \quad \forall w \in V_0.$$

The right-hand side of this relation is a bounded linear functional on V_0. Therefore, it determines a distribution, which is

$$f + \operatorname{div}(A\nabla v) \in H^{-1}.$$

Hence, we have the relation

$$\int_\Omega A\nabla(u - v) \cdot \nabla w \, dx = \langle f + \operatorname{div}(A\nabla v), w \rangle, \quad \forall w \in V_0.$$

Setting $w = u - v$, we derive the estimate

$$\|\nabla(u - v)\|_{2,\Omega} \le \lambda_{\min}^{-1} | f + \operatorname{div}(A\nabla v)|. \tag{4.1.26}$$

Next,

$$| f + \operatorname{div}(A\nabla u)| = \sup_{\varphi \in \overset{\circ}{H}{}^1(\Omega)} \frac{| \langle f + \operatorname{div}(A\nabla u), \varphi \rangle |}{\|\nabla\varphi\|_{2,\Omega}} =$$

$$= \sup_{\varphi \in \overset{\circ}{H}{}^1(\Omega)} \frac{| \int_\Omega A\nabla(u - v) \cdot \nabla\varphi \, dx |}{\|\nabla\varphi\|_{2,\Omega}} \le \lambda_{\max}\|\nabla(u - v)\|_{2,\Omega}. \tag{4.1.27}$$

Combining (4.1.26) and (4.1.27), we obtain

$$\lambda_{\max}^{-1}\, |R(v)| \leq \|\nabla(u-v)\|_{2,\Omega} \leq \lambda_{\min}^{-1}\, |R(v)|, \tag{4.1.28}$$

where

$$R(v) = f + \operatorname{div}(A\nabla v) \in H^{-1}(\Omega)$$

is a residual understood in the sense of distributions. The estimate (4.1.28) shows that upper and lower bounds of the error can be evaluated in terms of the negative norm of $R(v)$. It presents a certain analog of the estimate (4.1.8) for approximate solutions of elliptic type boundary-value problems.

There is a simple case in which the estimate of $|R(v)|$ is easy to find. Assume that $R(v) \in L^p(\Omega)$ with $p \geq 1$. Then

$$|R(v)| \leq \sup_{\varphi \in \overset{\circ}{H}^1(\Omega)} \frac{\|R(v)\|_{p,\Omega}\|\varphi\|_{q,\Omega}}{\|\nabla\varphi\|_{2,\Omega}}. \tag{4.1.29}$$

If the space $L^q(\Omega)$ with $q = \frac{p}{p-1}$ is embedded in $H^1(\Omega)$, then (4.1.29) implies the estimate

$$|R(v)| \leq C_q(\Omega)\|R(v)\|_{p,\Omega},$$

where $C_q(\Omega)$ is the constant in the inequality

$$\|\varphi\|_{q,\Omega} \leq C_q(\Omega)\|\nabla\varphi\|_{2,\Omega}, \quad \forall \varphi \in \overset{\circ}{H}^1(\Omega).$$

However, estimates of such a type have only a conventional meaning, because in the majority of cases approximate solutions obtained by finite element approximations cannot guarantee such a regularity of the residual.

4.2 Evaluation of negative norms of residuals

To use the estimate (4.1.28) one needs either to find $|R(v)|$ or to obtain a sharp estimate of this quantity. A wealth of papers devoted to the residual method is, in fact, focused on the way of overcoming the difficulties arising in the process. For continuous and piecewise smooth approximations, there are two main approaches to finding such estimates. One group encompasses the methods estimating $|R(v)|$ directly by using Galerkin orthogonality property and interpolation estimates for finite element patches. They provide explicit error bounds represented by sums of residuals on the elements and jumps over interelement boundaries.   These sums have weight coefficients that depend not only on the entire domain, but also on the shape of subdomains. In the other group of methods, the quantity $|R(v)|$ is estimated by solving auxiliary boundary-value problems whose right-hand side is formed by the residual. Below, we demonstrate ideas of both approaches with a paradigm of the problem (4.1.25).

4.2.1 Estimation of the residual in a one-dimensional problem

To clarify the idea of the first method, we consider a simple one-dimensional problem: find u such that

$$(\alpha u')' + f = 0, \tag{4.2.1}$$

$$u(0) = u(1) = 0, \tag{4.2.2}$$

where $\alpha = \alpha(x) > 0$ and $f = f(x)$ are given functions. Let I denote the interval $(0, 1)$, $f \in L^2(I)$, and $\alpha(x) \in C(\overline{I})$. A weak statement of the problem (4.2.1)–(4.2.1) is as follows: find $u \in V_0(I)$ such that

$$\int_I \alpha u' w' \, dx = \int_I f w' \, dx, \quad \forall w \in V_0(I) := \overset{\circ}{H}{}^1(I). \tag{4.2.3}$$

Divide I into a number of subintervals $I_i = (x_i, x_{i+1})$, where $x_0 = 0$, $x_{N+1} = 1$, and $|x_{i+1} - x_i| = h_i$. Assume that we have an aproximate solution $v \in \overset{\circ}{H}{}^1(I)$ that is smooth on any interval I_i. In this case,

$$|R(v)| = \sup_{w \in V_0(I),\ w \neq 0} \frac{\int_0^1 (-\alpha v' w' + fw)dx}{\|w'\|_{2,I}} =$$

$$= \sup_{w \in V_0(I),\ w \neq 0} \frac{\sum_{i=0}^N \int_{I_i} r_i(v) w \, dx + \sum_{i=1}^N \alpha(x_i) w(x_i) \mathrm{j}(v'(x_i))}{\|w'\|_{2,I}}, \tag{4.2.4}$$

where

$$\mathrm{j}(\phi(x)) := \phi(x + 0) - \phi(x - 0)$$

is the "jump–function" (note that $\mathrm{j}(\phi(x_0)) = 0$ if the function ϕ is continuous at x_0) and

$$r_i(v) = (\alpha v')' + f$$

is the residual on I_i. The supremum in (4.2.4) is taken over all functions from $V_0(I)$ other than zero function. To simplify further notation, we henseforth will not write this explicitly, but always assume that zero functions are eliminated in quotient type relations.

To find an upper bound of the right-hand side of (4.2.4), we restrict the class of admissible approximations and hereafter assume that v is the *Galerkin approximation* obtained on a finite–dimensional subspace V_{0h} formed by piecewise affine continuous functions. Let v is the Galerkin approximation $u_h \in V_{0h}$, which obeys the relation

$$\int_I \alpha u_h' w_h' \, dx = \int_I f w_h \, dx \quad \forall w_h \in V_{0h}. \tag{4.2.5}$$

Then,

$$|R(u_h)| = \sup_{w \in V_0(I)} \frac{\int_0^1 (-\alpha u_h'(w - \pi_h w)' + f(w - \pi_h w)) \, dx}{\|w'\|_{2,I}},$$

where $\pi_h : V_0 \to V_{0h}$ is the interpolation operator defined by the conditions $\pi_h v \in V_{0h}$, $\pi_h v(0) = \pi_h v(1) = 0$ and

$$\pi_h v(x_i) = v(x_i), \quad \forall x_i, \ i = 1, 2, ..., N. \tag{4.2.6}$$

Integrating by parts, we obtain

$$|R(u_h)| = \sup_{w \in V_0(I)} \left\{ \frac{\sum_{i=0}^N \int_{I_i} r_i(u_h)(w - \pi_h w) \, dx}{\|w'\|_{2,I}} + \right.$$
$$\left. + \frac{\sum_{i=1}^N \alpha(x_i)(w(x_i) - \pi_h w(x_i)) \mathrm{j}(u_h'(x_i))}{\|w'\|_{2,I}} \right\}.$$

By (4.2.6), $w(x_i) - \pi_h w(x_i) = 0$ so that the second sum vanishes. The first one admits the estimate

$$\sum_{i=0}^N \int_{I_i} r_i(u_h)(w - \pi_h w) \, dx \le \sum_{i=0}^N \|r_i(u_h)\|_{2,I_i} \|w - \pi_h w\|_{2,I_i}.$$

It can be represented in a more convenient form, provided that the estimates

$$\|w - \pi_h w\|_{2,I_i} \le c_i \|w'\|_{2,I_i} \tag{4.2.7}$$

hold with constants c_i dependent on I_i. Then, we obtain the estimate

$$\sum_{i=0}^N \int_{I_i} r_i(u_h)(w - \pi_h w) \, dx \le \sum_{i=0}^N c_i \|r_i(u_h)\|_{2,I_i} \|w'\|_{2,I_i} \le$$
$$\le \left(\sum_{i=0}^N c_i^2 \|r_i(u_h)\|_{2,I_i}^2 \right)^{1/2} \|w'\|_{2,I}, \tag{4.2.8}$$

which implies the desired upper bound

$$|R(u_h)| \leq \left(\sum_{i=0}^{N} c_i^2 \|r_i(u_h)\|_{2,I_i}^2 \right)^{1/2}. \tag{4.2.9}$$

This bound is the sum of local residuals with weights given by the interpolation constants c_i.

In this simple problem, the constants c_i can easily be found. Indeed, let γ_i be a constant that satisfies the condition

$$\inf_{w \in H^1(I_i)} \frac{\|w'\|_{2,I_i}^2}{\|w - \pi_h w\|_{2,I_i}^2} \geq \gamma_i. \tag{4.2.10}$$

Then,

$$\|w - \pi_h w\|_{2,I_i} \leq \gamma_i^{-1/2} \|w'\|_{2,I_i}$$

and one can set $c_i = \gamma_I^{-1/2}$. Note that

$$\int_{x_i}^{x_{i+1}} |w'|^2 \, dx = \int_{x_i}^{x_{i+1}} |(w - \pi_h w)' + (\pi_h w)'|^2 \, dx,$$

where $(\pi_h w)'$ is constant on (x_i, x_{i+1}). In view of (4.2.6),

$$\int_{x_i}^{x_{i+1}} (w - \pi_h w)'(\pi_h w)' \, dx = 0$$

and

$$\int_{x_i}^{x_{i+1}} |w'|^2 \, dx \geq \int_{x_i}^{x_{i+1}} |(w - \pi_h w)'|^2 \, dx.$$

Thus, we have

$$\inf_{w \in H^1(I_i)} \frac{\int_{x_i}^{x_{i+1}} |w'|^2 \, dx}{\int_{x_i}^{x_{i+1}} |w - \pi_h w|^2 \, dx} \geq$$

$$\geq \inf_{w \in H^1(I_i)} \frac{\int_{x_i}^{x_{i+1}} |(w - \pi_h w)'|^2 \, dx}{\int_{x_i}^{x_{i+1}} |w - \pi_h w|^2 \, dx} \geq$$

$$\geq \inf_{\eta \in \overset{\circ}{H}{}^1(I_i)} \frac{\int_{x_i}^{x_{i+1}} |\eta'|^2 \, dx}{\int_{x_i}^{x_{i+1}} |\eta|^2 \, dx} = \frac{\pi^2}{h_i^2},$$

so that $\gamma_i = \pi^2/h_i^2$ and $c_i = h_i/\pi$.

4.2.2 Estimation of the residual for linear elliptic problems

Consider the problem (4.1.25). Let Ω be represented as a union of elements T_i. For the sake of simplicity, we assume that Ω is a polygonal domain and $\overline{\Omega} = \cup_{i=1}^{N} \overline{T}_i$. Assume that V_{0h} consists of piecewise affine continuous functions. The respective Galerkin approximation u_h satisfies the relation

$$\int_{\Omega} A\nabla u_h \cdot \nabla w_h \, dx = \int_{\Omega} f w_h \, dx, \quad \forall w_h \in V_{0h}, \qquad (4.2.11)$$

where

$$V_{0h} = \{w_h \in V_0 \mid w_h \in P^1(T_i), \ T_i \in \mathcal{F}_h\}.$$

We have

$$|R(u_h)| = \sup_{w \in V_0} \frac{\int_{\Omega}(f w - A\nabla u_h \cdot \nabla w)\, dx}{\|\nabla w\|_{2,\Omega}} =$$

$$= \sup_{w \in V_0} \frac{\int_{\Omega}(f(w - \pi_h w) - A\nabla u_h \cdot \nabla(w - \pi_h w))\, dx}{\|\nabla w\|_{2,\Omega}},$$

where π_h is the Clement's interpolation operator (cf. (2.3.18)–(2.3.19)). Let $\sigma_h = A\nabla u_h$. Then, for any simplex T_i

$$\int_{T_i} \sigma_h \cdot \nabla(w - \pi_h w)\, dx =$$

$$= \int_{\partial T_i} (\sigma_h \cdot \nu)(w - \pi_h w)ds - \int_{T_i} \operatorname{div} \sigma_h (w - \pi_h w)\, dx,$$

where ν is the unit outward normal to ∂T_i. Since on the boundary $w - \pi_h w = 0$, we obtain

$$|R(u_h)| = \sup_{w \in V_0} \left\{ \frac{\sum_{i=1}^{N} \int_{T_i}(\operatorname{div}\sigma_h + f)(w - \pi_h w)\, dx}{\|\nabla w\|_{2,\Omega}} + \right.$$

$$\left. + \frac{\sum_{i=1}^{N}\sum_{j>i}^{N} \int_{E_{ij}} \mathrm{j}(\sigma_h \cdot \nu_{ij})(w - \pi_h w)\, ds}{\|\nabla w\|_{2,\Omega}} \right\}.$$

In the above relation, $\mathrm{j}(\sigma_h \cdot \nu_{ij})$ denotes the jumps of $\sigma_h \cdot \nu_{ij}$ on E_{ij} and ν_{ij} is the unit normal vector to E_{ij} outward to T_i if $i < j$. Note that

$$\int_{T_i}(\operatorname{div}\sigma_h + f)(w - \pi_h w)dx \leq \|\operatorname{div}\sigma_h + f\|_{2,T_i}\|w - \pi_h w\|_{2,T_i}$$

$$\leq c_{1i}\|\operatorname{div}\sigma_h + f\|_{2,T_i}\operatorname{diam}(T_i)\|w\|_{1,2,\omega_N(T_i)},$$

where $\omega_N(T_i)$ is the patch of elements having common nodes with T_i (see (2.3.20)–(2.3.21)). Then, we estimate the first sum as follows (cf. (2.1.2):

$$\sum_{i=1}^{N} \int_{T_i} (\operatorname{div}\sigma_h + f)(w - \pi_h w)dx \le$$

$$\le d_1 \left(\sum_{i=1}^{N} c_{1i}^2 \operatorname{diam}(T_i)^2 \|\operatorname{div}\sigma_h + f\|_{2,T_i}^2 \right)^{1/2} \|w\|_{1,2,\Omega}, \quad (4.2.12)$$

where the constant d_1 depends on the maximal number of elements in the set $\omega_N(T_i)$. For the second one, we have

$$\sum_{i=1}^{N}\sum_{j>i}^{N} \int_{E_{ij}} j(\sigma_h \cdot \nu_{ij})(w - \pi_h w)\, dx \le$$

$$\le \sum_{i=1}^{N}\sum_{j>i}^{N} \|j(\sigma_h \cdot \nu_{ij})\|_{2,E_{ij}} c_{2ij}|E_{ij}|^{1/2}\|w\|_{1,2,\omega_E(T_i)} \le$$

$$\le d_2 \left(\sum_{i=1}^{N}\sum_{j>i}^{N} c_{2ij}^2 |E_{ij}| \|j(\sigma_h \cdot \nu_{ij})\|_{2,E_{ij}}^2 \right)^{1/2} \|w\|_{1,2,\Omega}. \quad (4.2.13)$$

In the above relations, the constant d_2 depends on the maximal number of elements in the set $\omega_E(T_i)$. The relations (4.2.12) and (4.2.13) imply the estimate

$$|R(u_h)| \le C_0 \left(\left(\sum_{i=1}^{N} c_{1i}^2 \operatorname{diam}(T_i)^2 \|\operatorname{div}\sigma_h + f\|_{2,T_i}^2 \right)^{1/2} + \right.$$

$$\left. + \left(\sum_{i=1}^{N}\sum_{j>i}^{N} c_{2ij}^2 |E_{ij}| \|j(\sigma_h \cdot \nu_{ij})\|_{2,E_{ij}}^2 \right)^{1/2} \right). \quad (4.2.14)$$

Here $C_0 = \max\{d_1, d_2\}\sqrt{1 + C_\Omega^2}$, where C_Ω comes from (2.2.1). The right-hand side of (4.2.14) is the sum of locally defined quantities $\eta(T_i)$ multiplied by constants depending on properties of the chosen splitting $\mathcal{F}_h$ (e.g., on the maximal number of elements in a neighborhood of a node, the minimal angle of the triangulation, etc.). For quasi–uniform meshes all generic constants c_{1i} have approximately the same value and can be replaced by a single constant c_1. If the constants c_{2ij} are also estimated by a single constant c_2, then (4.2.14) implies an estimate

$$|R(u_h)| \le C_0 \left(\sum_{i=1}^{N} \eta^2(T_i) \right)^{1/2}, \quad (4.2.15)$$

where

$$\eta^2(T_i) = c_1^2 \operatorname{diam}(T_i)^2 \| \operatorname{div} \sigma_h + f \|_{2,T_i}^2 + \frac{c_2^2}{2} \sum_{E_{ij} \in \overline{T}_i} |E_{ij}| \| \mathrm{j}(\sigma_h \cdot \nu_{ij}) \|_{2,E_{ij}}^2.$$

$$(4.2.16)$$

Two parts of (4.2.16) are the residual norm on T_i and integrals related to the jumps on the edges. The multiplier $1/2$ arises, because any interior edge is common for two elements.

It is worth noting that for a strongly nonhomogeneous mesh such an estimate may lead to a considerable overestimation of $|R(u_h)|$. The sharper estimate (4.2.14) involves the constants c_{1i} and c_{2ij} whose evaluation requires the solution of the following variational problems:

$$\inf_{w \in V_0} \frac{\|w\|_{1,2,\omega_N(T_i)}}{\|w - \pi_h w\|_{2,T_i}} \operatorname{diam}(T_i) \qquad (4.2.17)$$

and

$$\inf_{w \in V_0} \frac{\|w\|_{1,2,\omega_E(T_i)}}{\|w - \pi_h w\|_{2,E_{ij}}} |E_{ij}|^{1/2}. \qquad (4.2.18)$$

4.2.3 Estimation of the residual norm by implicit methods

For a conforming approximation u_h, we have the basic relation

$$\int_\Omega A\nabla(u - u_h) \cdot \nabla w \, dx = \langle R(u_h), w \rangle, \quad \forall w \in V_0. \qquad (4.2.19)$$

Hence, the error $e = u - u_h$ is a solution of the problem

$$\operatorname{div} A\nabla e + R(u_h) = 0 \quad \text{in } \Omega, \qquad (4.2.20)$$

$$e = 0 \quad \text{on } \partial\Omega, \qquad (4.2.21)$$

where $R(u_h) = f + \operatorname{div} A\nabla u_h \in H^{-1}(\Omega)$ and the first equation is understood in the sense of distributions.

Formally, this idea can be extended to a wide class of linear problems. Indeed, if $\mathcal{A} : X \to Y$ is a linear operator for which we consider the problem

$$\mathcal{A}u + f = 0 \quad \text{in } \Omega, \qquad (4.2.22)$$

$$u = 0 \quad \text{on } \partial\Omega, \qquad (4.2.23)$$

then, for any $v \in X$,

$$\mathcal{A}e + R(v) = 0 \quad \text{in } \Omega, \qquad (4.2.24)$$

$$e = 0 \quad \text{on } \partial\Omega, \qquad (4.2.25)$$

where $R(v) = f + \mathcal{A}v$.

However, the evaluation of the error by directly using (4.2.24)–(4.2.25) may be a difficult task, because, in general, $R(v)$ is a distribution. Moreover, the accuracy of an approximate solution of (4.2.24)–(4.2.25) obviously affects the accuracy of the error estimation, so that a new error estimation problem arises.

In the literature focused on practical applications of this method, it is often suggested to split the functional $\langle R(u_h), w \rangle$ into a number of functionals defined as solutions of local subproblems. A version of such an approach is described below.

Let Ω be a union of nonoverlapping domains Ω_i, $i = 1, 2, ...N$. Assume that for each Ω_i we have found a function u_i such that

$$\langle R(u_h), w \rangle = \sum_{i=1}^{N} \int_{\Omega_i} A\nabla u_i \cdot \nabla w \, dx. \qquad (4.2.26)$$

Since A is a positive definite symmetric matrix, we supply $L^2(\Omega_i, \mathbb{R}^n)$ with a new norm

$$\| y \|_{\Omega_i} := \left(\int_{\Omega_i} Ay(x) \cdot y(x) \, dx \right)^{1/2},$$

which is equivalent to the standard one. Then, in view of the Cauchy inequality, we derive the estimate

$$| \langle R(u_h), w \rangle | = \sum_{i=1}^{N} \| \nabla u_i \|_{\Omega_i} \| \nabla w \|_{\Omega_i} \leq$$

$$\leq \| \nabla w \|_{\Omega} \left(\sum_{i=1}^{N} \| \nabla u_i \|_{\Omega_i}^2 \right)^{1/2} \leq c \|\nabla w\|_{2,\Omega} \left(\sum_{i=1}^{N} \| \nabla u_i \|_{\Omega_i}^2 \right)^{1/2}, \qquad (4.2.27)$$

where $c = \sqrt{\lambda_{\max}}$. Hence,

$$| R(u_h) | = \sup_{w \in V_0} \frac{| \langle R(u_h), w \rangle |}{\|\nabla w\|_{2,\Omega}} \leq c \left(\sum_{i=1}^{N} \| \nabla u_i \|_{\Omega_i}^2 \right)^{1/2} \qquad (4.2.28)$$

Now, the main task is to find the functions u_i. Let us define them as solutions of local subproblems with Neumann type boundary conditions on patches of elements Ω_i with a common node i. Denote the common edge of

Ω_i and Ω_j by Γ_{ij} and $\Gamma_{i0} := \partial\Omega_i \cap \partial\Omega$. For each Ω_i we solve the problem: find $u_i \in H^1(\Omega_i)$, such that $u_i = 0$ on Γ_{i0} and

$$\int_{\Omega_i} A\nabla u_i \cdot \nabla w dx = \int_{\Omega_i} fw dx +$$

$$+ \sum_{j=1}^{N} \int_{\Gamma_{ij}} \sigma_j^i g w ds - \int_{\Omega_i} A\nabla u_h \cdot \nabla w dx, \qquad \forall w \in V_0(\Omega_i). \quad (4.2.29)$$

In this relation, g is a function defined on all Γ_{ij} and it is assumed that the integral over Γ_{ij} is equal to zero if $\Gamma_{ij} = \emptyset$. The space $V_0(\Omega_i)$ is defined as follows. If $\Gamma_{0i} \neq \emptyset$ then

$$V_0(\Omega_i) = \{w \in H^1(\Omega_i) \mid w = 0 \text{ on } \Gamma_{i0}, \}.$$

If $\Gamma_{0i} = \emptyset$, then $V_0(\Omega_i)$ is the space $H^1(\Omega_i)$. The integer-valued function σ_j^i is defined by the rule

$$\sigma_j^i = \begin{cases} +1, & \text{if } i > j, \\ -1, & \text{if } j > i, \\ 0, & \text{if } i = j. \end{cases}$$

Usually, the function g is constructed on the basis of the information encompassed in u_h. For example, if the domains Ω_i coincide with the elements of a certain finite element sampling, then g can be determined by computing fluxes of a respective finite element solution.

Each internal boundary Γ_{ij} generates two integrals with equal absolute values and opposite signs. Therefore, the sum of all integrals does not contain such terms, and we obtain

$$\sum_{i=1}^{N} \int_{\Omega_i} A\nabla u_i \cdot \nabla w dx =$$

$$= \int_{\Omega} fw dx - \int_{\Omega} A\nabla u_h \cdot \nabla w dx = \langle R(u_h), w \rangle, \qquad \forall w \in V_0(\Omega). \quad (4.2.30)$$

Hence, (4.2.28) provides an upper bound of the error in terms of the norms of u_i.

Consider a function $\tilde{u} : \Omega \to \mathbb{R}$ that coincides with $u_i(x)$ if $x \in \Omega_i$. Assume that the functions u_i preserve continuity on the boundaries Γ_{ij} and the function $\tilde{u}(x)$ belongs to $H^1(\Omega)$. Then (4.2.30) reads

$$\int_{\Omega} A\nabla(\tilde{u} + u_h) \cdot \nabla w dx = \int_{\Omega} fw dx \qquad \forall w \in V_0(\Omega). \quad (4.2.31)$$

The relation (4.2.31) means that $u = u_i + u_h$ on Ω_i. Therefore, $u_i = u - u_h$ and by these functions we also obtain local errors.

However, this formally simple procedure contain certain technical difficulties. One of them is that for internal domains the function g (which defines the Neumann type boundary conditions of the local subproblems) cannot be taken arbitrarily. This follows from the fact that the Neumann problem may be unsolvable if the external data do not satisfy an additional condition. For the problem (4.2.29) this condition is as follows:

$$\int\limits_{\Omega_i} f\,dx + \sum_{j=1}^{N} \int\limits_{\Gamma_{ij}} \sigma_j^i g\,ds = 0. \qquad (4.2.32)$$

Therefore, it is required a special procedure (often called the "equilibration procedure") that transforms g in order to satisfy (4.2.32) on each element. After that, exact solutions of local problems must be found and then used to compute the upper bound of the error. In practice, the accuracy of such an estimate depends on the choice of g and on the accuracy of the computed approximations of u_i.

4.3 Error estimation methods using adjoint problems

Global error estimates give a general idea on the quality of an approximate solution and stopping criteria. However, from the viewpoint of engineering purposes, this information is not sufficient. In many cases, it is important to have special (problem–oriented) measures of the error associated with certain subdomains, lines, and points. Characteristics of such a type can be constructed by means of *specially selected linear functionals ℓ_s, $s = 1, 2, ...M$*. Having computable estimates of the quantities $\langle \ell_s, u - u_h \rangle$, one could judge the quality of a finite element approximation u_h in reference to the desired property. Moreover, such estimates also lead to integral type estimates of the error.

A sensible example of such a quantity is given by the functional

$$\langle \ell, v - u \rangle = \int\limits_{\Omega} \varphi_0 \left(v - u \right) dx,$$

where the weight function φ_0 is positive in $\omega \subset\subset \Omega$ and vanishes in $\Omega \setminus \omega$. Such a quantity gives information on the behavior of $v - u$ in ω. One can also construct other quantities characterizing $v - u$ along certain lines and also at some points of Ω. In other words, by choosing φ_0, one can create a wide spectrum of "quantities of interest" that complement the information obtained by global error estimates. This direction is often called *goal-oriented* error estimation methods (see, e.g., [3, 152]), because ℓ is chosen to fit a special goal.

Certainly, an estimate of the desired quantity is easy to obtain with the help of the obvious relation

$$|\langle \ell, u - u_h \rangle| \leq \|\ell\| \|u - u_h\|_V,$$

provided that the global error $\|v - u\|_V$ and the norm $\|\ell\|$ are estimated. However, in many cases, such an estimate will strongly overestimate the quantity of interest.

A posteriori estimates of the errors evaluated in terms of linear functionals are derived by attracting the *adjoint* boundary-value problem whose right-hand side is formed by the functional ℓ.

Let us represent this idea in the simplest form. Consider a system of linear simultaneous equations

$$Au = f,$$

where A is a positive definite matrix and f is a given vector. Let v be an approximate solution. We need to estimate the quantity $\ell \cdot (u - v)$, where ℓ is a given vector. Define u_ℓ by the relation

$$A^\star u_\ell = \ell$$

in which $A^\star$ is the matrix adjoint to A. Then,

$$\ell \cdot (u - v) = A^\star u_\ell \cdot u - \ell \cdot v = f \cdot u_\ell - \ell \cdot v = (f - Av) \cdot u_\ell$$

Hence, an estimate of $\ell \cdot (u - v)$ can be obtained fairly easily provided that u_ℓ is defined. Certainly, this quantity cannot completely characterize the error (e.g., $\ell \cdot (u - v)$ vanishes if the residual $R(v) = f - Av$ is orthogonal to u_ℓ). Therefore, it is desirable to obtain estimates for various functionals ℓ_s, which amounts to solving several adjoint problems.

In this section, we briefly describe a version of the approach to the a posteriori error estimation for finite element approximations based on the consideration of an adjoint problem. As before, we show this with a the paradigm of the problem (4.1.25). The reader interested in a deep investigation in this subject can find a thorough exposition of these methods in the book [26] (see also [27] and other publications given in the literature comments at the end of this chapter).

Consider again the problem (4.1.25). Now, we do not assume that A is a symmetric matrix and denote its adjoint matrix by $A^\star$. Also, it is convenient to denote the solution of (4.1.25) by u_f, i.e

$$\int_\Omega A\nabla u_f \cdot \nabla w\, dx = \int_\Omega fw\, dx, \quad \forall w \in V_0(\Omega).$$

The *adjoint* problem is to find $u_\ell \in V_0(\Omega)$ such that

$$\int_\Omega A^\star \nabla u_\ell \cdot \nabla w\, dx = \int_\Omega lw\, dx, \quad \forall w \in V_0(\Omega).$$

Let Ω be divided into a number of elements T_i, $i = 1, 2, ...N$. Given approximations on the elements, we define a finite-dimensional subspace $V_{0h} \in V_0(\Omega)$ and the Galerkin approximations u_{fh} and $u_{\ell h}$ as functions satisfying the relations

$$\int_\Omega A\nabla u_{fh} \cdot \nabla w_h\, dx = \int_\Omega fw_h dx, \quad \forall w_h \in V_{0h},$$

$$\int_\Omega A^\star \nabla u_{\ell h} \cdot \nabla w_h\, dx = \int_\Omega \ell w_h dx, \quad \forall w_h \in V_{0h}.$$

Since

$$\int_\Omega \ell(u_f - u_{fh})dx = \int_\Omega A^\star \nabla u_\ell \cdot \nabla(u_f - u_{fh})dx$$

and

$$\int_\Omega A^\star \nabla u_{\ell h} \cdot \nabla(u_f - u_{fh})dx = 0,$$

we arrive at the relation

$$\int_\Omega \ell(u_f - u_{fh})dx = \int_\Omega A^\star \nabla(u_\ell - u_{\ell h}) \cdot \nabla(u_f - u_{fh})dx \quad (4.3.1)$$

whose right-hand side is expressed in the form

$$\sum_{i=1}^{N} \int_{T_i} A\nabla(u_f - u_{fh}) \cdot \nabla(u_\ell - u_{\ell h})\, dx =$$

$$= \sum_{i=1}^{N} \left\{ -\int_{T_i} \text{div}\,(A\nabla(u_f - u_{fh}))\,(u_\ell - u_{\ell h})\, dx + \right.$$

$$\left. +\frac{1}{2}\int_{\partial T_i} \text{j}\,(\nu_i \cdot A\nabla(u_f - u_{fh}))\,(u_\ell - u_{\ell h})\, ds \right\}.$$

This relation implies the estimate

$$\int_\Omega l(u_f - u_{fh})dx = \sum_{i=1}^{N}\Big\{ \|\mathrm{div}\, A\nabla(u_f - u_{fh})\|_{2,T_i}\, \|u_\ell - u_{\ell h}\|_{2,T_i} +$$

$$+\tfrac{1}{2}\,\|\mathrm{j}(\nu_i \cdot A\nabla(u_f - u_{fh}))\|_{2,\partial T_i}\, \|u_\ell - u_{\ell h}\|_{2,\partial T_i}\Big\} =$$

$$= \sum_{i=1}^{N}\Big\{ \|f + \mathrm{div}\, A\nabla u_{fh}\|_{2,T_i}\, \|u_\ell - u_{\ell h}\|_{2,T_i} +$$

$$+\tfrac{1}{2}\,\|\mathrm{j}(\nu_i \cdot A\nabla(u_{fh}))\|_{2,\partial T_i}\, \|u_\ell - u_{\ell h}\|_{2,\partial T_i}\Big\}. \qquad (4.3.2)$$

We see that the principal terms of this estimate are the same as in (4.2.16), but the weights are given by the norms of $u_\ell - u_{\ell h}$. Their values depend on the properties of the adjoint problem. Assume that the latter has a sufficiently regular solution, and $u_{\ell h}$ is constructed by piecewise affine continuous approximations. Then the norms $\|u_\ell - u_{\ell h}\|_{T_i}$ and $\|u_\ell - u_{\ell h}\|_{2,\partial T_i}$ are estimated by the quantities $h^\alpha |u_\ell|_{2,2,T_i}$ with $\alpha = 1$ and $1/2$ and multipliers $\hat{c}_i$ and $\hat{c}_{ij}$, respectively. In this case, we obtain an estimate with constants defined by the standard $H^2 \to V_{0h}$ interpolation operator whose evaluation is easier than that of the constants arising in the $H^1 \to V_{0h}$ interpolation procedures. Moreover, if for any $\ell \in L^2(\Omega)$ the solution u_ℓ lies in H^2, then it is easy to derive an estimate of $\|u_f - u_{fh}\|_{2,\Omega}$. Indeed,

$$\|u_f - u_{fh}\|_{2,\Omega} = \sup_{\ell \in L^2} \frac{\int_\Omega \ell(u_f - u_{fh})}{\|\ell\|_{2,\Omega}}.$$

By (4.3.2), we estimate the numerator of this fraction as a sum of element residuals and interelement jumps with multipliers $\hat{c}_i$ and $\hat{c}_{ij}$ and the global multiplier $\|u_\ell\|_{2,2,\Omega}$. The latter norm is subject to the regularity estimate

$$\|u_\ell\|_{2,2,\Omega} \leq c_\Omega \|\ell\|_{2,\Omega},$$

which can also be viewed as a stability relation in the adjoint problem. Then, $\|\ell\|_{2,\Omega}$ occurs in the numerator and the fraction is reduced to an estimate of the form similar to (4.2.14), which, however, has other interpolation constants and includes a new "stability" constant. Estimates of such a type for various problems are analyzed in [112, 113, 114, 68] and other papers cited therein.

In [3, 152] and some other publications, it was suggested another way of evaluating errors in terms of linear functionals. It is based on the relation (4.3.1) and the algebraic identity

$$b(\sigma, \eta) = \tfrac{1}{4}b\left(\zeta\sigma + \tfrac{\eta}{\zeta}, \zeta\sigma + \tfrac{\eta}{\zeta}\right) - \tfrac{1}{4}b\left(\zeta\sigma - \tfrac{\eta}{\zeta}, \zeta\sigma - \tfrac{\eta}{\zeta}\right), \qquad (4.3.3)$$

which is valid for any symmetric bilinear form b and $\zeta \in \mathbb{R}$. By this identity, the right-hand side of (4.3.1) is expressed in the form

$$\frac{1}{4} \int_\Omega A^\star \nabla(\zeta e_\ell + \tfrac{e_f}{\zeta}) \cdot \nabla(\zeta e_\ell + \tfrac{e_f}{\zeta}) dx -$$

$$- \frac{1}{4} \int_\Omega A^\star \nabla(\zeta e_\ell - \tfrac{e_f}{\zeta}) \cdot \nabla(\zeta e_\ell - \tfrac{e_f}{\zeta}) dx,$$

where $e_\ell = u_\ell - u_{\ell h}$ and $e_f = u_f - u_{fh}$. Further, these two integrals are estimated by the equilibrated residual method based upon solving local problems with equilibrated Neumann type boundary conditions (cf. 2.3.2).

In Section 5, we present a new estimation method for goal–oriented functionals that uses the post-processing of approximate solutions u_ℓ and u_f. Further, in Chapter 6, these functionals will also be estimated by functional type a posteriori error bounds.

4.4   Methods based upon post-processing of finite element approximations

Finite element solutions are usually smooth at the interior points of elements, but in the entire domain they have a restricted regularity. For example, the solutions obtained by piecewise affine continuous approximations have piecewise constant gradients, i.e., they belong to $L^\infty(\Omega)$. In the vast majority of cases, this class of functions is much wider than the one that contains gradients of the exact solution. This fact is easy to demonstrate by considering the problem (4.1.25). Indeed, if $f \in L^2(\Omega)$, then the computed vector-valued function $\sigma_h = A\nabla u_h$ lies in $L^\infty(\Omega)$, but $\sigma = A\nabla u$ lies in $H(\Omega, \mathrm{div})$. Moreover, if $u \in H^k(\Omega)$, $k \geq 2$, and the coefficients of A are smooth, then $\sigma \in H^{k-1}$.

In more general terms, the situation that often arises in finite element methods is as follows. For a conforming approximation $u_h \in V_h$, the function Tu_h (where T is a certain linear operator) lies in the space U. However, a priori estimates of the solution guarantee that

$$Tu \in \overline{U} \subset U.$$

Assume that we have a computationally inexpensive continuous mapping G such that

$$G(Tv_h) \in \overline{U}, \qquad \forall v_h \in V_h.$$

Then, we may hope that the function $G(Tv_h)$ that possesses a priori regularity properties would be much closer to Tu than Tv_h. These arguments form

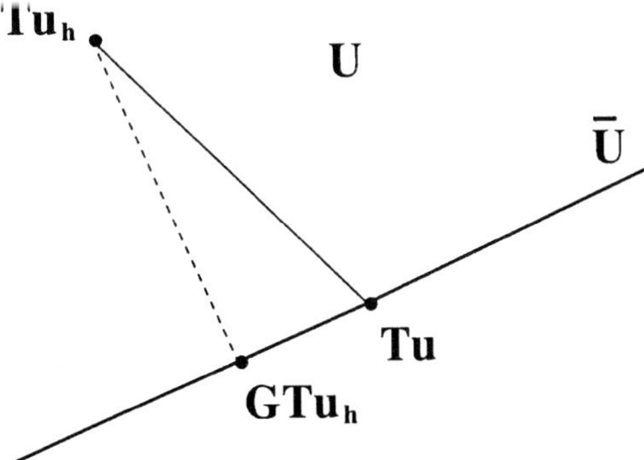

Figure 4.4.1: Post-processing operator and error estimation.

the basis of various *post-processing algorithms* that change a computed solution in accordance with some a priori knowledge of properties of the exact solution. Such algorithms are widely used in advanced numerical methods. If the error caused by violations of a priori regularity properties is dominant and the post-processing operator G is properly constructed, then

$$\|GTu_h - Tu\| \ll \|Tu_h - Tu\|.$$

Schematically, such a situation is illustrated in Fig. 4.4.1. In this case, the explicitly computable norm $\|GTu_h - Tu_h\|$ can be used to evaluate upper and lower bounds of the error. Indeed, assume that there is a positive number $\alpha < 1$ such that for the mapping T the estimate

$$\|GTu_h - Tu\| \leq \alpha \|Tu_h - Tu\| \tag{4.4.1}$$

holds. Then, we have

$$(1 - \alpha) \|Tu_h - Tu\| \leq \|Tu_h - Tu\| - \|GTu_h - Tu\| \leq$$
$$\leq \|GTu_h - Tu_h\| \leq \|GTu_h - Tu\| + \|Tu_h - Tu\| \leq$$
$$\leq (1 + \alpha) \|Tu_h - Tu\|. \tag{4.4.2}$$

Thus, we observe that if the image of Tu_h computed by a certain post-processing operator G is much closer to the exact solution than Tu_h (i.e., if $\alpha \ll 1$), then

$$\|Tu_h - Tu\| \simeq \|GTu_h - Tu_h\|. \tag{4.4.3}$$

In this case, the right-hand side of (4.4.3) can be used as an error indicator. The quality of this indicator is the better the smaller the value of α.

In practice, various types of post-processing operators are used. Below, we consider some of them for the problem (4.1.25), where T is either the gradient operator or the operator $A\nabla$.

4.4.1   Post-processing by averaging

Post-processing operators are often constructed by averaging Tu_h on finite element patches or on the entire domain Ω.

Integral averaging on patches

If Tu_h is a square summable function, then post-processing operators are obtained by various averaging procedures. Let Ω_i be a patch of finite elements, i.e.,

$$\overline{\Omega}_i = \bigcup T_{ij}, \quad j = 1, 2, ...M_i.$$

Let $P^k(\Omega_i, \mathbb{R}^n)$ be a subspace of $\overline{U}$ that consists of vector-valued polynomial functions of degrees less than or equal to k. Define $g_i \in P^k(\Omega_i, \mathbb{R}^n)$ as a solution of the following minimization problem:

$$\inf_{g \in P^k(\Omega_i, \mathbb{R}^n)} \int_{\Omega_i} |g - Tu_h|^2 \, dx. \qquad (4.4.4)$$

The minimizer g_i can be used to define the values of an averaged function at some points (nodes). Further, these values are utilized by a prolongation procedure that defines an averaged function

$$GTu_h : \Omega \to \mathbb{R}.$$

Consider the simplest case. Let T be the operator ∇ and u_h be a piecewise affine continuous function. Then,

$$\nabla u_h \in P^0(\Omega_i, \mathbb{R}^n) \quad \text{on } T_{ij}.$$

We denote the respective values of ∇u_h by $(\nabla u_h)_{ij}$. Set $k = 0$ and find $g_i \in P^0$ such that

$$\int_{\Omega_i} |g_i - \nabla u_h|^2 \, dx = \inf_{g \in P^0(\Omega_i)} \int_{\Omega_i} |g - \nabla u_h|^2 \, dx =$$

$$= \inf_{g \in P^0(\Omega_i)} \left\{ |g|^2 |\Omega_i| - 2g \cdot \sum_{j=1}^{M_i} (\nabla u_h)_{ij} |T_{ij}| + \sum_{j=1}^{M_i} |(\nabla u_h)_{ij}|^2 |T_{ij}| \right\}.$$

$$(4.4.5)$$

It is easy to see that g_i is given by a weighted sum of $(\nabla u_h)_{ij}$, namely,

$$g_i = \sum_{j=1}^{M_i} \frac{|T_{ij}|}{|\Omega_i|} (\nabla u_h)_{ij}. \tag{4.4.6}$$

Now, we define the value of $G(\nabla u_h)(x_i)$ as g_i. Repeat this procedure for all nodes and define the vector-valued function $G\nabla(u_h)$ by the piecewise affine prolongation of these values.

If the mesh is regular and all the quantities $|T_{ij}|$ are equal, then (4.4.6) reads

$$g_i = \sum_{j=1}^{M_i} \frac{1}{M_i} (\nabla u_h)_{ij}. \tag{4.4.7}$$

Various averaging formulas of this type are represented in the form

$$g_i = \sum_{j=1}^{M_i} \lambda_{ij} (\nabla u_h)_{ij}, \quad \sum_{j=1}^{M_i} \lambda_{ij} = 1, \tag{4.4.8}$$

where the quantities λ_{ij} are the weight factors. For internal nodes, they may be taken in accordance with (4.4.6) or defined by the rule

$$\lambda_{ij} = \frac{|\gamma_{ij}|}{2\pi},$$

where $|\gamma_{ij}|$ is the radian measure of the angle of T_{ij} associated with the node i. However, if a node belongs to the boundary, then it is better to choose special weights. Their values depend on the mesh and on the type of the boundary. The reader can find a detailed consideration of this question in [101].

Discrete averaging on patches

Consider the problem

$$\inf_{g \in \mathbb{P}^k(\Omega_i)} \sum_{s=1}^{m_i} |g(x_s) - Tu_h(x_s)|^2,$$

where the points x_s are specially selected in Ω_i. Usually, the points x_s are the so-called *superconvergent* points (see below). Let $g_i \in \mathbb{P}^k(\Omega_i)$ be the minimizer of this problem. Then, we define the values of GTu_h at some points and further use an inexpensive prolongation procedure to define GTu_h at any point.

If $k = 0$, then (cf. (4.4.6))

$$g_i = \frac{1}{m_i} \sum_{s=1}^{m_i} Tu_h(x_s). \tag{4.4.9}$$

Global averaging

In many cases, an efficient averaging operator is obtained if local minimization problems on patches are replaced by a global problem. Assume that Tu_h is a square summable function and solve the problem: find $\bar{g}_h \in V_h(\Omega) \subset \overline{U}$ such that

$$\|\bar{g}_h - Tu_h\|_\Omega^2 = \inf_{g_h \in V_h(\Omega)} \|g_h - Tu_h\|_\Omega^2. \tag{4.4.10}$$

The function $\bar{g}_h$ can be viewed as GTu_h. Very often $\bar{g}_h$ is a better image of Tu than the functions obtained by local procedures. Moreover, on can await that mathematical justifications of the methods based on advanced averaging procedures can be performed under weaker assumptions what makes them applicable to a wider class of problems (see, e.g., [43]).

4.4.2   Superconvergence

Let u_h be a Galerkin approximation of u computed on V_h. For piecewise affine approximations of the problem (4.1.25), we have the estimate

$$\|\nabla(u - u_h)\|_{2,\Omega} \le c_1 h. \tag{4.4.11}$$

Here, c_1 depends on the H^2-norm of the exact solution and on properties of the sampling $\mathcal{T}_h$. Under apropriate regularity assumptions similar estimates also hold for the maximum-norm. Such estimates show the maximal convergence rates that might be generally observed the type of approximations under consideration. However, it was discovered (see, e.g., [155, 216, 229]) that in many cases this rate may be higher. For example, the estimate

$$|u(x_s) - u_h(x_s)| \le Ch^{2+\sigma} \qquad \sigma > 0 \tag{4.4.12}$$

may hold at a certain point x_s, which is called a *superconvergent* point for the sequence $\{u_h\}$. Certainly, the location of superconvergent points (and their existence) strongly depends on the structure of $\mathcal{T}_h$.

Analogously, an operator G possesses a *superconvergence* property in $\omega \subset \Omega$ if

$$\|\nabla u - G\nabla u_h\|_{2,\omega} \le c_2 h^{1+\sigma}, \tag{4.4.13}$$

where the constant c_2 depends on higher norms of u and the structure of $\mathcal{T}_h$. For the problem (4.1.25) estimates of such a type can be found, e.g., in [101, 126, 127]. For example, let $\Omega \in \mathbb{R}^n$, $n = 2$, be a domain with piecewise smooth boundary, $A \in C^2(\Omega, \mathbb{M}^{n \times n})$, $u \in H^3(\Omega_1)$, where $\omega \subset\subset \Omega_1 \subset\subset \Omega$ and ω with its neighborhood is covered by congruent triangles. If the operator G is constructed in accordance with (4.4.6), then (4.4.13) holds with $\sigma = 1$ ([127]).

Nowadays it is a matter of common knowledge that convergent sequences of Galerkin approximations often possess certain superconvergence properties (see, e.g., [18, 213, 216, 219]). By exploiting these properties, one can usually construct a simple post-processing operator G satisfying (4.4.1). In this estimate, the value of α depends on h, which improves the quality of the error indication as h decreases. For this reason, error indicators of such a type are widely used in engineering computations (see, e.g., [17, 29, 30, 224, 226, 227, 228]).

4.4.3 Post-processing by equilibration

For a solution of the problem (4.1.25), it is a priori known that

$$\operatorname{div} \sigma + f = 0, \qquad (4.4.14)$$

where $\sigma = A\nabla u$. This suggests an idea to construct an operator G such that

$$\operatorname{div}(\mathrm{G}(A\nabla u_h)) + f = 0. \qquad (4.4.15)$$

If G possesses additional properties (linearity, boundedness), then we may hope that the function $GA\nabla u_h$ is closer to σ than $A\nabla u_h$ and use the quantity $\|A\nabla u_h - GA\nabla u_h\|$ as an error indicator.

This idea can be applied to an important class of problems

$$\Lambda^\star \mathrm{T}u + f = 0, \qquad \mathrm{T}u = \mathcal{A}\Lambda u, \qquad (4.4.16)$$

where $\mathcal{A}$ is a positive definite operator, Λ is a linear continuous operator, and $\Lambda^\star$ is the adjoint operator.

Note that in continuum mechanics, equations of the type (4.4.14) are referred to as the *equilibrium equations*. Therefore, it is natural to call an operator G (whose range consists of functions satisfying these equations) an *equilibration* operator. One way for constructing such an operator is as follows. Assume that we know the functions σ_i, $i = 0, 1, 2, ...M$, that satisfy the conditions

$$\operatorname{div} \sigma_0 + f = 0, \quad \operatorname{div} \sigma_i = 0, \quad i = 1, 2, .., M.$$

In this case, G may be defined as the projection to the subspace $\mathrm{Span}(\sigma_0, \sigma_1, ..., \sigma_M)$.

It should be outlined that by G we can also obtain a *guaranteed upper bound* of the error. This fact is very easy to establish in the simplest case of $A = \mathbb{I}$. Indeed,

$$\int_\Omega \nabla(u - u_h) \cdot \nabla(u - u_h)dx = \int_\Omega (\sigma - \nabla u_h) \cdot \nabla(u - u_h)dx,$$

where σ is a function in $H(\mathrm{div}, \Omega)$ satisfying (4.4.14). Then, we arrive at the estimate

$$\|\nabla(u - u_h)\| \leq \inf_{\sigma:\,\mathrm{div}\,\sigma + f = 0} \|\sigma - \nabla u_h\|, \tag{4.4.17}$$

which shows that

$$\|\nabla(u - u_h)\| \leq \|\mathrm{G}(A\nabla u_h) - \nabla u_h\|. \tag{4.4.18}$$

This type a posteriori error estimates have been proposed and investigated in [128, 129, 130] and some other papers. They lead to estimators that measure approximation errors by computing errors in constitutive relations. The major difficulty of this approach consists of finding a suitable projection a the set of functions exactly satisfying (4.4.14). In general, this may be an uneasy task, which makes a straightforward employment of (4.4.18) too expensive. In practice, various procedures that generate functions satisfying (4.4.15) or functions very close to equilibrated fields are used. Equilibrium finite element approximations in linear elasticity can be constructed with the help of the method exposed in [111].

4.5　Error estimation of goal-oriented quantities by gradient averaging techniques

4.5.1　General scheme

Post-processing procedures creates effective a posteriori estimates not only for integral error estimates, but also for errors evaluated by goal-oriented functionals (cf. Section 4.3). In the present section, we describe such a method suggested in [123, 124] for a class of linear elliptic type problems. Below, we consider it for the problem whose generalized solution is defined as a function $u \in V_0$ that meets the relation

$$(\mathcal{A}\Lambda u, \Lambda w) = \langle f, w \rangle, \quad \forall w \in V_0. \tag{4.5.1}$$

In (4.5.1), $\mathcal{A} \in \mathcal{L}(U,U)$, $\Lambda \in \mathcal{L}(V,U)$, V is a Banach space, V_0 is a subspace of V, $f \in V_0^*$, U is a Hilbert space equipped with scalar product $(\cdot,\cdot)$, and $\langle\cdot,\cdot\rangle$ denotes the duality pairing of V_0 and V_0^*. We assume that the operators $\mathcal{A}$ and Λ satisfy the conditions

$$c_1\|y\|_U^2 \le (\mathcal{A}y,y) \le c_2\|y\|_U^2, \forall y \in U, \tag{4.5.2}$$

$$c_3\|w\|_V \le \|\Lambda w\|_U, \quad \forall w \in V_0, \tag{4.5.3}$$

Henceforth, the problem (4.5.1) is called the *primal problem* (or Problem $\mathcal{P}$).

Let ℓ be an element of V_0^*. Our aim is to derive estimates of the quantity

$$\langle \ell, u - \widetilde{u} \rangle,$$

where $\widetilde{u}$ is an arbitrary element from V_0. For this purpose, we use the *adjoint problem* (or Problem $\mathcal{P}_a$): find $v \in V_0$ such that

$$(\mathcal{A}^\star \Lambda v, \Lambda w) = \langle \ell, w \rangle \quad \forall w \in V_0, \tag{4.5.4}$$

where $\mathcal{A}^\star$ is the operator adjoint to $\mathcal{A}$, i.e., $(\mathcal{A}y, z) = (y, \mathcal{A}^\star z)$ for all $y, z \in U$.

If $\mathcal{A}$ is a self-adjoint operator, then both problems are associated with a functional of the type

$$J(w) \;=\; \frac{1}{2}(\mathcal{A}\Lambda w, \Lambda w) + \langle \mu, w \rangle.$$

Under the conditions (4.5.2) and (4.5.3), this functional has a unique minimizer on V_0 for any $\mu \in V_0^*$.

Proposition 4.5.1. *Let u and v be solutions of problems $\mathcal{P}$ and $\mathcal{P}_a$, respectively. Then,*

$$\langle \ell, u - \widetilde{u} \rangle := E(\widetilde{u},\widetilde{v}) = E_0(\widetilde{u},\widetilde{v}) + E_1(\widetilde{u},\widetilde{v}), \tag{4.5.5}$$

where

$$E_0(\widetilde{u},\widetilde{v}) = \langle f, \widetilde{v} \rangle - (\mathcal{A}\Lambda\widetilde{u}, \Lambda\widetilde{v}) \tag{4.5.6}$$

and

$$E_1(\widetilde{u},\widetilde{v}) = (\mathcal{A}\Lambda(u - \widetilde{u}), \Lambda(v - \widetilde{v})). \tag{4.5.7}$$

Proof. By (4.5.4), we have

$$\langle \ell, u - \widetilde{u} \rangle = (\mathcal{A}^\star \Lambda v, \Lambda(u - \widetilde{u})).$$

Since

$$(\mathcal{A}^\star \Lambda v, \Lambda(u - \widetilde{u})) = (\mathcal{A}\Lambda(u - \widetilde{u}), \Lambda v),$$

where u satisfies (4.5.1), we find that

$$\langle \ell, u - \widetilde{u} \rangle = (\mathcal{A}\Lambda(u - \widetilde{u}), \Lambda(v - \widetilde{v})) + \langle f, \widetilde{v} \rangle - (\mathcal{A}\Lambda\widetilde{u}, \Lambda\widetilde{v}).$$

$$\square$$

In (4.5.5), the term $E_0(\widetilde{u}, \widetilde{v})$ is explicitly computable, whereas the term $E_1(\widetilde{u}, \widetilde{v})$ contains unknown solutions of the problems $(\mathcal{P})$ and $(\mathcal{P}_a)$. Let V_h and V_τ be two finite-dimensional subspaces of V_0, and let $\widetilde{u} = u_h$ and $\widetilde{v} = v_\tau$, where u_h and v_τ are solutions of the problems

$$(\mathcal{A}\Lambda u_h, \Lambda w_h) = \langle f, w_h \rangle \quad \forall w_h \in V_h, \tag{4.5.8}$$

$$(\mathcal{A}^\star \Lambda v_\tau, \Lambda w_\tau) = \langle \ell, w_\tau \rangle \quad \forall w_\tau \in V_\tau. \tag{4.5.9}$$

In the particular case $V_h \equiv V_\tau$, the relation (4.5.8) implies that $E_0(u_h, v_\tau) = 0$ so that there remains the term containing the product of the (unknown) energy errors. On the contrary, for noncoinciding spaces both terms E_0 and E_1 are present. Moreover, it is easy to show that the term E_0 dominates if $\widetilde{v}$ is close to v. Indeed, if $\widetilde{v} \to v$ in V, then $\Lambda(v - \widetilde{v}) \to 0$ in U, so that

$$E_1(\widetilde{u}, \widetilde{v}) \to 0.$$

However,

$$E_0(\widetilde{u}, \widetilde{v}) + E_1(\widetilde{u}, \widetilde{v}) = \langle \ell, u - \widetilde{u} \rangle \neq 0.$$

Hence, if $\widetilde{v}$ is sufficiently close to v, then the directly computable term $E_0(\widetilde{u}, \widetilde{v})$ contains the major part of the estimated quantity.

The case of nonhomogeneous boundary conditions in Problem $\mathcal{P}$ is reduced to the case under consideration. Indeed, assume that we find a function

$$u \in V_0 + u_0 = \{w \in V \mid w = w_0 + u_0,\ u_0 \in V,\ w_0 \in V_0\}$$

that satisfies (4.5.1). Set $\overset{\circ}{u} = u - u_0$ and represent Problem $\mathcal{P}$ as follows: find $\overset{\circ}{u} \in V_0$ such that

$$(\mathcal{A}\Lambda\overset{\circ}{u}, \Lambda w) = \langle \overset{\circ}{f}, w \rangle \quad \forall w \in V_0,$$

where

$$\langle \overset{\circ}{f}, w \rangle = \langle f, w \rangle - (\mathcal{A}\Lambda u_0, \Lambda w).$$

Let $\widetilde{u} \in V_0 + u_0$ be an approximation of u. Then, $\overset{\circ}{w} = \widetilde{u} - u_0 \in V_0$ is an approximation of $\overset{\circ}{u}$. By (4.5.5) we obtain

$$\langle \ell, u - \widetilde{u} \rangle = \langle \ell, \overset{\circ}{u} - \overset{\circ}{w} \rangle = \langle \overset{\circ}{f}, \widetilde{v} \rangle - (\mathcal{A}\Lambda \overset{\circ}{w}, \Lambda \widetilde{v}) + E_1(\overset{\circ}{w}, \widetilde{v}).$$

Certainly, finding a sharp approximation of the adjoint problem may require high computational costs. A more economical way is to use an approximate solution of the adjoint problem having approximately the same quality as the approximate solution of the primal one and to recover the unknown functions Λu and Λv by some post-processing techniques. Let G_h and G_τ be averaging operators defined on V_h and V_τ, respectively. Replace $E(u_h, v_\tau)$ by the directly computable functional

$$\widetilde{E}(u_h, v_\tau) := E_0(u_h, v_\tau) + \widetilde{E}_1(u_h, v_\tau), \qquad (4.5.10)$$

where

$$\widetilde{E}_1(u_h, v_\tau) = (\mathcal{A}(\mathrm{G}_h(\Lambda u_h) - \Lambda u_h), \mathrm{G}_\tau(\Lambda v_\tau) - \Lambda v_\tau).$$

If the operators G_h and G_τ perform a proper recovery of Λu_h and Λv_τ, then it is natural to expect that the difference between $E_1(u_h, v_\tau)$ and $\widetilde{E}(u_h, v_\tau)$ is given by higher order terms and, thus, the latter quantity can successfully be used instead of E_1. For a class of boundary-value problems, this observation was theoretically and numerically justified in [123]. Below, we present a concise summary of these results.

It is worth noting that the efficiency of the above approach strongly improves if one is interested not in a single solution of the primal problem but analyzes a set of approximate solutions obtained for various boundary conditions and right-hand sides (such a situation often arises in the engineering design). In this case, the adjoint problem must be solved only once, and its solution $\widetilde{v}$ can further be used in the estimate (4.5.5) for various primal problems.

4.5.2 Example

Let Ω be a bounded domain in $\mathbb{R}^2$ whose boundary consists of two measurable nonintersecting parts $\partial_1\Omega$ and $\partial_2\Omega$. Consider the problem

$$\begin{aligned} \mathrm{div}(A\nabla u) + f &= 0 && \text{in} \quad \Omega, \\ u &= 0 && \text{on} \quad \partial_1\Omega, \\ A\nabla u \cdot \nu &= g && \text{on} \quad \partial_2\Omega, \end{aligned}$$

where ν is the outward normal to $\partial\Omega$ and $A = \{a_{ij}(x)\}_{i,j=1}^2$ is a symmetric positive definite matrix with smooth entries.

Assum that

$$f \in L^2(\Omega), \qquad g \in L^2(\partial_2\Omega),$$

Now, the problem (4.5.1) has the form

$$\int_\Omega A\nabla u \cdot \nabla w dx = \int_\Omega fw dx + \int_{\partial_2\Omega} gw ds, \quad \forall w \in V_0. \qquad (4.5.11)$$

In this case, $U = L^2(\Omega, \mathbb{R}^2)$, $V = H^1(\Omega)$,

$$V_0 = \{v \in H^1(\Omega) \mid v = 0 \text{ on } \partial_1\Omega\},$$

and Problem $\mathcal{P}_a$ is to find $v \in V_0$ such that

$$\int_\Omega A\nabla v \cdot \nabla w dx = \langle \ell, w \rangle \quad \forall w \in V_0. \qquad (4.5.12)$$

By (4.5.5), we obtain

$$\ell(u - \widetilde{u}) \;=\; E_0(\widetilde{u}, \widetilde{v}) + E_1(\widetilde{u}, \widetilde{v}), \qquad (4.5.13)$$

where

$$E_0(\widetilde{u}, \widetilde{v}) = \int_\Omega f\widetilde{v} + \int_{\partial_2\Omega} g\widetilde{v}\, ds - \int_\Omega A\nabla\widetilde{v} \cdot \nabla\widetilde{u} dx$$

and

$$E_1(\widetilde{u}, \widetilde{v}) = \int_\Omega A\nabla(v - \widetilde{v}) \cdot \nabla(u - \widetilde{u}) dx.$$

Let V_h and V_τ be two finite-dimensional subspaces in V_0, constructed by piecewise affine finite element approximations on the triangulations $\mathcal{T}_h$ and $\mathcal{T}_\tau$, respectively, and u_h and v_τ be solutions of the following two problems:

$$\int_\Omega A\nabla u_h \cdot \nabla w_h dx = \int_\Omega fw_h dx + \int_{\partial_2\Omega} gw_h ds, \quad \forall w_h \in V_h, \qquad (4.5.14)$$

$$\int_\Omega A\nabla v_\tau \cdot \nabla w_\tau dx = \langle \ell, w_\tau \rangle, \quad \forall w_\tau \in V_\tau. \qquad (4.5.15)$$

Define

$$G_h : L^\infty(\Omega_h, \mathbb{R}^2) \to H^1(\Omega_h, \mathbb{R}^2)$$

and

$$G_\tau : L^\infty(\Omega_\tau, \mathbb{R}^2) \to H^1(\Omega_\tau, \mathbb{R}^2)$$

as gradient averaging operators on patches satisfying (4.4.6). Now, the functional (4.5.10) has the form

$$\widetilde{E}(w_h, w_\tau) = E_0(w_h, w_\tau) + \widetilde{E}_1(w_h, w_\tau), \tag{4.5.16}$$

where

$$\widetilde{E}_1(w_h, w_\tau) = \int_{\Omega_h} (\nabla w_h - G_h(\nabla w_h)) \cdot (\nabla w_\tau - G_\tau(\nabla w_\tau)) dx.$$

Let us qualitatively estimate the asymptotic order of the terms E_0, E_1, and $\widetilde{E}_1$. We have

$$|E_0(u_h, v_\tau)| = \left| \int_\Omega A(\nabla u - \nabla u_h) \cdot \nabla v_\tau \, dx \right| \le$$
$$\le \|A\|_{\infty,\Omega} \|\nabla v_\tau\|_{2,\Omega} \|\nabla(u - u_h)\|_{2,\Omega} \le$$
$$\le \|A\|_{\infty,\Omega} \left(\|\nabla(v - v_\tau)\|_{2,\Omega} + \|\nabla v\|_{2,\Omega} \right) \|\nabla(u - u_h)\|_{2,\Omega}.$$

Since $v_\tau \to v$ in V, the term in brackets is bounded and we see that

$$|E_0(u_h, v_\tau)| \le ch|u|_{2,2,\Omega}, \tag{4.5.17}$$

provided that $u \in H^2(\Omega)$. If the solutions of Problems $\mathcal{P}$ and $\mathcal{P}_a$ are sufficiently regular, then the terms $E_1(u_h, v_\tau)$ and $\widetilde{E}_1(u_h, v_\tau)$ are asymptotically smaller ones. Indeed,

$$|E_1(u_h, v_\tau)| \le ch\tau|u|_{2,2,\Omega}|v|_{2,2,\Omega}$$

and

$$|\widetilde{E}_1(u_h, v_\tau)| = \left| \int_{\Omega_h} A(\nabla u_h - G_h(\nabla u_h)) \cdot (\nabla v_\tau - G_\tau(\nabla v_\tau)) dx \right| \le$$
$$\le \|A\|_{\infty,\Omega} \|\nabla u_h - G_h(\nabla u_h)\|_{2,\Omega} \|\nabla v_\tau - G_\tau(\nabla v_\tau)\|_{2,\Omega}.$$

Since

$$\|\nabla u_h - G_h(\nabla u_h)\|_{2,\Omega} \le \|\nabla u - G_h(\nabla u_h)\|_{2,\Omega} + \|\nabla u_h - \nabla u\|_{2,\Omega},$$
$$\|\nabla u_\tau - G_\tau(\nabla u_\tau)\|_{2,\Omega} \le \|\nabla u - G_\tau(\nabla u_\tau)\|_{2,\Omega} + \|\nabla u_\tau - \nabla u\|_{2,\Omega},$$

the term $\widetilde{E}_1$ has the same order $h\tau$ if both gradient averaging operators possess superconvergence properties. In this case, $E_0(u_h, v_\tau)$ contains the major part of the quantity $\langle \ell, u - u_h \rangle$ (except the case $V_h \equiv V_\tau$, in which this term vanishes).

It remains to compare $E_1(u_h, v_\tau)$ with $\widetilde{E}_1(u_h, v_\tau)$. This question was considered in [123] under additional assumptions necessary to guarantee superconvergence properties of the gradient averaging operators. There it was shown that for sufficiently small h and τ,

$$|E_1(u_h, v_\tau) - \widetilde{E}_1(u_h, v_\tau)| \le$$
$$\le c(h\tau^{\frac{3}{2}} + \tau h^{\frac{3}{2}} + h\tau(h + \tau)^{\frac{1}{2} - \frac{1}{m}} + \tau^2) + \mu(h, \tau),$$

where m is any positive integer greater than 2 and $\mu(h, \tau)$ contains higher order terms.

The asymptotic behavior of the error indicator $\widetilde{E}$ suggests the following numerical strategy.

1. Define $\mathcal{T}_\tau$ that is sufficiently fine and accounts particular features of the functional ℓ. For example, if

$$\langle \ell, u - \widetilde{u} \rangle = \int_\Omega \varphi(u - \widetilde{u}) \, dx,$$

   where the weight φ is a locally supported function, i.e.,

$$\mathrm{supp}\varphi \subset \omega \subset\subset \Omega,$$

   then the mesh should have more degrees of freedom in a neighborhood of ω. Solve Problem $\mathcal{P}_a$ on V_τ and find v_τ;

2. Define V_h, solve the primal problem, and find u_h;

3. Compute $E_0(u_h, v_\tau)$ and use post-processed values of ∇u_h and ∇v_τ to evaluate $\widetilde{E}_1(u_h, v_\tau)$.

It is worth noting that if we have a series of approximate solutions u_h^k, $k = 1, 2, ...m$, computed for different right-hand sides and boundary conditions, then the operation of in part 1 must be performed only once.

4.5.3 Error estimates in terms of seminorms

Estimates of $u - \widetilde{u}$ computed on a set of linear functionals can be used for an evaluation of this difference in some seminorms.

Let ω be a chosen collection of simplexes and $\{\varphi_1, \varphi_2, ..., \varphi_d\}$ be a set of functions that vanish in $\Omega \setminus \omega$. By the method described above, we can estimate the quantities

$$I_{\varphi_s} := \int_\omega \varphi_s(x)(u(x) - \widetilde{u}(x)) \; dx.$$

However, we are also interested in estimating a certain norm of $u - \widetilde{u}$, for example, the norm

$$\|u - \widetilde{u}\|_{2,\omega}^2 = \int_\omega |u(x) - \widetilde{u}(x)|^2 \; dx.$$

Note that

$$\|u - \widetilde{u}\|_{2,\omega} = \sup_{\eta \in L^2(\omega)} \frac{\int_\omega \eta(u - \widetilde{u})dx}{\|\eta\|_{2,\omega}},$$

and the supremum is attained if $\eta = u - \widetilde{u}$. Thus, if the difference $u - \widetilde{u}$ is known to belong to a certain set $\Xi(\omega) \subset L^2(\omega)$, then the problem is reduced to the evaluation of the *seminorm*

$$|u - \widetilde{u}|_\Xi := \sup_{\eta \in \Xi(\omega)} \frac{\int_\omega \eta(u - \widetilde{u})dx}{\|\eta\|_{2,\omega}}.$$

In general, a seminorm does not provide complete information on the error, because the relation $|u - \widetilde{u}|_\Xi = 0$ does not mean that $u = \widetilde{u}$. Nevertheless, seminorms may give a useful unformation if Ξ contains sufficiently large amount of linearly independent functions. Let

$$\Xi = \mathrm{Span}\{\varphi_1, \varphi_2, ..., \varphi_d\}.$$

Then,

$$|u - \widetilde{u}|_\Xi = \sup_{a_i \in \mathbb{R}} \frac{\sum_{i=1}^{d} a_i \int_\omega \varphi_i(u - \widetilde{u})dx}{\left(\sum_{i,j=1}^{d} a_i a_j \int_\omega \varphi_i \varphi_j dx \right)^{\frac{1}{2}}}. \tag{4.5.18}$$

This problem is equivalent to the following one: find the quantity

$$\kappa^2(\omega, \Xi) = \inf_{a \in Q} B \, a \cdot a, \qquad (4.5.19)$$

where $a = (a_1, a_2, ..., a_d)$, $\boldsymbol{I} = (I_{\varphi_1}, I_{\varphi_2}, ..., I_{\varphi_d})$,

$$Q = \{a \in \mathbb{R}^d \mid a \cdot \boldsymbol{I} = 1\},$$

and

$$B = \{b_{ij}\}_{i,j=1,d}, \quad b_{ij} = \int_\omega \varphi_i \varphi_j \, dx.$$

The problem (4.5.19) has a minimax form

$$\kappa^2(\omega, \Xi) = \inf_{a \in \mathbb{R}^d} \sup_{\lambda \in \mathbb{R}} \{B \, a \cdot a + \lambda[(a \cdot \boldsymbol{I}) - 1]\} \geq$$

$$\geq \sup_{\lambda \in \mathbb{R}} \inf_{a \in \mathbb{R}^d} \{B \, a \cdot a + \lambda[(a \cdot \boldsymbol{I}) - 1]\}. \quad (4.5.20)$$

Assume that the functions φ_i are chosen in such a way that B is a positive definite matrix. Then (4.5.20) holds as equality. The minimization problem that stands on the right-hand side in (4.5.20) has a solution

$$a = -\frac{\lambda}{2} B^{-1} \boldsymbol{I}.$$

Therefore,

$$\kappa^2(\omega, \Xi) = \sup_{\lambda \in \mathbb{R}} \left\{ -\frac{\lambda^2}{4} B^{-1} \boldsymbol{I} \cdot \boldsymbol{I} - \lambda \right\} = \frac{1}{B^{-1} \boldsymbol{I} \cdot \boldsymbol{I}},$$

and

$$|u - \widetilde{u}|_\Xi = \frac{1}{\kappa(\omega, \Xi)} = (B^{-1} \boldsymbol{I} \cdot \boldsymbol{I})^{\frac{1}{2}}.$$

We see that the value of $|u - \widetilde{u}|_\Xi$ is not difficult to find, provided that the functions $\{\varphi_s\}$ are properly chosen and the respective quantities I_{φ_s} are defined.

Let us show that if $\widetilde{u} \in \Xi(\omega)$ and u is a sufficiently smooth function, then $|u - \widetilde{u}|_\Xi$ gives an estimate of $\|u - \widetilde{u}\|_{2,\omega}$. Let $\Xi(\omega)$ be a subspace of $L^2(\omega)$ and $\widehat{u}$ be an arbitrary function from $\Xi(\omega)$. Then

$$\|u - \widetilde{u}\|_{2,\omega} = \sup_{\eta \in L^2(\omega)} \frac{\int_\omega ((u - \widehat{u})\eta + (\widehat{u} - \widetilde{u})\eta) \, dx}{\|\eta\|_{2,\omega}} \leq$$

$$\leq \sup_{\eta \in L^2(\omega)} \frac{\int_\omega (\widehat{u} - \widetilde{u})\eta \, dx}{\|\eta\|_{2,\omega}} + \|u - \widehat{u}\|_{2,\omega}. \quad (4.11)$$

Since $\widehat{u} - \widetilde{u} \in \Xi(\omega)$, we have

$$\|u - \widetilde{u}\|_{2,\omega} \leq \sup_{\eta \in \Xi(\omega)} \frac{\int_\omega (\widehat{u} - \widetilde{u})\eta \, dx}{\|\eta\|_{2,\omega}} + \|u - \widehat{u}\|_{2,\omega}.$$

By rearranging again the numerator of the fraction on the right-hand side, we arrive at the estimate

$$\|u - \widetilde{u}\|_{2,\omega} = \sup_{\eta \in \Xi(\omega)} \frac{\int_\omega ((\widehat{u} - u)\eta + (u - \widetilde{u})\eta) \, dx}{\|\eta\|_{2,\omega}} + \|u - \widehat{u}\|_{2,\omega} \leq$$

$$\leq \sup_{\eta \in \Xi(\omega)} \frac{\int_\omega (u - \widetilde{u})\eta \, dx}{\|\eta\|_{2,\omega}} + 2\|u - \widehat{u}\|_{2,\omega} = |u - \widetilde{u}|_\Xi + 2\|u - \widehat{u}\|_{2,\omega},$$

in which $\widehat{u}$ is an arbitrary function from $\Xi(\omega)$. Hence, we see that

$$\|u - \widetilde{u}\|_{2,\omega} \leq |u - \widetilde{u}|_\Xi + 2\delta(u, \Xi(\omega)), \tag{4.5.21}$$

where

$$\delta(u, \Xi(\omega)) = \inf_{\widehat{u} \in \Xi(\omega)} \|u - \widehat{u}\|_{2,\omega}$$

is the distance between u and $\Xi(\omega)$.

If $\Xi(\Omega)$ is a set of polynomials and u is a smooth function then $\delta(u, \Xi(\omega))$ can be estimated with the help of well-known results of approximation theory (see, e.g., [53]).

By similar arguments, we obtain

$$\|u - \widetilde{u}\|_{2,\omega} = \sup_{\eta \in L^2(\omega)} \frac{\int_\omega ((u - \widehat{u})\eta + (\widehat{u} - \widetilde{u})\eta) \, dx}{\|\eta\|_{2,\omega}} \geq$$

$$\geq \sup_{\eta \in L^2(\omega)} \frac{\int_\omega (\widehat{u} - \widetilde{u})\eta \, dx}{\|\eta\|_{2,\omega}} - \|u - \widehat{u}\|_{2,\omega}.$$

Since $\widehat{u} - \widetilde{u}$ belongs to $\Xi(\omega)$, we find that

$$\|u - \widetilde{u}\|_{2,\omega} \geq \sup_{\eta \in \Xi(\omega)} \frac{\int_\omega ((\widehat{u} - u)\eta + (u - \widetilde{u})\eta) \, dx}{\|\eta\|_{2,\omega}} - \|u - \widehat{u}\|_{2,\omega} \geq$$

$$\geq \sup_{\eta \in \Xi(\omega)} \frac{\int_\omega (u - \widetilde{u})\eta \, dx}{\|\eta\|_{2,\omega}} - \sup_{\eta \in L^2(\omega)} \frac{\int_\omega (\widehat{u} - u)\eta \, dx}{\|\eta\|_{2,\omega}} - \|u - \widehat{u}\|_{2,\omega} =$$

$$= |u - \widetilde{u}|_\Xi - 2\|u - \widehat{u}\|_{2,\omega}, \quad \forall \widehat{u} \in \Xi(\omega).$$

From here, it follows that

$$\|u - \widetilde{u}\|_{2,\omega} \geq |u - \widetilde{u}|_{\Xi} - 2\delta(u, \Xi(\omega)). \qquad (4.5.22)$$

By (4.5.21) and (4.5.22), we conclude that the error arising if $\|u - u_h\|_{2,\Omega}$ is replaced by $|u - u_h|_{\Xi}$ depends on the regularity of u and approximation properties of $\Xi(\omega)$.

4.6 Notes for the Chapter

A posteriori error estimation for finite element approximations is the main subject of the books by M. Aintworth and T. Oden [3], I. Babuška and T. Stroboulis [18], W. Bangerth and R. Rannacher [26], K. Eriksson, D. Estep, P. Hansbo and C. Johnson [68], and R. Verfürth [213]. They present various approaches, results of numerical experiments and a wide list of references.

Below, we present a concise overview of some other publications related to the material considered in Chapter 4.

Explicit residual method was originated by I. Babuška and W. C. Rheinboldt [15, 16]. Later it was investigated and extended by many authors. Among other publications associated with residual based methods we refer to M. Ainsworth and J. T. Oden [1], C. Carstensen [42], C. Carstensen and S.A. Funken [44, 45] (in these papers the authors consider ways of computing bounds of the interpolation constants in the residual based error estimator), C. Carstensen and R. Verfürth [46], K. Eriksson and C. Johnson [69, 70], C. Johnson and P. Hansbo [110]. Estimates for problems with biharmonic operator are analysed in A. Charbonneau, K. Dossou and R. Pierre [47]. A posteriori estimates taking into account the influence of of the non-discretized part of the domain on the approximation error are considered in W. Dörfler and M. Rumpf [60]. Estimates in the L^{∞}-norm can be found, e.g., in S. W. Brady and A. R. Elcrat [31]. Adaptive methods for convection-diffusion problems are considered in C. Johnson [109].

A posteriori error estimators based on solving local Neumann type problems on each element were proposed in R.E. Bank and A. Weiser [24]. Its asymptotic properties were investigated in R. Duran and R. Rodriguez [63]. This error estimation method was in the focus of many researches. We present several publications, in which readers will find other references: I. Babuška, T. Strouboulis, C. S. Upadhyay and S. K. Gangaraj [19], I. Babuška, F. Ihlenburg, T. Strouboulis and S. K. Gangaraj [14], E. Stein and S. Ohnimus [202], E. Stein, F.J. Bartold, S. Ohnimus and M. Schmidt [203].

Probably earliest results on superconvergence were established in the papers by L. A. Oganesjan and L. A. Ruchovets [155] and M. Zlámal [229].

Also, we refer to I. Hlaváček and M. Křizek [101], M. Křižek and P. Neittaanmäki [127], P. Neittaanmäki and M. Křížek [148].

In the books L. Wahlbin [216] and M. Křižek, P. Neittaanmäki and R Stenberg [125] readers can find a detailed exposition of the subject and other references. Surveys on superconvergence are presented in M. Křižek and P. Neittaanmäki [126, 127] and J. R. Whiteman and G. Goodsell [219].

Simple and efficient error indicators based upon gradient averaging were suggested in O. C. Zienkiewicz and J. Z. Zhu [226]. This and other indicators were analyzed in many papers. Here, we refer to, e.g., M. Ainsworth, J. Z. Zhu, A. W. Craig and O. C. Zienkiewicz [5], I. Babuška and R. Rodriguez [17], R. D. Lazarov [136], R. Rodrigues [189], O. C. Zienkiewicz and J. Z. Zhu [227, 228], O. C. Zienkiewicz, B. Boroomand and J.Z. Zhu [224]. In C. Carstensen and S. Bartels [43], it is shown that each averaging technique leads to a certain a posteriori error estimate. Estimates based upon global averaging of gradients were considered in B.-O. Heimsund, X.-C. Tai and J. Wang [115]. In J. Wang [217], a generalization of the patch recovery technique of O. C. Zienkiewicz and J. Z. Zhu is considered. It consists of approximating an approximate solution data by a certain smooth function that is computed by the least–squares method.

A posteriori estimates that express approximation errors by errors in the equilibrium relations were suggested in P. Ladeveze and D. Leguillon [128] (see also P. Ladeveze, J.-P. Pelle and Ph. Rougeot [129], P. Ladeveze and Ph. Rougeot [130]). Post-processing procedures based upon equilibrating of recovered stresses were considered in B. Boroomand and O. C. Zienkiewicz [29, 30]. They result in stress fields that satisfy an equilibrium condition in a weak form. A method of equilibration is presented in P. Destuynder and B. Métivet [59]. This paper also contain numerical tests, in which a posteriori error estimates obtained by equilibration are compared with other estimates.

A posteriori error estimates are intended not only for the verification of the accuracy of numerical solutions. They must also provide a useful information for mesh adaptation algorithms. These questions are analyzed, e.g., in R. E. Bank and R. K. Smith [23], G. F. Carey and D. L. Humphrey [41], R. Löhner, K. Morgan and O. C. Zienkiewicz [139], J. T. Oden, L. Demkowicz, T. Strouboulis and P. Devloo [150], J. Peraire and A. T. Patera [159], R. Verfürth [213]. Methods of mesh adaptation are considered in the book by W. C. Rheinboldt [167].

The reader can find a detailed exposition of the a posteriori error estimation methods based upon using adjoint boundary-value problems in, e.g., W. Bangerth and R. Rannacher [26], R. Becker and R. Rannacher [27].

Applications of this techniques to elastic and elasto–plastic problems were considered in R. Rannacher and F. T. Suttmeier [165, 166], and to transport problems in P. Houston, R. Rannacher, E. Süli [103]. Applications to problems in the theory of optimal control were considered in R. Becker, H. Kapp and R. Rannacher [28].

Error estimation methods for goal-oriented quantities were considered, e.g., in M. Ainsworth and J .T .Oden [3] and J. T. Oden and S. Prudhomme [152].

Finally, it is necessary to say a few words about extensions of the residual based approach to nonlinear problems. In this case, the correlation between errors and residuals is not so obvious. This fact is easy to discover even in elementary examples. Let, for example a real-valued function $\phi(x)$ have an isolated root x_0. If ϕ is a smooth function, then

$$\phi(x) = \phi(x_0) + \phi'(x_0)(x - x_0) + \frac{1}{2}\phi''(x_0)(x - x_0)^2 + \ldots$$

and we see that

$$\|x - x_0\| \leq [\phi'(x_0)]^{-1}\phi(x) \tag{4.6.1}$$

provided that $\phi'(x_0) \neq 0$ and $\|x - x_0\|$ is so small that the third and other higher terms of the expansion are negligible with respect to the linear one. However, for a "distant" point $\widetilde{x}$ in which $\phi(\widetilde{x})$ is small the estimate (4.6.1) may be wrong because of the evident violation of the above locality condition. This simple example shows that, in general, error estimation methods based upon calculation of residuals require a deep investigation of the corresponding operator (mapping) at a neighborhood of the exact solution. Roughly speaking, a suitable estimate can be obtained if the operator considered admits a proper linearization at x_0. Such an estimate is valid in a ball centered at x_0 whose radius may be a priori unknown and, therefore, should be defined by an additional investigation. For these reasons, a posteriori error estimates for the abstract problem

$$F(x) = 0,$$

where $F : X \to Z$ is a continuous mapping of a Banach space X to a Banach space Z are usually derived under additional assumptions on F (see, e.g., J. Pousin and J. Rappaz [161, 162] and R. Verfürth [212, 213]). In particular, F must be a C^1 mapping, which possesses a unique solution u in a certain ball $\mathcal{B}(u, \delta)$. Moreover, the differential $DF(u)$ must be an isomorphism in a neighborhood of u. An extension of this approach to parabolic problems is exposed in R. Verfürth [213]).

Residual type a posteriori error estimates for finite element approximations of elliptic obstacle problems were obtained in Z. Chen and R. H.

Nochetto [48]. Their estimator contains terms of four types. Two of them are the same as in residual type estimates for linear elliptic problems and two others ("obstacle oscillations" and "obstacle jump residuals") reflect the behavior of an approximate solutions near the coincidence set. Various a posteriori error estimates for finite element approximations of variational inequalities were suggested by M. Ainsworth, J. T. Oden and C. Y. Lee [4], R. Hoppe and R. Kornhuber [102], R. Kornhuber [122], A. Veeser [210]. In J. Medina, M. Picasso and J. Rappaz [141], a priori and a posteriori error estimates are presented for nonlinear diffusion–convection problems. An aproaches to a posteriori error estimation for elasto–plastic problems is exposed in M. Schulz and W. L. Wendland [199, 200].

A posteriori error estimates for finite element approximations of problems in the theory of fluids are considered in, e.g., M. Ainsworth and J. T. Oden [2], M. Amara, M. Ben Younes and C. Bernardi [6], R. E. Bank and B. D. Welfert [25], C. Johnson and R. Rannacher [112], C. Johnson, R. Rannacher and M. Boman [113], J. T. Oden, W. Wu and M. Ainsworth [154], C. Padra [158], T. Strouboulis and J. T. Oden [205], R. Verfürth [211].

Chapter 5

Foundations of duality theory

A posteriori error estimates derived in Chapters 6–8 are based on duality theory in the calculus of variations. This chapter contains a concise exposition of a part of this theory that is used in subsequent chapters.

5.1 Convex sets and functionals

Let V be a linear space and $\mathbb{R}_+$ be the set of nonnegative real numbers.

Definition 5.1.1. *A set $K \subset V$ is called* convex *if*

$$\lambda_1 v_1 + \lambda_2 v_2 \in K \tag{5.1.1}$$

for all $v_1, v_2 \in K$ and all $\lambda_1, \lambda_2 \in \mathbb{R}_+$ such that $\lambda_1 + \lambda_2 = 1$.

We denote by $\mathbb{CS}(V)$ the set of all convex sets composed of elements of V. Note that $\emptyset \in \mathbb{CS}(V)$ and $V \in \mathbb{CS}(V)$. The minimal convex set containing K is called the *convex hull* (or *convex envelope*) of K. Conventional notation used for it is $\operatorname{conv} K$. Also, $\operatorname{conv} K$ can be defined as the set of all convex combinations of elements of K, i.e.,

$$\operatorname{conv} K = \left\{ v \in V \mid v = \sum_{i=1}^{m} \lambda_i v_i, \ v_i \in K, \ \sum_{i=1}^{m} \lambda_i = 1, \ \lambda_i \geq 0 \right\}. \tag{5.1.2}$$

It is obvious that $K = \operatorname{conv} K$ if and only if K is a convex set. Figure 5.1.1 depicts a convex set K_1, a nonconvex set K_2, and the set $\operatorname{conv} K_2$.

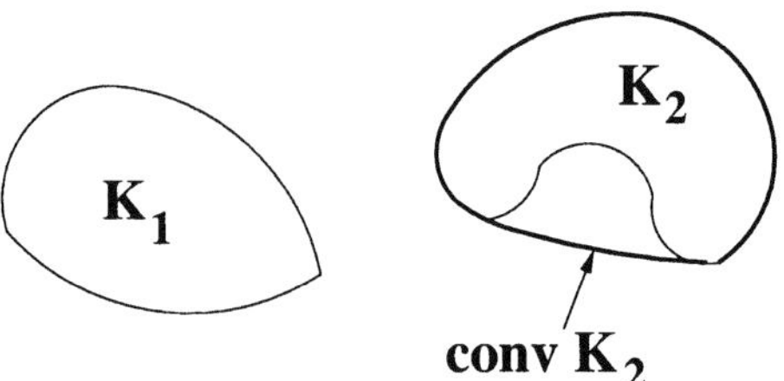

Figure 5.1.1: Convex and nonconvex sets.

If the condition $\lambda_i \geq 0$ in (5.1.2) is omitted, then we obtain the so-called *affine envelope* of K (denoted by aff K). For example, the affine envelopes of the above-described sets K_1 and K_2 coincide with the entire space $\mathbb{R}^2$.

A set K is called a *cone* if the inclusion $\lambda v \in K$ holds for any $\lambda \geq 0$ and any $v \in K$.

Definition 5.1.2. *Let K be a convex set. A functional $J : K \to \mathbb{R}$ is called* convex *if the inequality*

$$J(\lambda_1 v_1 + \lambda_2 v_2) \leq \lambda_1 J(v_1) + \lambda_2 J(v_2) \tag{5.1.3}$$

holds for all $v_1, v_2 \in K$ and all $\lambda_1, \lambda_2 \in \mathbb{R}_+$ such that $\lambda_1 + \lambda_2 = 1$.

We denote by $\mathbb{CF}(V)$ the set of all convex functionals defined on V. A functional J is called *strictly convex* if

$$J(\lambda_1 v_1 + \lambda_2 v_2) < \lambda_1 J(v_1) + \lambda_2 J(v_2) \tag{5.1.4}$$

for all $v_1, v_2 \in K$ and $\lambda \in (0, 1)$. A functional J is called *concave* (resp., *strictly concave*) if the functional $(-J)$ is convex (resp., strictly convex).

Example 5.1.1. The functional

$$\chi_K(v) = \begin{cases} 0 & \text{if } v \in K, \\ +\infty & \text{if } v \notin K \end{cases}$$

is called the *characteristic functional* of the set K. It is clear that it is convex if and only if the set K is convex.

In convex analysis, the functionals are often regarded as mappings to the extended set of real numbers

$$\overline{\mathbb{R}} := \{-\infty\} \cup \mathbb{R} \cup \{+\infty\}.$$

This is motivated by the following reasons. Assume that $J : K \to \mathbb{R}$ is a given functional. Define the extended functional $\widetilde{J} : V \to \overline{\mathbb{R}}$, setting

$$\widetilde{J}(v) = \begin{cases} J(v), & v \in K, \\ +\infty, & v \notin K. \end{cases}$$

The functional $\widetilde{J}$ is convex if and only if J is convex. Therefore, we can formally consider all the functionals as mappings from V to $\overline{\mathbb{R}}$. Such unification is useful in many instances.

Definition 5.1.3. *A functional* $J : V \to \overline{\mathbb{R}}$ *is called* proper *if* $J(v) > -\infty$ *for any* $v \in V$ *and* $J \not\equiv +\infty$.

Any functional $J : V \to \mathbb{R}$ is characterized by two sets:

$$\operatorname{dom} J := \{v \in V \mid J(v) < +\infty\},$$
$$\operatorname{epi} J := \{(v, \alpha) \mid v \in V, \ \alpha \in \mathbb{R}, \ J(v) \leq \alpha\}.$$

The first set contains elements (functions) that give finite values to J. The second one (called *supergraph* or *epigraph*) consists of "points" $(v, \alpha) \in V \times \overline{\mathbb{R}}$ that lie "above" the graph.

Proposition 5.1.1. *The set* $\operatorname{epi} J$ *is convex if and only if* J *is a convex functional.*

Proof. Let J be a convex functional (i.e., $J \in \mathbb{CF}(V)$) and

$$(v_1, \alpha_1) \in \operatorname{epi} J, \quad (v_2, \alpha_2) \in \operatorname{epi} J.$$

Then,

$$J(v_1) \leq \alpha_1, \quad J(v_2) \leq \alpha_2$$

and

$$J(\lambda_1 v_1 + \lambda_2 v_2) \leq \lambda_1 J(v_1) + \lambda_2 J(v_2) \leq \lambda_1 \alpha_1 + \lambda_2 \alpha_2,$$

where $\lambda_1, \lambda_2 \in \mathbb{R}_+$ are numbers such that $\lambda_1 + \lambda_2 = 1$. Hence,

$$\lambda_1 (v_1, \alpha_1) + \lambda_2 (v_2, \alpha_2) = (\lambda_1 v_1 + \lambda_2 v_2, \lambda_1 \alpha_1 + \lambda_2 \alpha_2) \in \operatorname{epi} J,$$

and, therefore,

$$\operatorname{epi} J \in \mathbb{CS}(V \times \overline{\mathbb{R}}).$$

Assume now that $\operatorname{epi} J$ is a convex set and $J(v_1) = \alpha_1$, $J(v_2) = \alpha_2$. Then $(v_1, \alpha_1) \in \operatorname{epi} J$, $(v_2, \alpha_2) \in \operatorname{epi} J$ and, consequently,

$$\lambda_1 (v_1, \alpha_1) + \lambda_2 (v_2, \alpha_2) \in \operatorname{epi} J.$$

The latter relation means that

$$J(\lambda_1\alpha_1 + \lambda_2\alpha_2) \leq \lambda_1\alpha_1 + \lambda_2\alpha_2 = \lambda_1 J(v_1) + \lambda_2 J(v_2).$$

$\square$

Proposition 5.1.2. *If $J \in \mathbb{CF}(V)$, then*

$$V_\alpha := \{v \in V \mid J(v) \leq \alpha, \ \alpha \in \mathbb{R}\} \in \mathbb{CS}(V).$$

Proof. The proof follows from the basic inequality (5.1.3). Let $v_1 \in V_\alpha$ and $v_2 \in V_\alpha$. Then $J(v_1) \leq \alpha$ and $J(v_2) \leq \alpha$ and for any $\lambda_1, \lambda_2 \in \mathbb{R}_+$ such that $\lambda_1 + \lambda_2 = 1$ we have

$$J(\lambda_1 v_1 + \lambda_2 v_2) \leq \lambda_1\alpha + \lambda_2\alpha = \alpha.$$

Thus, $\lambda_1 v_1 + \lambda_2 v_2 \in V_\alpha$. $\square$

Proposition 5.1.3. *Let $K_i \in \mathbb{CS}(V)$, $i = 1, 2, ..., m$. Then the intersection of these sets*

$$\overset{\cap}{K} = \{v \in V \mid v \in K_i , \quad \forall i = 1, 2, ..., m\}$$

is a convex set.

Proof. Let v_1 and v_2 be two elements of $\overset{\cap}{K}$. Then $\lambda_1 v_1 + \lambda_2 v_2 \in K_i$ for any $\lambda_1, \lambda_2 \in \mathbb{R}_+$ such that $\lambda_1 + \lambda_2 = 1$ and any $i = 1, 2, ..., m$. Hence, the intersection is a convex set. $\square$

Proposition 5.1.4. *Let $K_1 \in \mathbb{CS}(V)$, $\lambda \in \mathbb{R}_+$, and $v_0 \in V$. Then the set*

$$K_2 := \{v \in V \mid v = \lambda w + v_0, \ w \in K_1\}$$

is convex.

Proof. Assume that v_1 and v_2 are two elements of K_2. Then there exist $w_1, w_2 \in K_1$ such that $v_1 = \lambda w_1 + v_0$ and $v_2 = \lambda w_2 + v_0$. Since $K_1 \subset \mathbb{CS}(V)$, we have

$$\lambda_1 w_1 + \lambda_2 w_2 \in K_1,$$

where $\lambda_1, \lambda_2 \in \mathbb{R}_+$, $\lambda_1 + \lambda_2 = 1$. Then,

$$\lambda_1 v_1 + \lambda_2 v_2 = \lambda_1(\lambda w_1 + v_0) + \lambda_2(\lambda w_2 + v_0)$$
$$= \lambda(\lambda_1 w_1 + \lambda_2 w_2) + v_0 \in K_2,$$

and the assertion follows. $\square$

Proposition 5.1.5. *If $J_1 \in \mathbb{CF}(V)$ and $J_2 \in \mathbb{CF}(V)$, then for any $\mu_1, \mu_2 \in \mathbb{R}_+$ the functional*

$$G(v) := \mu_1 J_1(v) + \mu_2 J_2(v)$$

is convex.

Proof. From Definition 5.1.2, it follows that for any λ_1, λ_2 such that $\lambda_1 + \lambda_2 = 1$ and $v_1, v_2 \in V$

$$\lambda_1 J_1(v_1) + \lambda_2 J_1(v_2) \geq J_1(\lambda_1 v_1 + \lambda_2 v_2),$$
$$\lambda_1 J_2(v_1) + \lambda_2 J_2(v_2) \geq J_2(\lambda_1 v_1 + \lambda_2 v_2).$$

Multiplying the first inequality by μ_1 and the second one by μ_2 and adding them, we have

$$\lambda_1(\mu_1 J_1(v_1) + \mu_2 J_2(v_1)) + \lambda_2(\mu_1 J_1(v_2) + \mu_2 J_2(v_2))$$
$$\geq \mu_1 J_1(\lambda_1 v_1 + \lambda_2 v_2) + \mu_2 J_2(\lambda_1 v_1 + \lambda_2 v_2).$$

This means that

$$\lambda_1 G(v_1) + \lambda_2 G(v_2) \geq G(\lambda_1 v_1 + \lambda_2 v_2).$$

$\square$

Proposition 5.1.6. *If $J_1 \in \mathbb{CF}(V)$ and $J_2 \in \mathbb{CF}(V)$, then*

$$G(v) := \max\{J_1(v), J_2(v)\} \in \mathbb{CF}(V).$$

Proof. In view of Proposition 5.1.1, the sets $\text{epi}\, J_1$ and $\text{epi}\, J_2$ are convex. It is easy to see that

$$\text{epi}\, G = \text{epi}\, J_1 \cap \text{epi}\, J_2.$$

The intersection of convex sets is a convex set (see Proposition 5.1.3). Thus, we conclude that $\text{epi}\, J \in \mathbb{CS}(V \times \overline{\mathbb{R}})$ and, therefore, $J \in \mathbb{CF}(V)$. $\square$

The latter property remains valid for any amount of convex functionals, i.e., the upper bound taken over any set of convex functionals is a convex functional. This fact suggests the idea to represent all convex functionals as upper bounds computed on a special subclass of functionals. For this purpose, affine functionals are used. To present them, we first introduce the space V^* that contains all linear continuous functionals defined on V. The elements of this space are marked by stars and the value of a functional $v^* \in V^*$ on an element $v \in V$ is denoted by $\langle v^*, v \rangle$. The latter quantity is often called the *duality pairing* of the spaces V and V^*. Also, we define the set of *affine functionals* on V formed by the functionals $\langle v^*, v \rangle - \alpha$, where $v^* \in V^*$ and $\alpha \in \mathbb{R}$.

Definition 5.1.4. $\Gamma(V)$ *denotes the set of functionals that can be represented as upper bounds of affine functionals.*

From this definition it follows that

$$J \in \Gamma(V) \implies J \in \mathbb{CF}(V),$$

i.e., all functionals of the set $\Gamma(V)$ are convex. For any $J : V \to \overline{\mathbb{R}}$ define the set

$$\mathbb{AM}(J) = \{l(v) = \langle v^*, v \rangle - \alpha \mid l(v) \leq J(v) \ \forall v \in V\}.$$

We call it the set of *affine minorants* of J (formally, $\mathbb{AM}(J) \subset V^* \times \overline{\mathbb{R}}$). The procedure of taking the upper bound over this set gives rise to a new functional

$$\widehat{J}(v) = \sup_{l \in \mathbb{AM}(J)} \{l(v)\},$$

which is called the Γ-*regularization* of J. In general,

$$J(v) \geq \widehat{J}(v), \quad \forall v \in V.$$

However, there is one important case, in which J coincides with its Γ-regularization $\widehat{J}$. It will be considered in the next section. To complete this concise review of the theory of convex functionals, we prove another useful property. In the literature, it is known as Jensen's inequality (see [106, 188]).

Proposition 5.1.7. *Assume that $J \in \mathbb{CF}(V)$, v_i, $i = 1, 2, ..., n$, are given elements of V and $\lambda_i \in \mathbb{R}_+$ meet the condition*

$$\sum_{i=1}^{n} \lambda_i = 1.$$

Then

$$J\left(\sum_{i=1}^{n} \lambda_i v_i\right) \leq \sum_{i=1}^{n} \lambda_i J(v_i). \tag{5.1.5}$$

Proof. For $n = 2$, the inequality (5.1.5) coincides with Definition 5.1.2. Assume that (5.1.5) is proved for $n = k - 1$, i.e.,

$$J\left(\sum_{i=1}^{k-1} \lambda_i v_i\right) \leq \sum_{i=1}^{k-1} \lambda_i J(v_i), \tag{5.1.6}$$

where $\lambda_1, ..., \lambda_{k-1}$ are arbitrary nonnegative numbers such that $\lambda_1 + \lambda_2 + ... + \lambda_{k-1} = 1$. For any numbers $\mu_i \geq 0$, $i = 1, 2, ..., k$, that satisfy the condition $\mu_1 + \mu_2 + ... + \mu_k = 1$, we have

$$J((1 - \mu_k)w + \mu_k v_k) \leq (1 - \mu_k)J(w) + \mu_k J(v_k),$$

where w is an arbitrary element of V. Set

$$w = \frac{\mu_1}{1 - \mu_k}v_1 + \frac{\mu_2}{1 - \mu_k}v_2 + ... + \frac{\mu_{k-1}}{1 - \mu_k}v_{k-1}.$$

Then,

$$J(\mu_1 v_1 + \mu_2 v_2 + ... + \mu_k v_k) \leq (1 - \mu_k)J(w) + \mu_k J(v_k). \tag{5.1.7}$$

Let us introduce the numbers $\lambda_i = \frac{\mu_i}{1 - \mu_k}$, $i = 1, 2, ..., k-1$. Since $\sum_{i=1}^{k-1} \lambda_i = 1$, we apply (5.1.6) and obtain

$$J(w) \leq \sum_{i=1}^{k-1} \frac{\mu_i}{1 - \mu_k}J(v_i).$$

Combining this estimate with (5.1.7), we obtain the desired result

$$J\left(\sum_{i=1}^{k} \mu_i v_i\right) \leq \sum_{i=1}^{k} \mu_i J(v_i).$$

Arguing by induction, we can prove this estimate for any positive integer. $\qquad\square$

Remark 5.1.1. The inequality (5.1.5) also has an integral form. Let $J : \mathbb{R} \to \mathbb{R}$ be a convex function, $\Omega \subset \mathbb{R}^d$, $v : \Omega \to \mathbb{R}$ be a continuous function, and $\lambda : \Omega \to \mathbb{R}$ be an integrable function that satisfies the conditions

$$\int_\Omega \lambda(x)\, dx = 1, \quad \lambda(x) \geq 0 \quad \text{in } \Omega.$$

Then,

$$J\left(\int_\Omega \lambda(x)v(x)\, dx\right) \leq \int_\Omega \lambda(x)J(v(x))\, dx. \tag{5.1.8}$$

5.2   Dual functionals

5.2.1   Nomenclature

Let $v^* \in V^*$ and $\alpha \in \mathbb{R}$ be such that $\langle v^*, v \rangle - \alpha \in \mathbb{AM}(J)$, i.e.,

$$J(v) \geq \langle v^*, v \rangle - \alpha, \quad \forall v \in V. \tag{5.2.1}$$

In general, there are infinitely many different constants α that satisfy (5.2.1) with a fixed element v^*. Taking the supremum over all such constants, we define a certain number (finite or infinite) associated with v^*. This procedure defines a new functional $J^* : V^* \to \bar{\mathbb{R}}$, which is called *dual* (or *conjugate*) to J. From (5.2.1) it follows that

$$\alpha \geq \langle v^*, v \rangle - J(v)$$

and, consequently, J^* is defined as follows:

$$J^*(v^*) = \sup_{v \in V} \{ \langle v^*, v \rangle - J(v) \}. \tag{5.2.2}$$

Remark 5.2.1. If J is a smooth function that increases at infinity faster than any linear function, then J^* is the Legendre transform of J. The above definition was proposed by Young, Fenchel, and Moreau (see, e.g., [66, 74, 188]). The functional J^* is also called *polar* to J.

The functional

$$J^{**}(v) = \sup_{v^* \in V^*} \{ \langle v^*, v \rangle - J^*(v^*) \} \tag{5.2.3}$$

is called the *second conjugate* to J (or *bipolar*). Straightforwardly from the definition, it follows that J^* and J^{**} are convex functionals (they are defined as upper bounds of affine functionals).

Proposition 5.2.1. J^{**} *coincides with the Γ-regularization of J, i.e.,*

$$J^{**}(v) = \widehat{J}(v), \quad \forall v \in V. \tag{5.2.4}$$

Proof. $\widehat{J}(v)$ is the upper bound of the values of all affine functionals from the set $\mathbb{AM}(J)$. However, for any $v^* \in V^*$ the greatest minorant has the form $\langle v^*, v \rangle - J^*(v^*)$ (cf. the definition of $J^*(v^*)$). Therefore, the upper bound can be calculated by considering functionals only of such a type. Thus,

$$\widehat{J}(v) = \sup_{v^* \in V^*} \{ \langle v^*, v \rangle - J^*(v^*) \} = J^{**}(v).$$

$\square$

Corollary 5.2.1. If $J \in \Gamma(V)$, then $J(v) = \widehat{J}(v)$ and, therefore,

$$J^{**}(v) = J(v), \quad \forall v \in V.$$

Formally, one can also define the third conjugate functional

$$J^{***}(v^*) := \sup_{v \in V} \{ \langle v^*, v \rangle - J^{**}(v) \}. \tag{5.2.5}$$

However, this definition brings nothing new, because

$$J^{***}(v^*) = J^*(v^*), \quad \forall v^* \in V^*. \tag{5.2.6}$$

To prove this fact, we recall that

$$\widehat{J}(v) \le J(v), \quad \forall v \in V.$$

In view of (5.2.4), this means that

$$J^{**}(v) \le J(v), \quad \forall v \in V.$$

From this inequality it follows (see Property 5.2.1 below) that

$$J^{***}(v^*) \ge J^*(v^*), \quad \forall v^* \in V. \tag{5.2.7}$$

By definition, we have

$$J^{**}(v) \ge \langle v^*, v \rangle - J^*(v^*), \quad \forall v \in V.$$

Rewrite this relation as

$$J^*(v^*) \ge \langle v^*, v \rangle - J^{**}(v), \quad \forall v \in V,$$

and take the supremum over $v \in V$. We obtain

$$J^*(v^*) \ge J^{***}(v^*), \quad \forall v^* \in V^*. \tag{5.2.8}$$

Now (5.2.7) and (5.2.8) imply (5.2.6).

Definition 5.2.1. *Let* $J : V \to \overline{\mathbb{R}}$ *and* $G^* : V^* \to \overline{\mathbb{R}}$ *be two functionals defined on a Banach space* V *and its dual space* V^*, *respectively. These two functionals are called* mutually dual *(or* mutually conjugate*) if*

$$(G^*)^* = J \quad and \quad J^* = G^*. \tag{5.2.9}$$

5.2.2　Properties of dual functionals

Property 5.2.1. If $J : V \to \overline{\mathbb{R}}$ and $G : V \to \overline{\mathbb{R}}$ are such that

$$J(v) \geq G(v), \quad \forall v \in V, \tag{5.2.10}$$

then

$$J^*(v^*) \leq G^*(v^*), \quad \forall v^* \in V^*. \tag{5.2.11}$$

Proof. We have

$$J^*(v^*) = \sup_{v \in V}\{\langle v^*, v \rangle - J(v)\} \leq \sup_{v \in V}\{\langle v^*, v \rangle - G(v)\} = G^*(v^*).$$

$\square$

Property 5.2.2. For any $\lambda > 0$,

$$(\lambda J)^*(v^*) = \lambda J^*\left(\frac{v^*}{\lambda}\right). \tag{5.2.12}$$

Proof. This property is justified by direct calculations:

$$(\lambda J)^*(v^*) = \sup_{v \in V}\{\langle v^*, v \rangle - \lambda J(v)\} =$$

$$= \lambda \sup_{v \in V}\left\{\left\langle \frac{v^*}{\lambda}, v \right\rangle - J(v)\right\} = \lambda J^*\left(\frac{v^*}{\lambda}\right).$$

$\square$

Property 5.2.3. Let $J : V \to \overline{\mathbb{R}}$ and $J_\alpha(v) = J(v) + \alpha$, where $\alpha \in \mathbb{R}$. Then

$$J_\alpha^*(v^*) = J^*(v^*) - \alpha. \tag{5.2.13}$$

Proof. It follows from the obvious relation

$$\sup_{v \in V}\{\langle v^*, v \rangle - J(v) - \alpha\} = \sup_{v \in V}\{\langle v^*, v \rangle - J(v)\} - \alpha.$$

$\square$

Property 5.2.4. Let $v_0 \in V$ and $G(v) = J(v - v_0)$. Then

$$G^*(v^*) = J^*(v^*) + \langle v^*, v_0 \rangle. \tag{5.2.14}$$

Proof. Since

$$\sup_{v \in V}\{\langle v^*, v\rangle - J(v - v_0)\} = \sup_{w \in V}\{\langle v^*, w + v_0\rangle - J(w)\}$$
$$= \sup_{w \in V}\{\langle v^*, w\rangle - J(w)\} + \langle v^*, v_0\rangle = J^*(v^*) + \langle v^*, v_0\rangle,$$

we arrive at the relation (5.2.14). $\square$

Property 5.2.5. If
$$G(v) = \min_{i=1,\ldots,N}\{J_i(v)\},$$

then
$$G^*(v^*) = \max_{i=1,\ldots,N}\{J_i^*(v^*)\}. \tag{5.2.15}$$

Proof. We have

$$G^*(v^*) = \sup_{v \in V}\{\langle v^*, v\rangle - \min_{i=1,\ldots,N}\{J_i(v)\}\}$$
$$= \sup_{v \in V}\{\langle v^*, v\rangle + \max_{i=1,\ldots,N}\{-J_i(v)\}\}$$
$$= \sup_{v \in V} \max_{i=1,\ldots,N}\{\langle v^*, v\rangle - J_i(v)\}$$
$$= \max_{i=1,\ldots,N} \sup_{v \in V}\{\langle v^*, v\rangle - J_i(v)\} = \max_{i=1,\ldots,N}\{J_i^*(v^*)\}.$$

$\square$

To prove another property of conjugate functionals, we need to establish one subsidiary result.

Lemma 5.2.1. *Let $\Phi(x, y)$ be a functional defined on the elements of two nonempty sets X and Y. Then*

$$\sup_{x \in X} \inf_{y \in Y} \Phi(x, y) \le \inf_{y \in Y} \sup_{x \in X} \Phi(x, y). \tag{5.2.16}$$

Proof. It is easy to see that

$$\Phi(x, y) \ge \inf_{\xi \in X} \Phi(\xi, y), \quad \forall x \in X, \ y \in Y.$$

Taking the supremum over $y \in Y$, we obtain

$$\sup_{y \in Y} \Phi(x, y) \ge \sup_{y \in Y} \inf_{\xi \in X} \Phi(\xi, y), \quad \forall x \in X. \tag{5.2.17}$$

The left-hand side of (5.2.17) depends on x, while the right-hand side is a number. Thus, we have the inequality

$$\inf_{x \in X} \sup_{y \in Y} \Phi(x, y) \geq \sup_{y \in Y} \inf_{\xi \in X} \Phi(\xi, y),$$

which coincides with (5.2.16). $\qquad\square$

It should be emphasized that the inequality (5.2.16) has a general meaning. It holds for convex functionals, as well as for nonconvex ones, and does not depend on the nature of the sets X and Y. In many important cases, it holds as equality. This fact can be proved under additional assumptions on the structure of X, Y, and Φ.

Property 5.2.6. If

$$G(v) = \max_{i=1,\ldots,N} \{J_i(v)\},$$

then

$$G^*(v^*) \leq \min_{i=1,\ldots,N} \{J_i^*(v^*)\}. \qquad (5.2.18)$$

Proof. By definition, we have

$$G^*(v^*) = \sup_{v \in V} \{\langle v^*, v \rangle - \max_{i=1,\ldots,N} \{J_i(v)\}\}$$

$$= \sup_{v \in V} \{\langle v^*, v \rangle + \min_{i=1,\ldots,N} \{-J_i(v)\}\}$$

$$= \sup_{v \in V} \min_{i=1,\ldots,N} \{\langle v^*, v \rangle - J_i(v)\}\}.$$

Now we apply Lemma 5.2.1 to our case, in which $\langle v^*, v \rangle - J_i(v)$ stands for Φ. Then, (5.2.16) yields

$$G^*(v^*) \leq \min_{i=1,\ldots,N} \sup_{v \in V} \{\langle v^*, v \rangle - J_i(v)\} = \min_{i=1,\ldots,N} \{J_i^*(v^*)\}.$$

$$\square$$

5.2.3　Examples

To illustrate the definitions of conjugate functionals, we present below several examples for functionals defined on the Euclidean space E^d. In this case, V and V^* are isometrically isomorphic. Their elements are d-dimensional vectors denoted by ξ and ξ^*, respectively, and the quantity $\langle \xi^*, \xi \rangle$ is given by the scalar product $\xi^* \cdot \xi$.

Example 5.2.1. Let $A = \{a_{ij}\}$ be a real, positive definite matrix and

$$J(\xi) = \frac{1}{2}A\xi \cdot \xi = \frac{1}{2}a_{ij}\xi_i\xi_j.$$

Then

$$J^*(\xi^*) = \sup_{\xi \in E^d} \left\{ \xi^* \cdot \xi - \frac{1}{2}A\xi \cdot \xi \right\}.$$

This supremum is attained on an element ξ_0 such that

$$\xi^* = A\xi_0.$$

Therefore, we have the following pair of mutually conjugate functionals:

$$J(\xi) = \frac{1}{2}A\xi \cdot \xi \quad \text{and} \quad J^*(\xi^*) = \frac{1}{2}A^{-1}\xi^* \cdot \xi^*. \tag{5.2.19}$$

Example 5.2.2. Consider the functional

$$J(\xi) = \frac{1}{p}|\xi|^p,$$

where $p > 1$ and $|\xi| = (\xi \cdot \xi)^{1/2}$. It is easy to verify that the quantity

$$\xi^* \cdot \xi - \frac{1}{p}|\xi|^p$$

attains a supremum if $\xi = \xi_0$, where ξ_0 satisfies the relation

$$\xi^* - |\xi_0|^{p-2}\xi_0 = 0,$$

which yields

$$|\xi^*| = |\xi_0|^{p-1}, \quad \xi^* \cdot \xi_0 = |\xi_0|^p.$$

Therefore,

$$J^*(\xi^*) = \xi^* \cdot \xi_0 - \frac{1}{p}|\xi_0|^p = \left(1 - \frac{1}{p}\right)|\xi_0|^p = \frac{1}{p^*}|\xi^*|^{p^*},$$

where $p^* = \frac{p}{p-1}$. Thus, we obtain another pair of mutually conjugate functionals

$$J(\xi) = \frac{1}{p}|\xi|^p \quad \text{and} \quad J^*(\xi^*) = \frac{1}{p^*}|\xi^*|^{p^*}, \quad \frac{1}{p} + \frac{1}{p^*} = 1. \tag{5.2.20}$$

Remark 5.2.2. The relation (5.2.20) admits rather wide generalizations (see, e.g., [66]). Namely, let $\varphi : \mathbb{R} \to \mathbb{R}$ be a proper convex function that is, in addition, odd and let $\varphi^* : \mathbb{R} \to \mathbb{R}$ be its conjugate. Then

$$(\varphi(\|u\|_V))^* = \varphi^*(\|u^*\|_{V^*}). \tag{5.2.21}$$

Example 5.2.3. Let $J(\xi)$ be a linear functional, i.e.,

$$J(\xi) = l \cdot \xi, \quad l \in E^d.$$

It is easy to see that

$$J^*(\xi^*) = \sup_{\xi \in E^d} \{\xi^* \cdot \xi - l \cdot \xi\} = \begin{cases} 0 & \xi^* = l, \\ +\infty & \xi^* \neq l. \end{cases}$$

Denote by $\chi_{\{l\}}$ the characteristic functional of the set $\{l\} \subset E^d$. Then, another pair of mutually conjugate functionals is as follows:

$$J(\xi) = l \cdot \xi \quad \text{and} \quad J^*(\xi^*) = \chi_{\{l\}}(\xi^*). \tag{5.2.22}$$

Example 5.2.4. The case $J(\xi) = |\xi|$ also deserves a special consideration. The quantity

$$\sup_{\xi}\{\xi^* \cdot \xi - |\xi|\}$$

may be finite or infinite depending on the value of $|\xi^*|$. If $|\xi^*| > 1$, then, obviously, it is infinite. If $|\xi^*| \leq 1$, then, on the one hand,

$$\sup_{\xi}\{\xi^* \cdot \xi - |\xi|\} \leq \sup_{\xi}\{1 \cdot |\xi| - |\xi|\} = 0.$$

On the other hand,

$$\sup_{\xi}\{\xi^* \cdot \xi - |\xi|\} \geq \xi^* \cdot 0 - 0 = 0.$$

This means that $J^*(\xi^*) = 0$ if $|\xi^*| \leq 1$ and, thus,

$$J(\xi) = |\xi| \quad \text{and} \quad J^*(\xi^*) = \chi_{\mathcal{B}^*(0,1)}(\xi^*), \tag{5.2.23}$$

are mutually conjugate, where $\mathcal{B}^*(0,1) = \{\xi^* \in E^d \mid |\xi^*| \leq 1\}$. Now, in view of (5.2.12), we easily find the form of $[\lambda|\xi|]^*$:

$$[\lambda|\xi|]^*(\xi^*) = \lambda\,[|\xi|]^*\left(\frac{\xi^*}{\lambda}\right) = \chi_{\mathcal{B}^*(0,\lambda)}(\xi^*). \tag{5.2.24}$$

Example 5.2.5. Let K be a convex closed set in E^d and

$$J(\xi) = \chi_K(\xi).$$

The respective conjugate functional is defined as follows:

$$J^*(\xi^*) = \sup_{\xi \in E^d} \{\xi^* \cdot \xi - \chi_K(\xi)\} = \sup_{\xi \in K} \xi^* \cdot \xi.$$

This function is called the *support* function of K and is denoted by $\chi_K^*(\xi^*)$. For example, if $K = \mathcal{B}(0,1)$, then

$$\sup_{\xi \in K} \xi^* \cdot \xi = |\xi^*|,$$

so that $\chi_{\mathcal{B}(0,1)}^*(\xi^*) = |\xi^*|$.

Example 5.2.6. Another interesting example is given by the functional

$$J(\xi) = \frac{k}{2}|\xi|^2 + \chi_{\mathcal{B}(0,\lambda)}(\xi), \quad k > 0, \ \lambda > 0.$$

In this case,

$$J^*(\xi^*) = \sup_{\xi \in \mathcal{B}(0,\lambda)} \{\xi^* \cdot \xi - \frac{k}{2}|\xi|^2\}.$$

If ξ_0 meets the relation $\xi^* = k\xi_0$ and satisfies the condition $|\xi_0| \leq \lambda$, then it is the required maximizer. For such a ξ_0 we have

$$\xi^* \cdot \xi_0 - \frac{k}{2}|\xi_0|^2 = k|\xi_0|^2 - \frac{k}{2}|\xi_0|^2 = \frac{1}{2k}|\xi^*|^2.$$

If $|\xi_0| > \lambda$, then the maximizer ξ_m meets the conditions

$$|\xi_m| = \lambda, \quad \xi^* \cdot \xi_m \geq \xi^* \cdot \xi, \quad \forall \xi \in \mathcal{B}(0,\lambda),$$

which mean that $\xi_m = \lambda\frac{\xi^*}{|\xi^*|}$ and, consequently,

$$\xi^* \cdot \xi_m - \frac{k}{2}|\xi_m|^2 = \lambda|\xi^*| - \frac{k}{2}\lambda^2.$$

Thus, we obtain

$$J^*(\xi^*) = \begin{cases} \dfrac{1}{2k}|\xi^*|^2 & \text{if } |\xi^*| \leq k\lambda, \\[2mm] \lambda|\xi^*| - \dfrac{k}{2}\lambda^2 & \text{if } |\xi^*| > k\lambda. \end{cases} \tag{5.2.25}$$

In Examples 5.2.1–5.2.6, we considered proper convex functionals J such that J^{**} coincides with J. A different case is presented below.

Example 5.2.7. Let A_1 and A_2 be two positive definite symmetric matrices and

$$J(\xi) = \min\{J_1(\xi), J_2(\xi)\},$$

where

$$J_i(\xi) = \frac{1}{2} A_i(\xi - \xi_i) \cdot (\xi - \xi_i) + \mu_i, \quad i = 1, 2,$$

the ξ_i and μ_i are two given vectors and two positive numbers, respectively. The functional $J : E^d \to \mathbb{R}$ is nonconvex. Then, by Property 5.2.5,

$$J^*(\xi^*) = \max\{J_1^*(\xi^*), J_2^*(\xi^*)\}. \tag{5.2.26}$$

To find J_i^*, we recall Properties 5.2.3 and 5.2.4 and use (5.2.19). They lead to the relation

$$\begin{aligned}
J_i^*(\xi^*) &= \left(\frac{1}{2} A_i(\xi - \xi_i) \cdot (\xi - \xi_i) + \mu_i \right)^* \\
&= \left(\frac{1}{2} A_i(\xi - \xi_i) \cdot (\xi - \xi_i) \right)^* - \mu_i \\
&= \frac{1}{2} A_i^{-1} \xi^* \cdot \xi^* + \xi^* \cdot \xi_i - \mu_i.
\end{aligned}$$

In particular, assume that $\xi_1 = -\xi_0$, $\xi_2 = +\xi_0$, $\xi_0 > 0$, $\mu_1 = \mu_2 = 0$, and $A_i = \alpha_i I$, where the α_i are positive constants and I is the identity matrix. Then

$$J_1^*(\xi^*) = \frac{1}{2\alpha_1}|\xi^*|^2 - \xi^* \cdot \xi_0, \quad J_2^*(\xi^*) = \frac{1}{2\alpha_2}|\xi^*|^2 + \xi^* \cdot \xi_0.$$

Note that J^* is a nondifferentiable functional. Let us calculate the second conjugate functional J^{**} for the case $d = 1$ and $\alpha_1 = \alpha_2 = \alpha$. The quantity

$$\xi^*\xi - \max\left\{ \frac{1}{2\alpha}|\xi^*|^2 - \xi^*\xi_0, \frac{1}{2\alpha}|\xi^*|^2 + \xi^*\xi_0 \right\}$$

may attain a supremum in one of the following three cases:

1. $\xi^* = \xi_1^* := \alpha(\xi + \xi_0)$, $J_1^*(\xi_1^*) > J_2^*(\xi_1^*)$,

2. $\xi^* = \xi_2^* := \alpha(\xi - \xi_0)$, $J_1^*(\xi_2^*) < J_2^*(\xi_2^*)$,

3. $\xi^* = 0$.

Since the inequality $J_1^*(\xi_1^*) > J_2^*(\xi_1^*)$ is reduced to $\xi < -\xi_0$, and $J_1^*(\xi_2^*) < J_2^*(\xi_2^*)$ is equivalent to $\xi > \xi_0$, we conclude that

$$J^{**}(\xi) = \begin{cases} \alpha(\xi + \xi_0)^2/2 & \text{if } \xi < -\xi_0, \\ 0 & \text{if } |\xi| \le \xi_0, \\ \alpha(\xi - \xi_0)^2/2 & \text{if } \xi > +\xi_0. \end{cases}$$

It is clear that J^{**} does not coincide with J, which is

$$J(\xi) = \begin{cases} \alpha(\xi + \xi_0)^2/2 & \text{if } \xi < 0, \\ \alpha\xi_0^2/2 & \text{if } \xi = 0, \\ \alpha(\xi - \xi_0)^2/2 & \text{if } \xi > 0. \end{cases}$$

Example 5.2.8. Consider the functional

$$J(\xi) = \sqrt{1 + |\xi|^2},$$

which is of paramount importance for some variational problems having a geometrical meaning (e.g., for the nonparametric minimal surface problem). If $|\xi^*| > 1$, then the value of

$$\sup_{\xi \in E^d} \left\{ \xi^* \cdot \xi - \sqrt{1 + |\xi|^2} \right\}$$

is infinite. If $|\xi^*| \le 1$, then the maximizer ξ_0 satisfies the condition

$$\xi^* - \frac{\xi_0}{\sqrt{1 + |\xi_0|^2}} = 0,$$

which implies the relation

$$|\xi_0|^2 = \frac{|\xi^*|^2}{1 - |\xi^*|^2}.$$

Therefore, we obtain

$$J^*(\xi^*) = \begin{cases} -\sqrt{1 - |\xi^*|^2} & \text{if } |\xi^*| \le 1, \\ +\infty & \text{if } |\xi^*| > 1. \end{cases}$$

Several other examples of conjugate functions will be presented in Chapters 7 and 8.

5.3 Differentiation of convex functionals

5.3.1 Subdifferential

Let a functional $J : V \to \mathbb{R}$ take a finite value at $v_0 \in V$.

Definition 5.3.1. *The functional J is called subdifferentiable at v_0 if there exists an affine minorant $l \in \mathbb{AM}(J)$ such that $J(v_0) = l(v_0)$. A minorant with this property is called the exact minorant at v_0.*

Obviously, any affine minorant exact for J at v_0 has the form

$$l(v) = \langle v^*, v - v_0 \rangle + J(v_0), \quad l(v) \le J(v), \quad \forall v \in V. \tag{5.3.1}$$

The element v^* in (5.3.1) is called a *subgradient* of J at v_0. A functional may

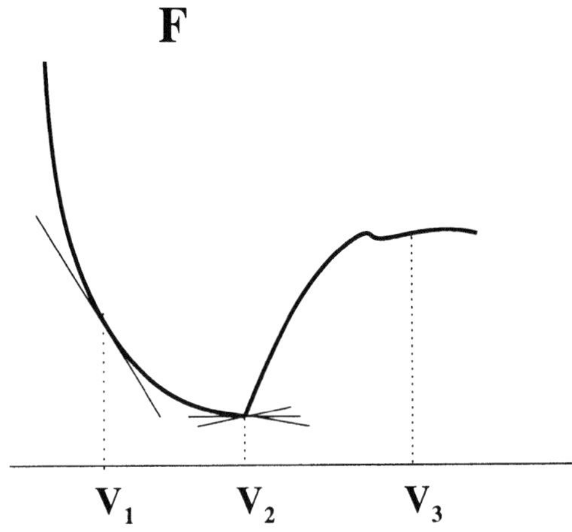

Figure 5.3.1: Affine minorants.

have only one subgradient at a given point (see the point v_1 in Fig. 5.3.1) or a collection of different subgradients (see the point v_2). Also, it may occur that a functional has no exact affine minorants at a given point (see v_3 in Fig. 5.3.1). In the latter case, we say that it has no subgradients. The set of all subgradients of J at v_0 forms a *subdifferential*, which is usually denoted by $\partial J(v_0)$. In view of the foregoing, $\partial J(v_0)$ may be empty or contain one element or infinitely many elements. An important property of convex functionals follows directly from the above definition. Assume that J is a convex functional and $v^* \in \partial J(v_0)$. Then there exists an affine minorant such that

$$\langle v^*, v \rangle - \alpha \leq J(v), \quad \forall v \in V,$$
$$\langle v^*, v_0 \rangle - \alpha = J(v_0).$$

Hence, we obtain

$$J(v) - J(v_0) \geq \langle v^*, v - v_0 \rangle. \tag{5.3.2}$$

The inequality (5.3.2) presents the basic incremental relation for convex functionals.

Proposition 5.3.1. *If $\partial J(v_0) \neq \emptyset$, then*

$$J(v_0) = J^{**}(v_0). \tag{5.3.3}$$

Proof. By assumption, $J(v_0) = l(v_0)$, where $l \in \mathbb{AM}(J)$. Hence, on the one hand,

$$J^{**}(v_0) \leq J(v_0) = l(v_0).$$

But J^{**} is the upper bound of all affine minorants. Thus, on the other hand,

$$J^{**}(v_0) \geq l(v_0).$$

This immediately implies (5.3.3). $\qquad\qquad\square$

To verify whether or not an element v^* belongs to the set $\partial J(v)$, we can use the criterion given below.

Proposition 5.3.2. *Let $J \in \Gamma(V)$. The following two statements are equivalent:*

$$J(v) + J^*(v^*) - \langle v^*, v \rangle = 0, \tag{5.3.4}$$

$$v^* \in \partial J(v) \text{ and } v \in \partial J^*(v^*). \tag{5.3.5}$$

Proof. Assume that $v^* \in \partial J(v)$. In accordance with (5.3.2), we have

$$J(w) \geq J(v) + \langle v^*, w - v \rangle, \quad \forall w \in V.$$

Hence,

$$\langle v^*, v \rangle - J(v) \geq \langle v^*, w \rangle - J(w), \quad \forall w \in V$$

and, consequently,

$$\langle v^*, v \rangle - J(v) \geq \sup_{w \in V} \left\{ \langle v^*, w \rangle - J(w) \right\} = J^*(v^*). \tag{5.3.6}$$

By the definition of J^*, the relation

$$J^*(v^*) + J(v) - \langle v^*, v \rangle \geq 0, \quad \forall v \in V, \ v^* \in V^*, \tag{5.3.7}$$

holds, so that (5.3.6) and (5.3.7) imply (5.3.4).

Assume that $v \in \partial J^*(v^*)$. Then

$$J^*(w^*) \geq J^*(v^*) + \langle w^* - v^*, v \rangle,$$

and we continue similarly to the previous case:

$$\langle v^*, v \rangle - J^*(v^*) \geq \langle w^*, v \rangle - J^*(w^*), \quad \forall w^* \in V^*,$$
$$\langle v^*, v \rangle - J^*(v^*) \geq J^{**}(v) = J(v).$$

Thus, we again arrive at (5.3.4).

Assume that (5.3.4) holds. Since

$$J^*(v^*) = \sup_{w \in V} \{ \langle v^*, w \rangle - J(w) \},$$

we obtain

$$0 = J(v) + J^*(v^*) - \langle v^*, v \rangle \geq J(v) - J(w) - \langle v^*, v - w \rangle,$$

where w is an arbitrary element of V. Rewrite this inequality in a more familiar form:

$$J(w) - J(v) \geq \langle v^*, w - v \rangle, \quad \forall w \in V,$$

which means that $J(v) + \langle v^*, v - w \rangle$ is an exact affine minorant of J (at v) and, consequently,

$$v^* \in \partial J(v).$$

The proof of the fact that $v^* \in \partial J^*(v^*)$ is quite analogous. $\square$

Remark 5.3.1. Proposition 5.3.2 plays an important role in the subsequent analysis (see Chapters 6, 7, and 8). In those chapters, we show that a posteriori error majorants of the functional type for approximate solutions of variational problems contain the functional

$$D_J(v, v^*) := J(v) + J^*(v^*) - \langle v^*, v \rangle.$$

This functional is defined on a pair of elements of two topologically dual spaces V and V^*. It is formed by a pair of mutually conjugate (dual) functionals J and J^*. For these reasons, henceforth we call it the *compound functional* and denote by the letter D. Proposition 5.3.2 states important property of a compound functional: it is nonnegative and vanishes only if the arguments satisfy the subdifferential relations, i.e.,

$$D_J(v, v^*) = 0 \quad \Longleftrightarrow \quad v \in \partial J^*(v^*) \text{ and } v^* \in \partial J(v). \qquad (5.3.8)$$

5.3.2 Properties of subdifferentials

Assume that J has a weak derivative $J'(v_0)$ in the sense of Gâteaux, i.e.,

$$\lim_{\lambda \to +0} \frac{J(v_0 + \lambda w) - J(v_0)}{\lambda} = \langle J'(v_0), w \rangle, \quad \forall w \in V.$$

Let $\partial J(v_0) \neq \emptyset$ and $v^* \in \partial J(v_0)$. Then

$$J(v) - J(v_0) \geq \langle v^*, v - v_0 \rangle, \quad \forall v \in V.$$

Setting $v = v_0 + \lambda w$, where $\lambda > 0$, we have

$$J(v_0 + \lambda w) - J(v_0) \geq \lambda \langle v^*, w \rangle, \quad \forall w \in V.$$

Therefore,

$$\langle J'(v_0), w \rangle = \lim_{\lambda \to +0} \frac{J(v_0 + \lambda w) - J(v_0)}{\lambda} \geq \langle v^*, w \rangle,$$

and we see that
$$\langle J'(v_0) - v^*, w \rangle \geq 0, \quad \forall w \in V.$$

Obviously, this inequality means that the Gâteaux derivative coincides with v^*. Thus, we conclude that

$$\partial J(v_0) = \{J'(v_0)\}. \tag{5.3.9}$$

In other words, if a functional is differentiable in the sense of Gâteaux, then the respective subdifferential contains only one element, which is the Gâteaux derivative.

Proposition 5.3.3. *If $J : V \to \overline{\mathbb{R}}$, then*

$$\partial(\lambda J)(v_0) = \lambda \partial J(v_0), \quad \forall v_0 \in V, \lambda \in \mathbb{R}_+. \tag{5.3.10}$$

Proof. If J has no exact affine minorants at v_0, then λJ also has no such minorants, so that both sets in (5.3.10) are empty. If there exists $v^* \in \partial J(v_0)$, then
$$J(v) \geq J(v_0) + \langle v^*, v - v_0 \rangle$$

and, consequently,

$$\lambda J(v) \geq \lambda J(v_0) + \langle \lambda v^*, v - v_0 \rangle.$$

Therefore, $\lambda v^* \in \partial(\lambda J)(v_0)$. $\qquad \square$

Let $J_i : V \to \overline{\mathbb{R}}$, $i = 1, 2$, be two functionals such that $\partial J_i(v_0) \neq \emptyset$. The important question is whether or not the classical additivity property of differentials remains valid for subdifferentials. Note that the inclusion

$$\partial J_1(v_0) + \partial J_2(v_0) \subset \partial(J_1 + J_2)(v_0) \tag{5.3.11}$$

is rather obvious. Indeed, if $v_1^* \in \partial J_1(v_0)$ and $v_2^* \in \partial J_2(v_0)$, then

$$J_1(v) \geq J_1(v_0) + \langle v_1^*, v - v_0 \rangle,$$
$$J_2(v) \geq J_2(v_0) + \langle v_2^*, v - v_0 \rangle.$$

Adding these inequalities, we obtain

$$(J_1 + J_2)(v) \geq (J_1 + J_2)(v_0) + \langle v_1^* + v_2^*, v - v_0 \rangle,$$

so that $v_1^* + v_2^* \in \partial(J_1 + J_2)(v_0)$. However, in general, $\partial(J_1 + J_2)$ may also contain elements other than $(v_1^* + v_2^*)$. The theorem below gives sufficient conditions for the additivity of subdifferentials. In the literature, it is known as the Moreau–Rockafellar theorem.

Proposition 5.3.4. *Assume that J_1 and J_2 are proper convex functionals and, additionally, there exists a $v_0 \in \mathrm{dom}\, J_1 \cap \mathrm{dom}\, J_2$ such that J_1 (or J_2) is continuous at v_0. Then*

$$\partial J_1(v) + \partial J_2(v) = \partial(J_1 + J_2)(v), \quad \forall v \in V. \tag{5.3.12}$$

The proof of this theorem and a more detailed presentation of properties of subdifferentials can be found, e.g., in [188]. Before closing this section, we establish one more property of convex functionals.

Proposition 5.3.5. *Assume that a convex functional J possesses the following two properties:*

$$J(0_V) = 0, \tag{5.3.13}$$

$$J(v) \leq C, \quad \forall v : \|v\|_V < \delta, \tag{5.3.14}$$

where 0_V is the zero element of V and δ is a positive number. Then J is continuous at 0_V.

Proof. For any $\varepsilon \in (0, 1]$, the elements $\pm v_\varepsilon$, where $v_\varepsilon = v/\varepsilon$, satisfy the condition $\|v_\varepsilon\| < \delta$ once v satisfies the condition $\|v\| < \varepsilon\delta$. Since J is a convex functional and

$$v = (1 - \varepsilon)0_V + \varepsilon v_\varepsilon,$$
$$(1 + \varepsilon)0_V = v + \varepsilon(-v_\varepsilon),$$

we have

$$J(v) \quad \le (1 - \varepsilon)J(0_v) + \varepsilon J(v_\varepsilon) \le \varepsilon C,$$

$$J(0_v) \quad \le \frac{1}{1 + \varepsilon} J(v) + \frac{\varepsilon}{1 + \varepsilon} J(-v_\varepsilon) \le$$

$$\le \frac{1}{1 + \varepsilon}\left(J(v) + \varepsilon C\right).$$

From these two inequalities it follows that

$$|J(v)| \le \varepsilon C \quad \text{for all } v \text{ such that } \|v\|_V < \varepsilon\delta. \qquad (5.3.15)$$

Let $\{v_k\}_{k=1}^{\infty}$ be a sequence converging to 0_v, i.e., $\|v_k\| \to 0$ as $k \to \infty$. Taking $\varepsilon_k = 1/\delta \, \|v_k\|$, we conclude from (5.3.15) that $|J(v_k)| \to 0$. Thus, J is continuous at 0_v. $\qquad \square$

Corollary 5.3.1. Let J be a convex functional, $J(v_0) \ne -\infty$, and $J(v) \le C$ for all $v \in \mathcal{B}(v_0, \delta)$. Then J is continuous at v_0.

Proof. Consider the functional

$$G(v) = J(v_0 - v) - J(v_0).$$

It is easy to see that $G(0_v) = 0$ and $G(v) \le C - J(v_0)$ if $v \in \mathcal{B}(0, \delta)$. By Proposition 5.3.5, we see that G is continuous at 0_v. Since $J(v_0 - v) = G(v) + J(v_0)$, the latter fact yields the continuity of J at v_0. $\qquad \square$

The above results enable us to conclude that a convex functional J is continuous at the interior points of the set dom J.

5.4 Lower semicontinuous functionals and existence of minimizers

Let $J : V \to \overline{\mathbb{R}}$ be a given functional. Consider the following problem: find $u \in V$ such that

$$J(u) \le J(v), \quad \forall v \in V. \qquad (5.4.1)$$

This problem is called the *primal variational problem* (or Problem $\mathcal{P}$). An investigation of such problems is motivated by the fact that many mathematical models in physics and mechanics have a variational form. An element $u \in V$ is said to be a *minimizer* to Problem $\mathcal{P}$ if

$$J(u) = \inf \mathcal{P} := \inf_{v \in V} J(v). \qquad (5.4.2)$$

The existence of a minimizer is the first question to be investigated. In general, the existence of a *minimizing sequence*, i.e., a sequence $\left\{v_k\right\}_{k=1}^{\infty}$ such that

$$J(v_k) \to \inf \mathcal{P}$$

does not guarantee the existence of a minimizer, because $\left\{v_k\right\}_{k=1}^{\infty}$ may have no limit in V. Therefore, the first subject to be investigated is the existence and uniqueness of minimizers. Below we briefly present the main facts of the existence theory for variational problems with convex functionals (the reader can find a detailed exposition of this theory in, e.g., [66, 105]).

5.4.1   Lower semicontinuous functionals

Recall that the functional $J : V \to \mathbb{R}$ is called *continuous* at v_0 if for any sequence $\left\{v_k\right\}_{k=1}^{\infty}$ converging to v_0 in V, we have

$$\lim_{k \to \infty} J(v_k) = J(v_0). \tag{5.4.3}$$

Also, J is said to be continuous at v_0 if for any α and β such that $\alpha < J(v_0) < \beta$ there exists a number $\varepsilon > 0$ such that

$$\alpha < J(v) < \beta, \quad \forall\, v \in \mathcal{B}(v_0, \varepsilon).$$

Let $\left\{\xi_k\right\}_{k=1}^{\infty}$ be a sequence of real numbers. By $\varliminf_{k \to \infty} \xi_k$ we denote the lower limit of this sequence, which may be finite or infinite.

Definition 5.4.1. *We say that $J : V \to \overline{\mathbb{R}}$ is lower semicontinuous at $v_0 \in V$ if*

$$\varliminf_{k \to \infty} J(v_k) \geq J(v_0) \tag{5.4.4}$$

for any sequence $\left\{v_k\right\}_{k=1}^{\infty}$ converging to v_0 in V.

Another definition of lower semicontinuous functionals is as follows.

Definition 5.4.2. *A functional J is said to be lower semicontinuous at $v_0 \in V$ if the inequality $J(v_0) > \alpha$ implies that this inequality also holds in a ball centered at v_0, i.e.,*

$$J(v) > \alpha, \quad \forall\, v \in \mathcal{B}(v_0, \varepsilon),$$

where ε is a positive number.

The relationship between these definitions is quite transparent. Indeed, assume that J is lower semicontinuous in the sense of Definition 5.4.2 and $v_k \to v_0$ in V. Take α such that $J(v_0) > \alpha$ and choose a respective $\mathcal{B}(v_0, \varepsilon)$. Then, $v_k \in \mathcal{B}(v_0, \varepsilon)$, provided that k is greater than a certain positive integer $N(\varepsilon)$. Consequently, $J(v_k) > \alpha$ and $\underline{\lim} J(v_k) \geq J(v_0)$. If V is a space with a countable base, then both definitions are equivalent. In general topological spaces, Definition 5.4.2 should be used.

Definition 5.4.3. *A functional that is lower semicontinuous at any point is called lower semicontinuous or an l.s.c. functional.*

Definition 5.4.4. *A functional G is called upper semicontinuous if $G = -J$, where J is a lower semicontinuous functional.*

Note that a functional is continuous if and only if it is simultaneously lower and upper semicontinuous.

Proposition 5.4.1. *The sets*

$$V_\alpha := \{v \in V \mid J(v) \leq \alpha, \quad \alpha \in \mathbb{R}\}$$

are closed if and only if J is an l.s.c. functional.

Proof. Let J be a lower semicontinuous functional. Then the set

$$V \setminus V_\alpha = \{v \in V \mid J(v) > \alpha\}$$

is open. Indeed, this fact follows from Definition 5.4.2, which says that any point v_0 belongs to $V \setminus V_\alpha$ with a certain neighborhood. Therefore, V_α is a closed set. If V_α is closed for any α, then $V \setminus V_\alpha$ is open and, consequently, any $v_0 \in V \setminus V_\alpha$ has a neighborhood $\mathcal{B}(v_0, \varepsilon) \subset V \setminus V_\alpha$. This means that $J(v) > \alpha$ for all $v \in \mathcal{B}(v_0, \varepsilon)$, i.e., J is a lower semicontinuous functional. $\square$

The theorem below shows the meaning of lower semicontinuity in proving existence theorems for variational problem.

Proposition 5.4.2. *Assume that $J : V \to \overline{\mathbb{R}}$ is an l.s.c. functional, the set V_α is compact for some $\alpha > \inf \mathcal{P}$, and*

$$\inf_{v \in V} J(v) = \inf \mathcal{P} > -\infty.$$

Then there exists an element $u \in V$ such that $J(u) = \inf \mathcal{P}$.

Proof. Let $\{v_k\}_{k=1}^{\infty}$ be a minimizing sequence. Then, $v_k \in V_\alpha$ for $k > N(\alpha)$. Since V_α is compact, we can extract a subsequence $\{v_{k_s}\}$ such that $v_{k_s} \to u \in V_\alpha$. Then

$$\inf \mathcal{P} \leq J(u) \leq \varliminf_{s \to \infty} J(v_{k_s}) = \inf \mathcal{P},$$

which shows that $J(u) = \inf \mathcal{P}$. $\qquad\square$

However, this theorem is difficult to use, because in the majority of variational models that are of practical importance, the sets V_α are not compact. This suggests to impose stronger conditions on J with less demanding conditions for V_α. Along this way, the lower semicontinuity condition is replaced by the *weak lower semicontinuity* condition.

Definition 5.4.5. *A functional* $J : V \to \overline{\mathbb{R}}$ *is said to be weakly lower semicontinuous at* $v_0 \in V$ *if*

$$\varliminf_{k \to \infty} J(v_k) \geq J(v_0)$$

for any sequence $\{v_k\}_{k=1}^{\infty}$ *that weakly converges to* v_0.

The concept of weak lower semicontinuity plays an important role in the calculus of variations. First, we need a suitable criterion for verifying whether or not a functional possesses such a property. In general, this is not an easy task. However, the weak lower semicontinuity of convex functionals is found to follow from the lower semicontinuity, which makes the verification of it rather simple. Below we prove this fact for Hilbert spaces.

Proposition 5.4.3 (Banach, Saks, and Mazur). *Let* $\{v_k\}$ *be a sequence of elements in the Hilbert space* H, *which weakly converges to* $v_0 \in H$. *Then one can find a subsequence* $\{v_{k_i}\}$ *such that the sequence* $w_m = \dfrac{1}{m} \sum_{i=1}^{m} v_{k_i}$ *strongly converges to* v_0 *in* H.

Proof. Without loss of generality, we may consider only the case $v_0 = 0$. Let us construct a subsequence $\{v_{k_i}\}$ by the following procedure. Set $v_{k_1} = v_1$. Since $(v_{k_1}, v) \to 0$ as $k \to \infty$, we find v_{k_2} such that $|(v_{k_1}, v_{k_2})| < 1$. Assume that $v_{k_1}, \ldots, v_{k_i}$ are determined. Find $v_{k_{i+1}}$ such that

$$\left|(v_{k_1}, v_{k_{i+1}})\right| < \frac{1}{i}, \ \left|(v_{k_2}, v_{k_{i+1}})\right| < \frac{1}{i}, \ \ldots, \ \left|(v_{k_i}, v_{k_{i+1}})\right| < \frac{1}{i}.$$

The sequence $\{v_k\}$ is bounded and, therefore, $\|v_{k_i}\| \leq c < +\infty$. Note that the amount of products (v_{k_i}, v_{k_j}), where $i < j$, is $2(j-1)$ (the multiplier 2 arises due to the symmetry) and the value of such a product is no greater than $1/(j-1)$. In view of this fact, we have

$$\|w_m\|^2 = \frac{1}{m^2}\left(\sum_{i=1}^m v_{k_i}, \sum_{j=1}^m v_{k_j}\right) \le \frac{1}{m^2}\left(mc^2 + 2(m-1)\right).$$

The right-hand side of this estimate tends to zero as $m \to +\infty$. Hence, the required statement is proved. $\qquad\square$

Remark 5.4.1. Proposition 5.4.3 implies the following result: any convex closed set K is weakly closed. Indeed, if $\{v_k\} \in K$ and v_k weakly converges to $v_0 \in H$ (i.e., $v_k \rightharpoonup v_0$), then one can find a sequence of "averaged" elements $\{w_k\}$ that converges to v_0. Since K is closed, we see that $v_0 \in K$ and, therefore, K is weakly closed.

Proposition 5.4.4. *Let $J : H \to \overline{\mathbb{R}}$ be a lower semicontinuous functional. If J is convex, then it is weakly lower semicontinuous.*

Proof. Our aim is to show that

$$\varliminf_{k\to\infty} J(v_k) \ge J(v_0)$$

for any sequence $\{v_k\}$ such that $v_k \rightharpoonup v_0$. By the definition of the lower limit, the sequence $\{v_k\}$ contains a subsequence $\{v_{k_s}\}$ such that

$$\varliminf_{k\to\infty} J(v_k) = \lim_{s\to\infty} J(v_{k_s}).$$

Obviously, v_{k_s} weakly converges to v_0. Thus, we may apply Proposition 5.4.3, which shows that $\{v_{k_s}\}$ contains a subsequence $\{v_i\}$ such that $w_m = 1/m \sum_{i=1}^m v_i$ strongly converges to v_0. By the convexity of J, we obtain

$$\frac{1}{m}\sum_{i=1}^m J(v_i) \ge J\left(\frac{1}{m}\sum_{i=1}^m v_i\right) = J(w_m).$$

Since J is lower semicontinuous,

$$\varliminf_{m\to\infty} J(w_m) \ge J(v_0).$$

we see that

$$\varliminf_{k\to\infty} J(v_k) = \lim_{m\to\infty} \frac{1}{m}\sum_{i=1}^m J(v_i) \ge J(v_0).$$

$\qquad\square$

Remark 5.4.2. Propositions 5.4.3 and 5.4.4 admit of wide generalizations. They remain valid for reflexive Banach spaces. After the replacement of weak convergence by the so-called $*$–weak convergence, they are also valid for topological vector spaces (see, e.g., [190]).

5.4.2 Existence of minimizers

Let us now return to the main question and formulate sufficient conditions for the existence of a minimizer to Problem $\mathcal{P}$.

Definition 5.4.6. *The functional $J : V \to \overline{\mathbb{R}}$ is called coercive if*

$$J(v_k) \to +\infty \tag{5.4.5}$$

for any sequence $\{v_k\}_{k=1}^{\infty}$ such that $\|v_k\| \to +\infty$.

Proposition 5.4.5. *If $J : H \to \overline{\mathbb{R}}$ is coercive, then the sets*

$$V_\alpha := \{v \in V \,|\, J(v) \leq \alpha\}$$

are bounded.

Proof. We prove this proposition by contradiction. Assume that V_α is unbounded, i.e., it cannot be embedded into any ball $\mathcal{B}(0, d)$. Then, for any $k \in \mathbf{N}$ we can find $\{v_k\} \in V_\alpha$ such that $\|v_k\| > k$, so that from (5.4.5) we conclude that $J(v_k) \to +\infty$. But this is impossible, because $J(v_k) < \alpha$. We arrive at the contradiction, which shows that the assumption is wrong. $\square$

Now we proceed to proving the main result in this section.

Theorem 5.4.1. *Let $J : V \to \overline{\mathbb{R}}$ be a proper convex functional that is coercive and lower semicontinuous on the reflexive Banach space V. In addition, assume that*

$$\inf \mathcal{P} > -\infty.$$

Then, Problem $\mathcal{P}$ has a solution. Moreover, if J is strictly convex, then the solution is unique.

Proof. Take a sequence $\{v_k\}_{k=1}^{\infty}$ such that

$$J(v_1) \geq J(v_2) \geq \ldots J(v_k) \to \inf \mathcal{P}.$$

Then,

$$\{v_k\} \in V_\alpha, \qquad \alpha = J(v_1).$$

The set V_α is convex and closed (see Propositions 5.1.2 and 5.4.1). Thus, this set is also weakly closed, i.e., it contains the limits of all weakly converging sequences (see Remark 5.4.1). The set V_α is bounded (see Proposition 5.4.5). Therefore, the sequence $\{v_k\}$ contains a weakly converging subsequence $\{v_{k_i}\}$, whose weak limit u belongs to the set V_α. The functional J is weakly lower semicontinuous (see Proposition 5.4.4) and, consequently,

$$\inf \mathcal{P} = \varliminf_{i \to \infty} J(v_{k_i}) \geq J(u).$$

Since $J(u) \geq \inf \mathcal{P}$, we conclude that $J(u) = \inf \mathcal{P}$, which means that a minimizer exists.

Assume that u_1 and u_2 are two different minimizers of Problem $\mathcal{P}$, i.e.,

$$\inf \mathcal{P} = J(u_1) = J(u_2).$$

If the functional is strictly convex, then

$$J\left(\lambda_1 u_1 + \lambda_2 u_2\right) < \lambda_1 J(u_1) + \lambda_2 J(u_2),$$

or

$$J\left(\lambda_1 u_1 + \lambda_2 u_2\right) < \inf \mathcal{P},$$

and we arrive at a contradiction, which shows that the minimizer is unique. $\qquad\square$

5.4.3 Dual variational problems

In addition to V and V^*, introduce another pair of mutually dual reflexive spaces Y and Y^* with duality pairing $\langle\!\langle \cdot, \cdot \rangle\!\rangle$. By K and K^* we denote nonempty convex sets in V and Y^*, respectively. Let us define a functional $L : K \times K^* \to \mathbb{R}$ that possesses the following properties:

(i) for any $y^* \in K^*$, the functional $v \mapsto L(v, y^*)$ is convex and lower semicontinuous;

(ii) for any $v \in K$, the functional $y^* \mapsto L(v, y^*)$ is concave and upper semicontinuous.

The functional L is called *Lagrangian.* It gives rise to two variational problems. To define them, we introduce two functionals:

$$J(v) := \sup_{y^* \in K^*} L(v, y^*) \tag{5.4.6}$$

and

$$I^*(y^*) := \inf_{v \in K} L(v, y^*). \tag{5.4.7}$$

The functional J is defined as an upper bound of convex l.s.c. functionals and, therefore, is a convex l.s.c. functional. Analogously, I^* is a concave and upper semicontinuous functional. The functionals J and I^* generate two variational problems.

Problem $\mathcal{P}$.   Find $u \in K$ such that

$$J(u) = \inf \mathcal{P} := \inf_{v \in K} J(v). \tag{5.4.8}$$

Problem $\mathcal{P}^*$.   Find $p^* \in K^*$ such that

$$I^*(p^*) = \sup \mathcal{P}^* := \sup_{y^* \in K^*} I^*(y^*). \tag{5.4.9}$$

Henceforth, Problems $\mathcal{P}$ and $\mathcal{P}^*$ are called *primal* and *dual*, respectively. They are closely related to a minimax problem, which is called the *saddle-point* problem for the Lagrangian L.

Problem $\mathcal{L}$.   Find $(\overline{u}, \overline{p}^*) \in K \times K^*$ such that

$$L(\overline{u}, y^*) \leq L(\overline{u}, \overline{p}^*) \leq L(v, \overline{p}^*), \quad \forall v \in K, y^* \in K^*. \tag{5.4.10}$$

The pair of elements $(\overline{u}, \overline{p}^*)$ is called a saddle point of L on $K \times K^*$. In the remainder of this section, we analyze properties of $\overline{u}$ and $\overline{p}^*$ and their relations with solutions of Problems $\mathcal{P}$ and $\mathcal{P}^*$. First, we note that

$$\sup \mathcal{P}^* \leq \inf \mathcal{P}. \tag{5.4.11}$$

Since

$$\sup \mathcal{P}^* = \sup_{y^* \in K^*} \inf_{v \in K} L(v, y^*) \quad \text{and} \quad \inf \mathcal{P} = \inf_{v \in K} \sup_{y^* \in K^*} L(v, y^*),$$

the relation (5.4.11) follows from (5.2.16). Clearly, in some cases left- or right-hand side of (5.4.11) (or both of them) may be infinite. Let us prove that if Problem $\mathcal{L}$ has a solution, then its components are solutions of Problems $\mathcal{P}$ and $\mathcal{P}^*$ and (5.4.11) holds as equality. First, we present a simple saddle-point criterion.

Proposition 5.4.6. *If there exist a constant α such that*

$$L(u, y^*) \leq \alpha, \quad \forall y^* \in K^*, \tag{5.4.12}$$

and

$$L(v, p^*) \geq \alpha, \quad \forall v \in K, \tag{5.4.13}$$

then (u, p^*) is a saddle point. Moreover, we have the relation

$$\alpha = \inf_{v \in K} \sup_{y^* \in K^*} L(v, y^*) = \sup_{y^* \in K^*} \inf_{v \in K} L(v, y^*). \qquad (5.4.14)$$

Proof. From (5.4.12) and (5.4.13), we obtain

$$L(u, p^*) \leq \alpha \leq L(u, p^*).$$

Therefore, $L(u, p^*) = \alpha$ and

$$L(u, y^*) \leq L(u, p^*) \leq L(v, p^*), \quad \forall v \in K, \forall y \in K^*,$$

which means that (u, p^*) is a saddle point. Since

$$\sup_{y^* \in K^*} L(u, y^*) = L(u, p^*) = \alpha$$

and

$$\inf_{v \in K} L(v, p^*) = L(u, p^*) = \alpha,$$

we have

$$\inf \mathcal{P} = \inf_{v \in K} \sup_{y^* \in K^*} L(v, y^*) \leq \sup_{y^* \in K^*} L(u, y^*) = \alpha,$$
$$\sup \mathcal{P}^* = \sup_{y^* \in K^*} \inf_{v \in K} L(v, y^*) \geq \inf_{v \in K} L(v, p^*) = \alpha.$$

In view of (5.4.11), we arrive at (5.4.14) . $\qquad\square$

Proposition 5.4.7. *The set of all saddle points of the Lagrangian L has the form $K_0 \times K_0^*$, where K_0 and K_0^* are convex subsets of K and K^*, respectively.*

Proof. Let (u_1, p_1^*) and (u_2, p_2^*) be two different saddle points. Then

$$L(u_1, y^*) \leq L(u_1, p_1^*) = \alpha = L(u_2, p_2^*) \leq L(v, p_1^*),$$
$$L(u_2, y^*) \leq L(u_1, p_1^*) = \alpha = L(u_2, p_2^*) \leq L(v, p_2^*),$$

where v and y^* are arbitrary elements of the sets K and K^*, respectively. From the first relation, we obtain

$$L(u_2, p_1^*) \geq \alpha,$$

and from the second one we have

$$L(u_2, p_1^*) \leq \alpha.$$

Now, Proposition 5.4.6 implies that (u_2, p_1^*) is a saddle point. The same conclusion is obviously true for (u_1, p_2^*). Therefore, if (u_i, p_i^*), $i = 1, 2, \ldots$, are saddle points, then the pairs (u_i, p_j^*), $i = 1, 2, \ldots, j = 1, 2, \ldots$, are also saddle points. This means that the set of all saddle points has a structure $K_0 \times K_0^*$, where K_0 and K_0^* are some subsets of K and K_0^*, respectively. Let u_1 and u_2 be two different elements of K_0. Then,

$$L(u_1, y^*) \leq L(u_1, p_1^*) = \alpha, \quad \forall y^* \in K^*,$$
$$L(u_2, y^*) \leq L(u_2, p_1^*) = \alpha, \quad \forall y^* \in K^*.$$

By assumption (i), the functional $v \mapsto L(v, y^*)$ is convex. Therefore, for any λ_1, λ_2 such that $\lambda_1 + \lambda_2 = 1$ we have

$$L\left(\lambda_1 u_1 + \lambda_2 u_2, y^*\right) \leq \lambda_1 L(u_1, y^*) + \lambda_2 L(u_2, y^*) \leq \alpha.$$

Since

$$L(v, p_1^*) \geq \alpha, \quad \forall v \in K,$$

we deduce the opposite inequality

$$L\left(\lambda_1 u_1 + \lambda_2 u_2, p_1^*\right) \geq \alpha.$$

It remains to conclude that

$$L\left(\lambda_1 u_1 + \lambda_2 u_2, p_1^*\right) = \alpha.$$

Hence, $\lambda_1 u_1 + \lambda_2 u_2 \in K_0$. The convexity of K_0^* can be proved in the same fashion. $\qquad\square$

Now we state the main theorem that establishes a link between the solutions of Problems $\mathcal{P}$ and $\mathcal{P}^*$ and the saddle points of Problems $\mathcal{L}$.

Theorem 5.4.2. *The following two statements are equivalent:*

1. there exists a pair of elements $u \in K$ and $p^* \in K^*$ such that

$$J(u) = \inf \mathcal{P}, \tag{5.4.15}$$

$$I^*(p^*) = \sup \mathcal{P}^*, \tag{5.4.16}$$

$$\inf \mathcal{P} = \sup \mathcal{P}^* \tag{5.4.17}$$

2. (u, p^*) is a saddle point of the Lagrangian L on $K \times K^*$.

Moreover, any of the above two assertions implies the principal relation

$$I^*(p^*) = L(u, p^*) = J(u). \tag{5.4.18}$$

Proof. Let the first assumption be true. Then

$$L(u, y^*) \leq \sup_{y^* \in K^*} L(u, y^*) = J(u) = \alpha, \quad \forall y^* \in K^*,$$

$$L(v, p^*) \geq \inf_{v \in K} L(v, p^*) = I^*(p^*) = \alpha, \quad \forall v \in K.$$

According to Proposition 5.4.6, (u, p^*) is a saddle point. Let (u^*, p^*) be a saddle point, i.e.,

$$L(u, y^*) \leq L(u, p^*) \leq L(v, p^*), \quad \forall v \in K, y^* \in K^*.$$

From this double inequality we obtain

$$J(u) = \sup_{y^* \in K^*} L(u, y^*) \leq L(u, p^*) \leq$$

$$\leq L(v, p^*) \leq \sup_{y^* \in K^*} L(v, y^*) = J(v), \quad \forall v \in K,$$

and

$$I^*(p^*) = \inf_{v \in K} L(v, p^*) \geq L(u, p^*) \geq$$

$$\geq L(u, y^*) \geq \inf_{v \in K} L(v, y^*) = I^*(y^*), \quad \forall y^* \in K^*.$$

Hence, $u \in K$ and $p^* \in K^*$ are solutions of Problems $\mathcal{P}$ and $\mathcal{P}^*$, respectively. Furthermore,

$$L(u, p^*) \leq \sup_{y^* \in K^*} L(u, y^*) = J(u) \leq L(u, p^*),$$

$$L(u, p^*) \geq \inf_{v \in K} L(v, p^*) = I^*(p^*) \geq L(u, p^*),$$

and the relation (5.4.18) follows. $\qquad\square$

Before closing this concise review of saddle-point theory, we shall present two assertions that may be useful in checking the correctness of particular saddle-point problems.

Theorem 5.4.3. *If assumptions (i) and (ii) imposed on L hold and the sets K and K^* are bounded, then L possesses at least one saddle point and* $\inf \mathcal{P} = \sup \mathcal{P}^*$.

Theorem 5.4.4. *Let (i) and (ii) hold and there exist elements $p^* \in K^*$ and $u \in K$ such that*

$$L(v_k, p^*) \to +\infty \qquad \textit{for any sequence } \{v_k\} \in K$$
$$\textit{such that } \|v_k\|_V \to +\infty, \qquad (5.4.19)$$

$$L(u, y_k^*) \to -\infty \qquad \textit{for any sequence } \{y_k^*\} \in K^*$$
$$\textit{such that } \|y_k^*\|_{Y^*} \to +\infty. \qquad (5.4.20)$$

Then, L has at least one saddle point.

Combining the conditions of the above two theorems, one can prove, for example, that a saddle point exists if K is bounded and the coercivity condition (5.4.20) holds (or K^* is bounded and (5.4.19) holds). It is also worth noting that the basic relation $\inf \mathcal{P} = \sup \mathcal{P}^*$ is true even if only one of the coercivity conditions hold (provided that (i) and (ii) also hold). Proofs of all these results and a more detailed exposition of saddle-point theory can be found in [66]. Consider a particular but important class of Lagrangians

$$L(v, y^*) := \langle y^*, \Lambda v \rangle - G^*(y^*) - \langle\langle l^*, v \rangle\rangle,$$

where $\Lambda : V \to Y$ is a linear bounded operator, $G^* : Y^* \to \mathbb{R}$ is a convex l.s.c. functional, and $l^* \in V^*$. Assume that L has a saddle point $(u, p^*) \in K \times K^*$, i.e.,

$$L(u, y^*) \leq L(u, p^*) \leq L(v, p^*), \quad \forall v \in K, y^* \in K^*.$$

The inequality on the right-hand side means that

$$\langle p^*, \Lambda(u - v) \rangle \leq \langle\langle l^*, u - v \rangle\rangle, \quad \forall v \in K, \tag{5.4.21}$$

and the inequality on the left-hand side leads to the relation

$$\langle p^* - y^*, \Lambda u \rangle \geq G^*(p^*) - G^*(y^*), \quad \forall y^* \in K^*. \tag{5.4.22}$$

Let $\Lambda^* : Y^* \to V^*$ denote the operator conjugate to Λ, i.e., $\langle y^*, \Lambda v \rangle = \langle\langle \Lambda^* y^*, v \rangle\rangle$. Then, (5.4.21) has the form

$$\langle\langle \Lambda^* p^* - l^*, u - v \rangle\rangle \leq 0, \quad \forall v \in K, \tag{5.4.23}$$

which yields *a necessary condition* for the maximizer of Problem $\mathcal{P}^*$. If K is a linear manifold, then (5.4.23) implies the relation

$$\Lambda^* p^* = l^*. \tag{5.4.24}$$

By (5.4.22), we obtain

$$G^*(y^*) \geq G^*(p^*) + \langle y^* - p^*, \Lambda u \rangle, \quad \forall y^* \in K^*.$$

If K^* coincides with Y^*, then this relation shows that

$$\Lambda u \in \partial G^*(p^*). \tag{5.4.25}$$

By virtue of Proposition 5.3.2, this also means that

$$p^* \in \partial G(\Lambda u). \tag{5.4.26}$$

Hence, we observe that the solutions of Problems $\mathcal{P}$ and $\mathcal{P}^*$ are connected by some relations called henceforth the *duality relations*.

5.5 Uniformly convex functionals

In this section, we consider a special class of convex functionals. As before, Y denotes a reflexive Banach space and Y^* is its topologically dual counterpart. Consider a proper l.s.c. functional $\Upsilon : Y \to \overline{\mathbb{R}}$ subject to the following conditions:

$$\Upsilon(y) \geq 0, \qquad \forall y \in Y, \tag{5.5.1}$$

$$\Upsilon(y) = 0 \iff y = 0_Y. \tag{5.5.2}$$

The set of all such functionals is denoted by $\Gamma_0^+(Y)$.

Proposition 5.5.1. *If $\Upsilon \in \Gamma_0^+(Y)$, then $\Upsilon^* \in \Gamma_0^+(Y^*)$.*

Proof. For any $y^* \in Y^*$, we have

$$\Upsilon^*(y^*) = \sup_{y \in \mathbb{R}}\{\langle y^*, y \rangle - \Upsilon(y)\} \geq \langle y^*, 0_Y \rangle - \Upsilon(0_Y) = 0.$$

Further, since

$$\Upsilon^*(0_{Y^*}) = \sup_{y \in Y}\{-\Upsilon(y)\},$$

where $\Upsilon(y) \geq 0$, we see that $\Upsilon^*(0_{Y^*}) = 0$. $\square$

Definition 5.5.1. *A convex functional $I : Y \to \overline{\mathbb{R}}$ is called uniformly convex in $\mathcal{B}(0_Y, \delta)$ if there exists a functional $\Upsilon_\delta \in \Gamma_0^+(Y)$ such that for all $y_1, y_2 \in \mathcal{B}(0_Y, \delta)$ the following inequality holds:*

$$I\left(\frac{y_1 + y_2}{2}\right) + \Upsilon_\delta(y_1 - y_2) \leq \frac{1}{2}\left(I(y_1) + I(y_2)\right). \tag{5.5.3}$$

The functional Υ_δ enforces standard convexity inequality. For this reason, it is called a *forcing* functional. Henceforth, we assume that Υ_δ is even, i.e.,

$$\Upsilon_\delta(y) = \Upsilon_\delta(-y) \quad \forall y \in \mathcal{B}(0_Y, \delta).$$

Definition 5.5.1 leads to the following statement.

Proposition 5.5.2. *Assume that I is uniformly convex in $\mathcal{B}(0_Y, \delta)$. Then, for any λ_1, λ_2 such that $\lambda_1 + \lambda_2 = 1$, there exists a functional $\Upsilon_\delta^\lambda \in \Gamma_0^+(Y)$ such that the relation*

$$I(\lambda_1 y_1 + \lambda_2 y_2) + \Upsilon_\delta^\lambda(y_1 - y_2) \leq \lambda_1 I(y_1) + \lambda_2 I(y_2) \tag{5.5.4}$$

holds for any $y_1, y_2 \in \mathcal{B}(0_Y, \delta)$.

Proof. Let $z \in \mathcal{B}(0_Y, \delta)$. Then,

$$I\left(\frac{y_1 + z}{2}\right) + \Upsilon_\delta(y_1 - z) \leq \frac{1}{2}\left(I(y_1) + I(z)\right),$$

$$I\left(\frac{y_2 + z}{2}\right) + \Upsilon_\delta(y_2 - z) \leq \frac{1}{2}\left(I(y_2) + I(z)\right).$$

Multiply the first inequality by λ_1 and the second one by λ_2 and add them. We have

$$\lambda_1 I\left(\frac{y_1 + z}{2}\right) + \lambda_2 I\left(\frac{y_2 + z}{2}\right) +$$

$$+ \lambda_1 \Upsilon_\delta(y_1 - z) + \lambda_2 \Upsilon_\delta(y_2 - z) \leq$$

$$\leq \frac{\lambda_1}{2} I(y_1) + \frac{\lambda_2}{2} I(y_2) + \frac{1}{2} I(z). \quad (5.5.5)$$

Take $z = \lambda_1 y_1 + \lambda_2 y_2 \in \mathcal{B}(0_Y, \delta)$ and note that

$$\lambda_1 I\left(\frac{y_1 + z}{2}\right) + \lambda_2 I\left(\frac{y_2 + z}{2}\right) \geq I\left(\frac{\lambda_1 y_1 + \lambda_2 y_2 + z}{2}\right).$$

Then, we obtain

$$\frac{1}{2} I\left(\lambda_1 y_1 + \lambda_2 y_2\right) + \lambda_1 \Upsilon_\delta\left(\lambda_2(y_1 - y_2)\right) +$$

$$+ \lambda_2 \Upsilon_\delta\left(\lambda_1(y_2 - y_1)\right) \leq \frac{\lambda_1}{2} I(y_1) + \frac{\lambda_2}{2} I(y_2). \quad (5.5.6)$$

The relation (5.5.6) is equivalent to (5.5.4) for

$$\Upsilon_\delta^\lambda(y_1 - y_2) = 2\lambda_1 \Upsilon_\delta\left(\lambda_2(y_1 - y_2)\right) + 2\lambda_2 \Upsilon_\delta\left(\lambda_1(y_2 - y_1)\right).$$

$\square$

From (5.5.4) it is clear that any uniformly convex functional is convex in $\mathcal{B}(0_Y, \delta)$. Now we establish two important inequalities that hold for uniformly convex functionals.

Proposition 5.5.3. *If $I : Y \to \overline{\mathbb{R}}$ is uniformly convex and Gâteaux differentiable in $\mathcal{B}(0_Y, \delta)$, then for any $y, z \in \mathcal{B}(0_Y, \delta)$ the following relations hold:*

$$I(z) \geq I(y) + \langle I'(y), z - y \rangle + 2\Upsilon_\delta(z - y) \quad (5.5.7)$$

and

$$\langle I'(z) - I'(y), z - y \rangle \geq 2\Upsilon_\delta(z - y) + 2\Upsilon_\delta(y - z). \quad (5.5.8)$$

Proof. From the definition, it follows that

$$\Upsilon_\delta(z - y) \le \frac{1}{2}I(z) + \frac{1}{2}I(y) - I\left(\frac{z + y}{2}\right).$$

Since I is convex and differentiable (cf. (5.3.9)), we have

$$I\left(\frac{z + y}{2}\right) \ge I(y) + \left\langle I'(y), \frac{z - y}{2}\right\rangle,$$

and, therefore,

$$2\Upsilon_\delta(z - y) \le I(z) - I(y) - \left\langle I'(y), z - y\right\rangle.$$

The latter inequality coincides with (5.5.7). We can rewrite it replacing z by y

$$2\Upsilon_\delta(y - z) \le I(y) - I(z) + \left\langle I'(z), z - y\right\rangle$$

and obtain (5.5.8). $\qquad\square$

Assume that $y_\bullet \in \mathcal{B}(0_Y, \delta)$ is a minimizer of I, so that $I'(y_\bullet) = 0_{Y^*}$. Then, Proposition 5.5.3 implies the relation

$$\Upsilon_\delta(z - y_\bullet) \le \frac{1}{2}\left(I(z) - I(y_\bullet)\right), \quad \forall z \in \mathcal{B}(0_Y, \delta). \tag{5.5.9}$$

Estimate (5.5.9) can also be extended to the entire class of uniformly convex functionals.

Proposition 5.5.4. *Let a functional I be uniformly convex and $y_\bullet \in \mathcal{B}(0_Y, \delta)$ be the minimizer of I. Then (5.5.9) holds.*

Proof. Since $I\left(\frac{y_\bullet + z}{2}\right) \ge I(y_\bullet)$, we obtain

$$\Upsilon_\delta(z - y_\bullet) \le \frac{1}{2}I(y_\bullet) + \frac{1}{2}I(z) - I\left(\frac{y_\bullet + z}{2}\right) \le$$

$$\le \frac{1}{2}\left(I(z) - I(y_\bullet)\right).$$

$$\square$$

Estimate (5.5.9) is the first step in deriving a posteriori error estimates of the functional type. It shows that deviations from the minimizer (measured in terms of the functional Υ_δ) are controlled by the difference of the functionals.

Corollary 5.5.1. Let I be differentiable. Rewrite (5.5.9) in the form

$$\Upsilon_\delta(z - y_\bullet) + I\left(\frac{y_\bullet + z}{2}\right) - I(y_\bullet) \le \frac{1}{2}\left(I(z) - I(y_\bullet)\right).$$

By virtue of (5.5.7), we have

$$I\left(\frac{y_\bullet + z}{2}\right) - I(y_\bullet) \ge 2\Upsilon_\delta\left(\frac{z - y_\bullet}{2}\right)$$

and, therefore, we arrive at the strengthened estimate

$$\Upsilon_\delta(z - y_\bullet) + 2\Upsilon_\delta\left(\frac{z - y_\bullet}{2}\right) \le \frac{1}{2}\left(I(z) - I(y_\bullet)\right). \qquad (5.5.10)$$

Assume that I is twice differentiable in the vicinity of $y_\bullet$ and satisfies the finite increment relation

$$I\left(\frac{y_\bullet + z}{2}\right) = \quad I(y_\bullet) + \left\langle I'(y_\bullet), \frac{z - y_\bullet}{2}\right\rangle +$$
$$+ \frac{1}{2}\left\langle I''\left(y_\bullet + \xi\frac{z + y_\bullet}{2}\right)\frac{z - y_\bullet}{2}, \frac{z - y_\bullet}{2}\right\rangle,$$

where $\xi \in (0, 1)$. Since $I'(y_\bullet) = 0_{Y^*}$, we have another estimate:

$$\Upsilon_\delta(z - y_\bullet) + \frac{1}{8}\left\langle I''\left((1 + \tfrac{\xi}{2})y_\bullet + \tfrac{\xi}{2}z\right)(z - y_\bullet), z - y_\bullet\right\rangle \le$$
$$\le \frac{1}{2}\left(I(z) - I(y_\bullet)\right). \qquad (5.5.11)$$

Example 5.5.1. Consider a self-adjoint operator $A \in \mathcal{L}(H, H)$ defined on a Hilbert space H with scalar product $(\cdot, \cdot)$. Assume that it satisfies the condition

$$\alpha_1 \|y\|^2 \le (Ay, y) \le \alpha_2 \|y\|^2, \qquad \forall y \in H.$$

Define the functional

$$I(y) = \frac{1}{2}(Ay, y) + (l, y), \qquad l \in H. \qquad (5.5.12)$$

Since

$$\frac{1}{2}I(y) + \frac{1}{2}I(z) - I\left(\frac{y + z}{2}\right) =$$
$$= \frac{1}{4}(Ay, y) + \frac{1}{4}(Az, z) - \frac{1}{8}(A(y + z), y + z) =$$
$$= \frac{1}{8}(A(z - y), (z - y)),$$

the functional I is uniformly convex in any ball with

$$\Upsilon(z - y) = \frac{1}{8}(A(z - y), (z - y)).$$

Let $y_\bullet$ be a minimizer of I, i.e., $I(y_\bullet) \leq I(y)$ for any $y \in H$. Then (5.5.11) gives the estimate

$$\frac{1}{2}(A(z - y_\bullet), (z - y_\bullet)) \leq I(z) - I(y_\bullet). \tag{5.5.13}$$

Note that, in this case, we have

$$I(z) - I(y_\bullet) = (Ay_\bullet + l, z - y_\bullet) + \frac{1}{2}(A(z - y_\bullet), z - y_\bullet).$$

and

$$(Ay_\bullet + l, y) = 0, \qquad \forall y \in Y. \tag{5.5.14}$$

Therefore, (5.5.13) holds as equality.

Proposition 5.5.5. *Let I_1 and I_2 be uniformly convex in $\mathcal{B}(0_Y, \delta)$ with functionals $\Upsilon_{1\delta}$ and $\Upsilon_{2\delta}$, respectively. Then the functional $\mu_1 I_1 + \mu_2 I_2$, where $\mu_1, \mu_2 \geq 0$, is uniformly convex in $\mathcal{B}(0_Y, \delta)$ with $\Upsilon_\delta = \mu_1 \Upsilon_{1\delta} + \mu_2 \Upsilon_{2\delta}$.*

Proof. The proposition follows directly from Definition 5.5.1. $\square$

Example 5.5.2. Consider the functional

$$I(y) = \frac{1}{2}(Ay, y) + (l, y) + \Psi(y), \tag{5.5.15}$$

where $\Psi(y)$ is a convex and l.s.c. functional. Applying Proposition 5.5.5 with $\mu_1 = \mu_2 = 1$, $I_1(y) = \frac{1}{2}(Ay, y) + (l, y)$, and $I_2(y) = \Psi(y)$, we see that I is uniformly convex with functional Υ defined in Example 5.5.1.

Proposition 5.5.6. *Let I_1 and I_2 be uniformly convex in $\mathcal{B}(0_Y, \delta)$ with functionals $\Upsilon_{1\delta}$ and $\Upsilon_{2\delta}$, respectively. Then the functional $I(y) = \max\{I_1(y), I_2(y)\}$ is uniformly convex with $\Upsilon_\delta = \min\{\Upsilon_{1\delta}, \Upsilon_{2\delta}\}$.*

Proof. We have

$$\frac{1}{2}I(y) + \frac{1}{2}I(z) - I\left(\tfrac{y+z}{2}\right) = \frac{1}{2}\max\{I_1(y), I_2(y)\}+$$

$$+ \frac{1}{2}\max\{I_1(z), I_2(z)\} - \max\left\{I_1\left(\tfrac{y+z}{2}\right), I_2\left(\tfrac{y+z}{2}\right)\right\}.$$

Assume that

$$\max\left\{I_1\left(\tfrac{y+z}{2}\right), I_2\left(\tfrac{y+z}{2}\right)\right\} = I_1\left(\tfrac{y+z}{2}\right).$$

Then

$$\frac{1}{2}\left(I(y)+I(z)\right) - I\left(\tfrac{y+z}{2}\right) \geq \frac{1}{2}\left(I_1(y)+I_1(z)\right) - I_1\left(\tfrac{y+z}{2}\right) \geq \Upsilon_{1\delta}(z-y).$$

If

$$\max\left\{I_1\left(\tfrac{y+z}{2}\right), I_2\left(\tfrac{y+z}{2}\right)\right\} = I_2\left(\tfrac{y+z}{2}\right),$$

then

$$\frac{1}{2}I(y) + \frac{1}{2}I(z) - I\left(\tfrac{y+z}{2}\right) \geq \Upsilon_{2\delta}(z-y).$$

Thus, in both cases the lower bound is given by the functional

$$\min\{\Upsilon_{1\delta}(z-y), \Upsilon_{2\delta}(z-y)\}.$$

$\square$

5.6   Integral functionals

In many applications, variational problems are stated in terms of integral functionals. For example, consider the functional

$$G(y) = \int_{\Omega} g(x, y(x))\, dx, \tag{5.6.1}$$

where $g : \Omega \times \mathbb{R}^n \to \mathbb{R}$ is a given function and Ω is a bounded open domain in $\mathbb{R}^d$. The functional G is an integral functional with *integrand g*. It is of significance that the main properties of integral functionals (convexity, uniform convexity, lower semicontinuity) depend on the properties of their integrands. Owing to this fact, those properties can be verified by investigating the function g. Moreover, for the cases analyzed in the book, the determination of Fenchel conjugate functionals and Gâteaux derivatives is also reduced to operations with integrands. Here, we present a concise overview of the respective results. For a more detailed exposition and proofs, we refer the reader to [66, 105] and the references therein.

5.6.1   Lower semicontinuity

We restrict ourselves to the case in which the integrand is subject to the following conditions: $g(x, \xi)$ is nonnegative, it is measurable and integrable with respect to x and continuous with respect to ξ for almost all $x \in \Omega$.

Proposition 5.6.1. *Let $\{y_k\}_{k=1}^{\infty}$ be a sequence of measurable functions converging to y_0 almost everywhere. Then*

$$\lim_{k\to\infty} \int_{\Omega} g(x, y_k(x))\, dx \geq \int_{\Omega} g(x, y_0(x))\, dx. \qquad (5.6.2)$$

Proof. The proof is based upon the Fatou Lemma: *if a sequence $\{f_k(x)\}_{k=1}^{\infty}$ of measurable nonnegative functions converges to $f_0(x)$ almost everywhere in Ω and $\int_{\Omega} f_k(x)\, dx \leq C$, then f_0 is integrable and $\int_{\Omega} f_0(x)\, dx \leq C$.*

We have a sequence $f_k(x) = g(x, y_k(x))$ that meets the conditions of this lemma. For almost all $x \in \Omega$, $y_k(x) \to y_0(x)$, whence $f_k(x) \to f_0(x) = g(x, y_0(x))$. Take any C such that $C \geq \lim_{k\to\infty} \int_{\Omega} f_k(x)\, dx$. Then $C \geq \int_{\Omega} f_0(x)\, dx$ and, therefore, the relation (5.6.2) holds. $\qquad\square$

Remark 5.6.1. Let $\{y_k(x)\}_{k=1}^{\infty}$ converge to $y_0(x)$ in $L^p(\Omega)$ $(p \geq 1)$. Such a sequence contains a subsequence $\{y_{k_i}(x)\}_{i=1}^{\infty}$ that converges to $y_0(x)$ for almost all $x \in \Omega$. Since $\int_{\Omega} g(x, y_{k_i}(x))$ tends to the same value as the whole sequence, we conclude that the integral functional preserves the lower semicontinuity property.

5.6.2 Convexity

It is quite obvious that the functional G inherits the convexity property of its integrand.

Proposition 5.6.2. *If a function $g(x, \xi)$ is convex with respect to ξ for almost all $x \in \Omega$, then the functional G is convex on Y.*

Proof. Integrating the inequality

$$g(x, \lambda_1 y_1(x) + \lambda_2 y_2(x)) \leq \lambda_1 g(x, y_1(x)) + \lambda_2 g(x, y_2(x))$$

over all $x \in \Omega$, we arrive at the desired result. $\qquad\square$

The assertion below shows a way for establishing the uniform convexity of an integral functional that consists of proving this property for its integrand.

Proposition 5.6.3. *Assume that $g(x, \xi)$ is uniformly convex with respect to ξ, i.e., for almost all $x \in \Omega$ and all $\xi_1, \xi_2 \in \mathbb{R}^n$ such that $|\xi_1|$ and $|\xi_2|$ are less than $\delta > 0$, the following inequality holds:*

$$g\left(x, \frac{\xi_1 + \xi_2}{2}\right) + \varphi_\delta(x, \xi_1 - \xi_2) \leq \frac{1}{2}g(x, \xi_1) + \frac{1}{2}g(x, \xi_2), \qquad (5.6.3)$$

where $\varphi_\delta(x,\xi)$ is a nonzero function from $\in \Gamma_0^+(\mathbb{R}^n)$. Then the functional $G : Y \to \overline{\mathbb{R}}$ is uniformly convex in the ball $\|y\|_{\infty,\Omega} \leq \delta$ with

$$\Upsilon_\delta(y) = \int_\Omega \varphi_\delta(x, y(x))\, dx.$$

Proof. Take two functions $y_1, y_2 \in Y$ that belong to $\operatorname{dom} G$. By assumption, we have

$$g\left(x, \frac{y_1(x) + y_2(x)}{2}\right) + \varphi_\delta(x, y_1(x) - y_2(x)) \leq$$

$$\leq \frac{1}{2}g(x, y_1(x)) + \frac{1}{2}g(x, y_2(x))$$

for almost all $x \in \Omega$, provided that $|y_i(x)| \leq \delta$, $i = 1, 2$. Integrate this inequality over Ω. Note that the integral of the right-hand side exists. Hence, the integral of the left-hand side also exists and

$$G\left(\frac{y_1 + y_2}{2}\right) + \Upsilon_\delta(y_1 - y_2) \leq \frac{1}{2}G(y_1) + \frac{1}{2}G(y_2).$$

$$\square$$

5.6.3   Conjugate functionals

The main result is that in the vast majority of cases the determination of G^* is reduced to the determination of g^*, i.e.,

$$G^*(y^*) = \int_\Omega g^*(x, y^*(x))\, dx, \qquad (5.6.4)$$

where

$$g^*(x, \xi^*) = \sup_{\xi \in \mathbb{R}^n} \{\xi^* \cdot \xi - g(x, \xi)\}, \quad \forall \xi^* \in \mathbb{R}^n.$$

In this concise overview, we have no room to prove (5.6.4) in the most general form. Instead of that, we consider three particular cases, which are frequently encountered in various applied problems.

Example 5.6.1. Let $g(x, \xi) = \frac{1}{2}A(x)\xi{\cdot}\xi$, where A is a positive definite matrix such that

$$\alpha_1 |\xi|^2 \leq A(x)\xi \cdot \xi \leq \alpha_2 |\xi|^2, \quad \forall \xi \in \mathbb{R}^n. \qquad (5.6.5)$$

Then, G is defined on the space $Y = L^2(\Omega, \mathbb{R}^n)$. It is easy to check that $g^*(x, \xi^*) = \frac{1}{2}A^{-1}(x)\xi^* \cdot \xi^*$, where

$$\alpha_2^{-1} |\xi^*|^2 \leq A^{-1}\xi^* \cdot \xi^* \leq \alpha_1^{-1} |\xi^*|^2, \quad \forall \xi^* \in \mathbb{R}^n. \qquad (5.6.6)$$

The conjugate functional G^* is defined on $Y^* = (L^2(\Omega; \mathbb{R}^n))^* = Y$. By definition,

$$G^*(y^*) \;=\; \sup_{y \in L^2(\Omega;\mathbb{R}^n)} \left\{ \int_\Omega y^*(x) \cdot y(x)\, dx - \int_\Omega g(x, y(x))\, dx \right\} \le$$

$$\le \int_\Omega \sup_{\xi \in \mathbb{R}^n} \{y^*(x) \cdot \xi - g(x, \xi)\} dx = \int_\Omega g^*(x, y^*(x)) dx. \tag{5.6.7}$$

However, taking the function $\tilde{y}(x) = \frac{1}{2} A^{-1}(x) y^*(x) \in L^2(\Omega; \mathbb{R}^n)$, we obtain

$$G^*(y^*) \;\ge\; \int_\Omega \left\{ y^*(x) \cdot \tilde{y}(x) - \frac{1}{2} A\tilde{y}(x) \cdot \tilde{y}(x) \right\} dx =$$

$$= \int_\Omega g^*(x, y^*(x)) dx. \tag{5.6.8}$$

The relations (5.6.7) and (5.6.8) show that

$$G^*(y^*) = \int_\Omega g^*(x, y^*(x))\, dx.$$

Example 5.6.2. Let $g(x, \xi) = \alpha(x)\frac{|\xi|^p}{p}$, where $\alpha_1 \le \alpha(x) \le \alpha_2$ and $1 < p < +\infty$. In this case, $Y = L^p(\Omega; \mathbb{R}^n)$ and, therefore, $Y^* = L^{p^*}(\Omega; \mathbb{R}^n)$, where $p^* = \frac{p}{p-1}$. For any function $y^* \in Y^*$, the problem

$$\sup_{\xi \in \mathbb{R}^n} \left\{ y^*(x) \cdot \xi - \alpha(x) \frac{|\xi|^p}{p} \right\}, \quad x \in \Omega,$$

has a solution $\tilde{\xi}$ satisfying the relation $y^*(x) = \alpha(x)\tilde{\xi}\left|\tilde{\xi}\right|^{p-2}$. From this it follows that $\left|\tilde{\xi}\right|^p = (1/\alpha)^{p^*} |y^*(x)|^{p^*}$. The function

$$\tilde{y}(x) = \left(\frac{1}{\alpha}\right)^{\frac{1}{p-1}} y^* |y^*|^{\frac{2-p}{p-1}}$$

satisfies the above relation at any point $x \in \Omega$. Since

$$\int_\Omega |\tilde{y}(x)|^p\, dx = \int_\Omega \left(\frac{1}{\alpha}\right)^{p^*} |y^*|^{p^*}\, dx < +\infty,$$

this function belongs to Y. Now, we use the same arguments as in Example 5.6.1 and obtain

$$G^*(y^*) = \int_\Omega g^*(x,\xi)\,dx, \quad \text{where } g^*(x,\xi) = \frac{1}{p^*}\alpha^{1-p^*}\,|y^*|^{p^*}.$$

In particular, this means that the functional conjugate to $\frac{1}{p}\|y\|_{p,\Omega}^p$ is $\frac{1}{p^*}\|y^*\|_{p^*,\Omega}^{p^*}$.

Example 5.6.3. Let $y(x) \in Y = L^1(\Omega)$ and $G(y) = \int_\Omega |y(x)|\,dx$. In this case, $Y^* = L^\infty(\Omega)$. We are aimed at showing that

$$G^*(y^*) = \int_\Omega \chi(y^*(x))\,dx, \quad \forall y^* \in Y^*, \tag{5.6.9}$$

where χ is the support function of the interval $[-1,1]$. Consider two cases: 1) $|y^*| \le 1$ for almost all $x \in \Omega$ and 2) there exists a set $\omega \subset \Omega$ such that $|\omega| > 0$ and $|y^*(x)| > 1$ for almost all $x \in \omega$. In the first case,

$$G^*(y^*) = \sup_{y \in L^1} \int_\Omega \{y^*(x)y(x) - |y(x)|\}\,dx = 0 \tag{5.6.10}$$

and we note that (5.6.9) holds. In the second case, we take the following sequence of functions:

$$\widetilde{y}_k(x) = \begin{cases} 0 & \text{if } x \in \Omega \setminus \omega, \\ k\operatorname{sign} y^*(x) & \text{if } x \in \omega. \end{cases}$$

Substituting the sequence $\{\widetilde{y}_k\}$ in (5.6.10), we see that

$$G^*(y^*) \ge k \int_\omega (|y^*| - 1)\,dx \to +\infty.$$

It remains to note that the relation (5.6.9) also holds for vector-valued functions.

5.6.4   Differentiation

Let G and G^* be integral functionals introduced above. By Proposition 5.3.2, a function y^* belongs to the set $\partial G(y)$ and a function y belongs to the set $\partial G^*(y^*)$ if and only if the compound functional

$$D_G(y,y^*) := G(y) + G^*(y^*) - \langle y^*, y \rangle$$

vanishes. In the cases considered, $Y = L^p(\Omega; \mathbb{R}^n)$ and $Y^* = L^{p^*}(\Omega; \mathbb{R}^n)$. The product is defined by the integral

$$\langle y^*, y \rangle = \int_\Omega y^* \cdot y \, dx$$

and

$$D_G(y, y^*) = \int_\Omega \mathcal{D}_g(x, y(x), y^*(x)) \, dx,$$

where

$$\mathcal{D}_g(x, \xi, \xi^*) = g(x, \xi) + g^*(x, \xi^*) - \xi \cdot \xi^*, \quad x \in \Omega, \ \xi^* \in \mathbb{R}^n.$$

It is easy to see that $D_G(y, y^*) = 0$ if and only if

$$\mathcal{D}_g(x, y(x), y^*(x)) = 0 \quad \text{for almost all } x \text{ in } \Omega. \tag{5.6.11}$$

Hence, we conclude that $y^* \in \partial G(y)$ and $y \in \partial G^*(y^*)$ if and only if for almost all $x \in \Omega$, the following relations hold:

$$y^*(x) = \partial g(x, y(x)), \text{ and } y(x) = \partial g(x, y^*(x)). \tag{5.6.12}$$

Chapter 6

Two-sided a posteriori estimates for linear elliptic problems

6.1 Basic relations

In this chapter, we consider problems that can be represented in the following general form: to find $u \in V_0 + u_0$ such that

$$(\mathcal{A}\Lambda u, \Lambda w) + \langle l, w \rangle = 0 \quad \forall w \in V_0. \tag{6.1.1}$$

Here V_0 is a subspace of a reflexive Banach space V, Λ is a linear bounded operator acting from V to a Hilbert space U with scalar product $(\cdot, \cdot)$, $l \in V_0^*$, and $\mathcal{A} \in \mathcal{L}(U, U)$ is a self-adjoint operator. We assume that V is compactly embedded in U and the operators Λ and $\mathcal{A}$ satisfy the relations

$$c_1 \|y\|^2 \leq (\mathcal{A}y, y) \leq c_2 \|y\|^2, \qquad \forall y \in H, \tag{6.1.2}$$

and

$$\|\Lambda w\| \geq c_3 \|w\|_V, \quad \forall w \in V_0, \tag{6.1.3}$$

with positive constants c_1, c_2, and c_3. We take the functional l in the form $\langle l, w \rangle = (f, w) + (g, \Lambda w)$, where f and g are some elements of U. Such a functional is well defined and finite on the elements of V.

For our analysis, it is convenient to introduce two more spaces. The quantity $(\mathcal{A}y, y)^{1/2}$ determines a new norm $\|\| y \|\|$, which is equivalent to the original norm $\|y\| = (y, y)^{1/2}$. Another equivalent norm is $\|\| y \|\|_* =$

$(\mathcal{A}^{-1}y, y)^{1/2}$, where $\mathcal{A}^{-1}$ is the operator inverse to $\mathcal{A}$. The spaces Y and Y^* contain elements of U equipped with the norms $\|\cdot\|$ and $\|\cdot\|_*$, respectively.

It is easy to see that (6.1.1) is the Euler relation for the following problem.

Problem $\mathcal{P}$. Find $u \in V_0 + u_0$ such that

$$J(u) = \inf_{v \in V_0 + u_0} J(u) := \inf \mathcal{P},$$

where

$$J(v) = \frac{1}{2} \| \Lambda v \|^2 + \langle l, v \rangle.$$

On the set $(V_0 + u_0) \times Y^*$, we define the Lagrangian

$$L(v, y^*) = (y^*, \Lambda v) - \frac{1}{2} \| y^* \|^2 + \langle l, v \rangle$$

and the functional

$$I^*(y^*) = \inf_{v \in K} L(v, y^*) = \begin{cases} (y^*, \Lambda u_0) - \frac{1}{2} \| y^* \|_*^2 + \langle l, u_0 \rangle, & y^* \in Q_l^*, \\ -\infty, & y^* \notin Q_l^*, \end{cases}$$

where

$$Q_l^* := \left\{ y^* \in Y^* \big| (y^*, \Lambda w) + \langle l, w \rangle = 0, \quad \forall w \in V_0 \right\}.$$

The problem dual to $\mathcal{P}$ is as follows.

Problem $\mathcal{P}^*$. Find $p^* \in Q_l^*$ such that

$$I^*(p^*) = \sup_{y^* \in Q_l^*} I^*(y^*) := \sup \mathcal{P}^*.$$

The functionals J and $(-I^*)$ are evidently convex and coercive on V and Y^*, respectively. The sets $V_0 + u_0$ and Q_l^* are closed affine manifolds. By Theorem 5.4.1, we conclude that there exist $u \in V_0 + u_0$ and $p^* \in Q_l^*$ such that

$$J(u) = \inf \mathcal{P}, \qquad I^*(p^*) = \sup \mathcal{P}^*. \tag{6.1.4}$$

The minimizer u obeys (6.1.1) and the maximizer p^* satisfies the relation

$$(\Lambda u_0 - \mathcal{A}^{-1} p^*, y^*) = 0, \quad \forall y^* \in Q_0^*, \tag{6.1.5}$$

where

$$Q_0^* := \left\{ y^* \in Y^* \,|\, (y^*, \Lambda w) = 0, \quad \forall w \in V_0 \right\}.$$

Note that $\mathcal{A}\Lambda u \in Q_l^*$, so that, setting $u_0 = u$, we obtain

$$I^*(\mathcal{A}\Lambda u) = (\mathcal{A}\Lambda u, \Lambda u) - \frac{1}{2} \parallel \mathcal{A}\Lambda u \parallel_*^2 + \langle l, u \rangle \le \sup \mathcal{P}^*. \qquad (6.1.6)$$

Since $\parallel \mathcal{A}\Lambda u \parallel_*^2 = \parallel \Lambda u \parallel^2$, we see that

$$I^*(\mathcal{A}\Lambda u) = J(u) = \inf \mathcal{P} \qquad (6.1.7)$$

and (cf. (5.4.11))

$$\sup \mathcal{P}^* \equiv \inf \mathcal{P}.$$

The latter relation means that

$$(p^*, \Lambda u) - \frac{1}{2} \parallel p^* \parallel_*^2 + \langle l, u \rangle = \frac{1}{2} \parallel \Lambda u \parallel^2 + \langle l, u \rangle,$$

which is equivalent to the relation

$$D(\Lambda u, p^*) = \frac{1}{2} \parallel \Lambda u \parallel^2 + \frac{1}{2} \parallel p^* \parallel_*^2 - (p^*, \Lambda u) = 0, \qquad (6.1.8)$$

or simply

$$p^* = \mathcal{A}\Lambda u \quad \text{a.e in } \Omega. \qquad (6.1.9)$$

This is the *duality relation* for the pair (u, p^*).

Let $v \in V_0 + u_0$ and $y^* \in Y^*$ be some approximations of u and p^*, respectively. Our goal is to obtain two-sided estimates of the quantities $\parallel \Lambda(v - u) \parallel$ and $\parallel y^* - p^* \parallel$ that are norms of *deviations* from the exact solutions u and p^*. We will call them *deviation estimates*. Deviation estimates are derived without referring to a particular numerical method used for finding approximate solutions. For this reason, they lead to a posteriori estimates valid for conforming approximations of all types. Below, we show how such estimates are derived by methods of the calculus of variations. First, we establish the following basic result.

Proposition 6.1.1. *For any $v \in V_0 + u_0$ and $q^* \in Q_l^*$, hold the following relations:*

$$\parallel \Lambda(v - u) \parallel^2 + \parallel q^* - p^* \parallel_*^2 = 2(J(v) - I^*(q^*)), \qquad (6.1.10)$$

$$\parallel \Lambda(v - u) \parallel^2 + \parallel q^* - p^* \parallel_*^2 = 2D(\Lambda v, q^*). \qquad (6.1.11)$$

Proof. In view of (6.1.1) and (6.1.5), we have

$$\frac{1}{2} \parallel \Lambda(v-u) \parallel^2 = J(v) - J(u) + (\mathcal{A}\Lambda u, \Lambda(v-u)) +$$

$$+ \langle l, v-u \rangle = J(v) - J(u),$$

and

$$\frac{1}{2} \parallel q^* - p^* \parallel^2_* \;=\; I^*(p^*) - I^*(q^*) + (\Lambda u_0 - \mathcal{A}^{-1}p^*, p^* - q^*) =$$

$$=\; I^*(p^*) - I^*(q^*).$$

Since

$$J(u) = I^*(p^*),$$

these relations imply (6.1.10). For $q^* \in Q_l^*$ the difference $J(v) - I^*(q^*)$ is equal to $D(\Lambda v, q^*)$, so that (6.1.11) follows from (6.1.10). $\square$

The estimates (6.1.10) and (6.1.11) are valid only for $q^* \in Q_l^*$, which poses some difficulties. Below it is shown how we can overcome this drawback. First, we establish one subsidiary result.

Proposition 6.1.2. *Let* $q^* \in Q_l^*$, $v \in V_0 + u_0$, $\beta \in \mathbb{R}_+$, *and* $y^* \in Y^*$. *Then*

$$J(v) - I^*(q^*) \leq$$

$$\leq (1+\beta)D(\Lambda v, y^*) + \left(1 + \frac{1}{\beta}\right)\frac{1}{2} \parallel q^* - y^* \parallel^2_* . \quad (6.1.12)$$

Proof. For any $y^* \in Y^*$, we have

$$J(v) - I^*(q^*) = \frac{1}{2}\Big(\parallel \Lambda v \parallel^2 + \parallel y^* \parallel^2_*\Big) +$$

$$+ \frac{1}{2}\Big(\parallel q^* \parallel^2_* - \parallel y^* \parallel^2_*\Big) - (\Lambda u_0, q^*) + \langle l, v - u_0 \rangle.$$

Since

$$\langle l, v - u_0 \rangle = (q^*, \Lambda(u_0 - v)),$$

we find that

$$J(v) - I^*(q^*) \;=\; \frac{1}{2}\left(\parallel \Lambda v \parallel^2 + \parallel y^* \parallel^2_*\right) - (q^*, \Lambda v) +$$

$$+ \frac{1}{2}\left(\parallel q^* \parallel^2_* - \parallel y^* \parallel^2_*\right) =$$

$$= D(\Lambda v, y^*) + \left(y^* - q^*, \Lambda v - \mathcal{A}^{-1}y^*\right) + \frac{1}{2}\parallel q^* - y^* \parallel^2_* .$$

This relation yields (6.1.12) if we note that

$$2\left(y^* - q^*, \Lambda v - \mathcal{A}^{-1}y^*\right) \le \beta \parallel \Lambda v - \mathcal{A}^{-1}y^* \parallel^2 + \beta^{-1} \parallel y^* - q^* \parallel_*^2 .$$

$\square$

Introduce the quantity

$$d_l^2(y^*) := \inf_{q^* \in Q_l^*} \parallel q^* - y^* \parallel_*^2 .$$

Then, (6.1.10) and (6.1.12) imply the estimate

$$\frac{1}{2} \parallel \Lambda(v - u) \parallel^2 \le (1 + \beta)D(\Lambda v, y^*) + \left(1 + \frac{1}{\beta}\right)\frac{1}{2}d_l^2(y^*) \qquad (6.1.13)$$

with $v \in V_0 + u_0$ and $y^* \in Y^*$. From (6.1.13), it follows that

$$\frac{1}{2} \parallel \Lambda(v - u) \parallel^2 \le \mathcal{M}(v, \beta), \quad \forall v \in V_0 + u_0, \quad \beta \in \mathbb{R}_+,$$

where

$$\mathcal{M}(v, \beta) := \inf_{y^* \in Y^*} \left\{ (1 + \beta)D(\Lambda v, y^*) + \left(1 + \frac{1}{\beta}\right)\frac{1}{2}d_l^2(y^*) \right\} .$$

Let us show that this relation holds as equality.

Proposition 6.1.3. *For any $\beta \in \mathbb{R}_+$,*

$$\frac{1}{2} \parallel \Lambda(v - u) \parallel^2 = \mathcal{M}(v, \beta). \qquad (6.1.14)$$

Proof. Set $y^* = \lambda p^* + (1 - \lambda)\mathcal{A}\Lambda v$. Then

$$D(\Lambda v, y^*) = \frac{1}{2}\lambda^2 \parallel \Lambda(v - u) \parallel^2 .$$

Since

$$d_l^2(y^*) \le \parallel p^* - y^* \parallel_*^2 =$$
$$= (1 - \lambda)^2 \parallel p^* - \mathcal{A}\Lambda v \parallel_*^2 = (1 - \lambda)^2 \parallel \mathcal{A}\Lambda(u - v) \parallel_*^2 =$$
$$= (1 - \lambda)^2 \parallel \Lambda(u - v) \parallel^2,$$

we obtain

$$\mathcal{M}(v, \beta) \le \frac{1}{2}\left((1 + \beta)\lambda^2 + \left(1 + \frac{1}{\beta}\right)(1 - \lambda)^2\right) \parallel \Lambda(v - u) \parallel^2 .$$

The right-hand side attains its minimal value at $\lambda = 1/(1+\beta)$, which leads to the estimate

$$\frac{1}{2} \parallel \Lambda(v - u) \parallel^2 \geq \mathcal{M}(v, \beta), \quad \forall v \in V_0 + u_0, \quad \beta \in \mathbb{R}_+ .$$

Recalling that the inverse inequality has already been established, we arrive at (6.1.14). $\qquad\square$

Proposition 6.1.4.

$$\frac{1}{2} d_l^2(y^*) = \sup_{w \in V_0} \left\{ -\frac{1}{2} \parallel \Lambda w \parallel^2 - \langle l, w \rangle - (y^*, \Lambda w) \right\} . \tag{6.1.15}$$

Proof. This assertion is a consequence of the basic duality relation $\inf \mathcal{P} = \sup \mathcal{P}^*$. Indeed,

$$\frac{1}{2} d_l^2(y^*) = - \sup_{\eta^* \in Q_l^*} \left\{ -\frac{1}{2} \parallel y^* - \eta^* \parallel_*^2 \right\} = - \sup_{\eta^* \in Q_l^* - y^*} \left\{ -\frac{1}{2} \parallel \eta^* \parallel_*^2 \right\},$$

where

$$Q_l^* - y^* := \left\{ \eta^* \in Y^* \, \big| \, \eta^* = \ae^* - y^*, \quad \ae^* \in Q_l^* \right\} .$$

In other words, $\eta^* \in Q_l^* - y^*$ if

$$(\eta^*, \Lambda w) = -\langle l, w \rangle - (y^*, \Lambda w), \quad \forall w \in V_0 .$$

The right-hand side of this relation is a linear continuous functional. We denote it by l_y and rewrite the relation as follows:

$$(\eta^*, \Lambda w) + \langle l_y, w \rangle = 0 \quad \forall w \in V_0 .$$

Then, $Q_l^* - y^* = Q_{l_y}^*$ and

$$\frac{1}{2} d_l^2(y^*) = - \sup_{\eta^* \in Q_{l_y}^*} \left\{ -\frac{1}{2} \parallel \eta^* \parallel_*^2 \right\} .$$

This maximization problem coincides with Problem $\mathcal{P}^*$ if we set $u_0 = 0$ and $l = l_y$. From this fact, we conclude that

$$\begin{aligned}
\frac{1}{2} d_l^2(y^*) &= - \inf_{w \in V_0} \left\{ \frac{1}{2} \parallel \Lambda w \parallel^2 + \langle l_y, w \rangle \right\} = \\
&= - \inf_{w \in V_0} \left\{ \frac{1}{2} \parallel \Lambda w \parallel^2 + \langle l, w \rangle + (y^*, \Lambda w) \right\} = \\
&= \sup_{w \in V_0} \left\{ -\frac{1}{2} \parallel \Lambda w \parallel^2 - \langle l, w \rangle - (y^*, \Lambda w) \right\} .
\end{aligned}$$

$$\qquad\square$$

Corollary 6.1.1. Propositions 6.1.3 and 6.1.4 lead to the conclusion that the functional $\mathcal{M}(v, \beta)$ has a minimax form

$$\mathcal{M}(v, \beta) =$$
$$\inf_{y \in Y^*} \sup_{w \in V_0} \left\{ (1 + \beta) D(\Lambda v, y^*) + \left(1 + \frac{1}{\beta} \right) \left(-(y^*, \Lambda w) - J(w) \right) \right\}.$$
$$(6.1.16)$$

This relation is crucial for deriving upper and lower bounds of deviations.

6.2 Two-sided estimates of deviations

6.2.1 Upper estimates

In view of (6.1.16), we have

$$\mathcal{M}(v, \beta) \leq (1 + \beta) D(\Lambda v, y^*) +$$
$$+ \left(1 + \frac{1}{\beta} \right) \sup_{w \in V_0} \left\{ -\frac{1}{2} \parallel \Lambda w \parallel^2 - \langle l, w \rangle - (y^*, \Lambda w) \right\},$$

where $y^* \in Y^*$. Let Λ^* denote the operator conjugate to Λ, which is defined by the relation

$$(y^*, \Lambda w) = \langle \Lambda^* y^*, w \rangle, \qquad \forall w \in V_0.$$

Then

$$\langle l, w \rangle + \langle y^*, \Lambda w \rangle = \langle l + \Lambda^* y^*, w \rangle \leq | l + \Lambda^* y^* | \, \|\Lambda w\|,$$

where

$$| l + \Lambda^* y^* | := \sup_{w \in V_0} \frac{\langle l + \Lambda^* y^*, w \rangle}{\parallel \Lambda w \parallel}. \tag{6.2.1}$$

Note that

$$\langle l + \Lambda^* y^*, w \rangle \quad \leq \|l\|_{V_0^*} \|w\|_V + \|y^*\| \|\Lambda w\| \leq$$
$$\leq \left(c_3^{-1} \|l\|_{V_0^*} + \|y^*\| \right) \|\Lambda w\| \leq$$
$$\leq c_1^{-1/2} \left(c_3^{-1} \|l\|_{V_0^*} + \|y^*\| \right) \parallel \Lambda w \parallel.$$

Hence, the value of the supremum in (6.2.1) is finite and

$$
\begin{aligned}
\sup_{w\in V_0} &\left\{ -\frac{1}{2}\parallel \Lambda w \parallel^2 -\langle l, w\rangle - (y^*, \Lambda w) \right\} \leq \\
&\leq \sup_{w\in V_0} \left\{ -\frac{1}{2}\parallel \Lambda w \parallel^2 +|\, l + \Lambda^* y^*|\|\Lambda w\| \right\} \leq \\
&\leq \sup_{t>0} \left\{ -\frac{1}{2}t^2 +|\, l + \Lambda^* y^*|\, t \right\} = \frac{1}{2}|\, l + \Lambda^* y^*|^2.
\end{aligned}
\tag{6.2.2}
$$

Define the functional

$$
\mathcal{M}_{\oplus}(v, \beta, y^*) := (1+\beta)D(\Lambda v, y^*) + \left(1+\frac{1}{\beta}\right)\frac{1}{2}|\, l + \Lambda^* y^*|^2.
$$

By (6.1.16) we see that

$$
\boxed{\ \tfrac{1}{2}\parallel \Lambda(v-u)\parallel^2 = \mathcal{M}(v,\beta) \leq \mathcal{M}_{\oplus}(v,\beta,y^*).\ }
\tag{6.2.3}
$$

Thus, for any $y^* \in Y^*$ and $\beta > 0$. the functional $\mathcal{M}_{\oplus}(v,\beta,y^*)$ is a *majorant* that provides an upper bound for the deviation $v-u$ evaluated in the natural energy norm.

6.2.2   Lower estimates

Since

$$
\mathcal{M}(v,\beta) \geq \sup_{w\in V_0}\ \inf_{y^*\in Y^*}\ \Big\{ (1+\beta)D(\Lambda v, y^*)-
$$

$$
- \left(1+\frac{1}{\beta}\right)\left(\frac{1}{2}\parallel \Lambda w \parallel^2 +\langle l, w\rangle + (y^*, \Lambda w)\right)\Big\},
$$

we have

$$
\mathcal{M}(v,\beta) \geq
$$

$$
\inf_{y^*\in Y^*}\left\{ (1+\beta)\left(\frac{1}{2}\parallel y^* \parallel_*^2 -(y^*, \Lambda v)\right) - \left(1+\frac{1}{\beta}\right)(y^*, \Lambda w)\right\} +
$$

$$
+ (1+\beta)\frac{1}{2}\parallel \Lambda v \parallel^2 - \left(1+\frac{1}{\beta}\right)\left(\frac{1}{2}\parallel \Lambda w \parallel^2 +\langle l, w\rangle\right),
$$

where w is an arbitrary element of V_0. Evidently, this estimate is also valid for βw, which yields

$$\mathcal{M}(v,\beta) \geq (1+\beta) \inf_{y^* \in Y^*} \left\{ \frac{1}{2} \parallel y^* \parallel_*^2 - (y^*, \Lambda(v+w)) \right\} +$$
$$+ (1+\beta) \left(\frac{1}{2} \parallel \Lambda v \parallel^2 - \frac{\beta}{2} \parallel \Lambda w \parallel^2 - \langle l, w \rangle \right).$$

Note that

$$\inf_{y^* \in Y^*} \left\{ \frac{1}{2} \parallel y^* \parallel_*^2 - (y^*, \Lambda(v+w)) \right\} \geq$$
$$\geq \inf_{y^* \in Y^*} \left\{ \frac{1}{2} \parallel y^* \parallel_*^2 - \parallel y^* \parallel_* \parallel \Lambda(v+w) \parallel \right\} = -\frac{1}{2} \parallel \Lambda(v+w) \parallel^2.$$

Thus, we obtain

$$\mathcal{M}(v,\beta) \geq (1+\beta) \left\{ -\frac{1}{2} \parallel \Lambda(v+w) \parallel^2 + \right.$$
$$+ \frac{1}{2} \parallel \Lambda v \parallel^2 - \frac{\beta}{2} \parallel \Lambda w \parallel^2 - \langle l, w \rangle \right\} =$$
$$= (1+\beta) \left\{ -(\mathcal{A}\Lambda v, \Lambda w) - \frac{1+\beta}{2} \parallel \Lambda w \parallel^2 - \langle l, w \rangle \right\}.$$

Here, w is an arbitrary function in V_0. Therefore, we may replace w by $w/(1+\beta)$. Such a replacement leads to the estimate

$$\boxed{\ \tfrac{1}{2} \parallel \Lambda(v-w) \parallel^2 \geq \mathcal{M}_\ominus(v,w), \quad w \in V_0,\ }\qquad (6.2.4)$$

where the *minorant* is defined by the relation

$$\mathcal{M}_\ominus(v,w) := -\frac{1}{2} \parallel \Lambda w \parallel^2 - (\mathcal{A}\Lambda v, \Lambda w) - \langle l, w \rangle.$$

Remark 6.2.1. Assume that v coincides with u. In this case,

$$\mathcal{M}_\ominus(u,w) = -\frac{1}{2} \parallel \Lambda w \parallel^2 - (\mathcal{A}\Lambda u, \Lambda w) - \langle l, w \rangle = -\frac{1}{2} \parallel \Lambda w \parallel^2$$

and, therefore,

$$\sup_{w \in V_0} \mathcal{M}_\ominus(u,w) = 0.$$

The same is true for the majorant. Indeed, set $\widehat{y}^* = \mathcal{A}\Lambda u$. Then,

$$\mathcal{M}_\oplus(u,\beta,\widehat{y}^*) = (1+\beta)D(\Lambda u,\widehat{y}^*) + \left(1+\frac{1}{\beta}\right)\frac{1}{2}|\,l+\Lambda^*\widehat{y}^*\,|^2,$$

where the first term is zero and

$$|\,l+\Lambda^*\widehat{y}^*\,| = \sup_{w\in V_0}\frac{\langle l,w\rangle + (\mathcal{A}\Lambda u,\Lambda w)}{\|\,\Lambda w\,\|},$$

so that it is zero as well. Thus,

$$\inf_{y^*\in Y^*}\mathcal{M}_\oplus(u,\beta,y^*) = 0.$$

6.2.3　Estimates of deviations in terms of the dual variable

Let $y^* \in Y^*$ be an approximation of p^*. For any $q^* \in Q_l^*$, we have

$$\|\,y^*-p^*\,\|_*^2 \le (1+\gamma)\,\|\,y^*-q^*\,\|_*^2 + \left(1+\frac{1}{\gamma}\right)\|\,q^*-p^*\,\|_*^2, \qquad (6.2.5)$$

where γ is a positive number. From (6.2.5), (6.1.10), and (6.1.12), we obtain

$$\begin{aligned}
\|\,y^*-p^*\,\|_*^2 \;&\le\; (1+\gamma)\,\|\,y^*-q^*\,\|_*^2 + \\
&\quad +2\left(1+\frac{1}{\gamma}\right)\left(J(v)-I^*(q^*)\right) = \\
&= (1+\gamma)\left(1+\frac{1}{\gamma}+\frac{1}{\beta\gamma}\right)\|\,y^*-q^*\,\|_*^2 + \\
&\quad +2(1+\beta)\left(1+\frac{1}{\gamma}\right)D(\Lambda v,y^*).
\end{aligned}$$

Therefore,

$$\begin{aligned}
\tfrac{1}{2}\,\|\,y^*-p^*\,\|_*^2 \;&\le\; (1+\gamma)\left(1+\frac{1}{\gamma}+\frac{1}{\beta\gamma}\right)\frac{1}{2}d_l^2(y^*) + \\
&\quad +(1+\beta)\left(1+\frac{1}{\gamma}\right)D(\Lambda v,y^*) \qquad (6.2.6)
\end{aligned}$$

and

$$\begin{aligned}
\tfrac{1}{2}\,\|\,y^*-p^*\,\|_*^2 \;&\le\; (1+\gamma)\left(1+\frac{1}{\gamma}+\frac{1}{\beta\gamma}\right)\frac{1}{2}|\,l+\Lambda^*y^*\,|^2 + \\
&\quad +(1+\beta)\left(1+\frac{1}{\gamma}\right)D(\Lambda v,y^*). \qquad (6.2.7)
\end{aligned}$$

Rewrite this estimate as follows:

$$\tfrac{1}{2} \ \| \ y^* - p^* \ \|_*^2 \leq \mathcal{M}_\oplus^*(y^*, \beta, \gamma, v), \qquad (6.2.8)$$

where $\mathcal{M}_\oplus^*$ denotes the right-hand side of (6.2.7). This estimate holds for any $y^* \in Y^*$, positive parameters β, γ, and any $v \in V_0 + u_0$.

It is important to outline a certain duality of the estimates (6.2.2) and (6.2.8). In (6.2.2), the majorant contains a "free" function y^* that belongs to the space Y^* of the dual problem. The minimization of $\mathcal{M}_\oplus$ over Y^* gives the sharpest upper bound of the deviation evaluated in the energy norm of the primal problem. In (6.2.7), the majorant $\mathcal{M}_\oplus^*$ contains a "free" function v, which belongs to the set $V_0 + u_0$ of the primal problem. Minimization over this set leads to the sharpest upper bound of the deviation evaluated in the energy norm of the dual problem.

Estimates of the deviations in the dual energy norm may be useful in problems arising in continuum mechanics, where the dual variable often has an important physical meaning (e.g., in solid mechanics it is a stress function).

6.3 Properties of two-sided estimates

6.3.1 Exactness

Values of $\mathcal{M}_\ominus(v, w)$ depend on the choice of w and values of $\mathcal{M}_\oplus(v, \beta, y^*)$ depend on β and y^*. Let us show that w, β, and y^* can always be chosen such that the minorant coincides with the majorant and, consequently, coincides with the energy norm of $(v - u)$.

Since

$$|\, l + \Lambda^* p^* \,| = \sup_{w \in V_0} \frac{\langle l, w \rangle + (A\Lambda u, \Lambda w)}{\|\, \Lambda w \,\|} = 0,$$

we find that

$$\mathcal{M}_\oplus(v, \beta, p^*) = (1 + \beta) D(\Lambda v, p^*) =$$

$$= (1 + \beta) \left(\frac{1}{2} \ \| \ \Lambda v \ \|^2 + \frac{1}{2} \ \| \ \Lambda u \ \|^2 - (A\Lambda u, \Lambda v) \right) =$$

$$= \frac{1 + \beta}{2} \ \| \ \Lambda(v - u) \ \|^2$$

and, therefore,

$$\inf_{\substack{y^* \in Y^* \\ \beta \in \mathbb{R}_+}} \mathcal{M}_\oplus(v, \beta, y^*) = \frac{1}{2} \, \| \, \Lambda(v - u) \, \|^2 \, .$$

Similarly,

$$\mathcal{M}_\ominus(v, u - v) =$$
$$= -\frac{1}{2} \, \| \, \Lambda(u - v) \, \|^2 - (\, \mathcal{A}\Lambda v, \Lambda(u - v) \,) - \langle l, u - v \rangle =$$
$$= -\frac{1}{2} \, \| \, \Lambda(u - v) \, \|^2 + (\, \mathcal{A}\Lambda(u - v), \Lambda(u - v) \,) = \frac{1}{2} \, \| \, \Lambda(u - v) \, \|^2 \, .$$

Thus, the estimate (6.2.4) is exact the in sense that

$$\sup_{w \in V_0} \mathcal{M}_\ominus(v, w) = \frac{1}{2} \, \| \, \Lambda(u - v) \, \|^2 \, .$$

Let $\{Y_i^*\}_{i=1}^\infty$ and $\{V_{0i}\}_{i=1}^\infty$ be two sets of finite-dimensional subspaces. We assume that they are dense in Y^* and V_0, respectively. The latter means that, given $\varepsilon > 0$ and arbitrary elements $y^* \in Y^*$ and $w \in V_0$, one can find a natural number k_ε such that

$$\inf_{\tilde{w} \in V_{0i}} \|\tilde{w} - w\|_V \le \varepsilon, \quad \inf_{\tilde{y}^* \in Y_i^*} \| \, \tilde{y}^* - y^* \, \| \le \varepsilon, \quad \forall i \ge k_\varepsilon. \tag{6.3.1}$$

The methods of creating sequences of finite-dimensional subspaces with above-mentioned property are well known in numerical analysis. In particular, they are represented by regular finite element approximations. Let us show that two-sided bounds can be evaluated precisely by minimizing the majorant on $\{Y_i^*\}$ and maximizing the minorant on $\{V_{0i}\}$.

Take a small $\varepsilon > 0$,. Then there exists a number k and elements $w_k \in V_{0k}$ and $p_k^* \in Y_{0k}^*$ satisfying the conditions

$$\|w_k - (u - v)\|_V \le \varepsilon, \quad \| \, p_k^* - p^* \, \|_* \le \varepsilon. \tag{6.3.2}$$

Define two quantities

$$\mathcal{M}_\oplus^k = \inf_{\substack{y_k^* \in Y_k^* \\ \beta \in \mathbb{R}_+}} \mathcal{M}_\oplus(v, \beta, y_k^*) \tag{6.3.3}$$

and

$$\mathcal{M}_\ominus^k = \sup_{w_k \in V_{0k}} \mathcal{M}_\ominus(v, w_k). \tag{6.3.4}$$

From (6.3.3) and (6.3.4), it follows that

$$\mathcal{M}_\oplus^k \leq \mathcal{M}_\oplus(v, \beta, p_k^*), \quad \mathcal{M}_\ominus^k \geq \mathcal{M}_\ominus(v, w_k). \tag{6.3.5}$$

Let us show that $\mathcal{M}_\oplus^k$ and $\mathcal{M}_\ominus^k$ tend to $\frac{1}{2}\|\Lambda(v - u)\|^2$ as $k \to \infty$. First, we the upper estimate first. We have

$$\mathcal{M}_\oplus(v, \beta, p_k^*) = (1 + \beta)D(\Lambda v, p_k^*) + \left(1 + \frac{1}{\beta}\right)\frac{1}{2}|l + \Lambda^* p_k^*|^2. \tag{6.3.6}$$

Here

$$|l + \Lambda^* p_k^*| = \sup_{\varphi \in V_0} \frac{(p^* - p_k^*, \Lambda\varphi)}{\|\Lambda\varphi\|} \leq \varepsilon \tag{6.3.7}$$

and

$$D(\Lambda v, p_k^*) = \frac{1}{2}(\Lambda v - \mathcal{A}^{-1}p_k^*, \mathcal{A}\Lambda v - p_k^*) =$$

$$= \frac{1}{2}\left(\Lambda(v - u) - \mathcal{A}^{-1}(p_k^* - p^*), \mathcal{A}\Lambda(v - u) - (p_k^* - p^*)\right) =$$

$$= \frac{1}{2}\|\Lambda(v - u)\|^2 + \|p_k^* - p^*\|_*^2 - (\Lambda(v - u), p_k^* - p^*).$$

From the latter estimate we see that

$$D(\Lambda v, p_k^*) \leq \frac{1}{2}\|\Lambda(v - u)\|^2 + \varepsilon\|\Lambda(v - u)\| + \frac{1}{2}\varepsilon^2. \tag{6.3.8}$$

Combining (6.3.6), (6.3.7), and (6.3.8), we obtain

$$\mathcal{M}_\oplus^k \leq \mathcal{M}_\oplus(v, \varepsilon, p_k^*) =$$

$$= (1 + \varepsilon)\left(\frac{1}{2}\|\Lambda(v - u)\|^2 + \varepsilon\|\Lambda(v - u)\| + \frac{1}{2}\varepsilon^2\right) +$$

$$+ \frac{1}{2}\varepsilon(1 + \varepsilon) = \frac{1}{2}\|\Lambda(v - u)\|^2 + c_4\varepsilon + o(\varepsilon), \tag{6.3.9}$$

where $c_4 = \frac{1}{2}\left(1 + \|\Lambda(v - u)\|\right)^2$. Recall that for any small ε there exists a number k such that the estimates (6.3.2) hold. Thus, we conclude that

$$\mathcal{M}_\oplus^k \longrightarrow \frac{1}{2}\|\Lambda(v - u)\|^2 \quad \text{as} \quad k \to \infty. \tag{6.3.10}$$

Remark 6.3.1. Note that the constant c_4 depends on the norm of $(v - u)$, so that for a good approximation this convergence is faster than for a bad one.

Now, let us focus our attention on lower estimates. It is easy to see that

$$
\mathcal{M}_\ominus(v, w_k) = -\frac{1}{2} \parallel \Lambda w_k \parallel^2 - (\mathcal{A}\Lambda v, \Lambda w_k) - \langle l, w_k \rangle =
$$

$$
= -\frac{1}{2} \parallel \Lambda w_k \parallel + (\mathcal{A}\Lambda(u - v), \Lambda w_k) =
$$

$$
= \frac{1}{2} \parallel \Lambda(u - v) \parallel^2 - \frac{1}{2} \parallel \Lambda(w_k - (u - v)) \parallel^2 \geq
$$

$$
\geq \frac{1}{2} \parallel \Lambda(u - v) \parallel^2 - \frac{1}{2} c_2 \parallel \Lambda(w_k - (u - v)) \parallel^2 .
$$

This implies the estimate

$$
\frac{1}{2} \parallel \Lambda(u - v) \parallel^2 \geq \mathcal{M}_\ominus^k \geq \frac{1}{2} \parallel \Lambda(u - v) \parallel^2 - c_5 \varepsilon^2, \tag{6.3.11}
$$

where c_5 is a positive constant dependent on the norm of Λ. The estimate (6.3.11) shows that

$$
\mathcal{M}_\ominus^k \to \frac{1}{2} \parallel \Lambda(u - v) \parallel^2 \quad \text{as } k \to \infty. \tag{6.3.12}
$$

Remark 6.3.2. Having $\mathcal{M}_\oplus^k$ and $\mathcal{M}_\ominus^k$, one can define the number

$$
\eta_k := \frac{\mathcal{M}_\oplus^k}{\mathcal{M}_\ominus^k} \geq 1, \tag{6.3.13}
$$

which gives an idea of the quality of the error estimation. From (6.3.10) and (6.3.12) it follows that

$$
\eta_k \to 1, \qquad \text{as } k \to +\infty.
$$

6.3.2   Computability

By *computability* we mean that upper and lower estimates can be computed with any a priori given accuracy. The quantities $\mathcal{M}_\ominus^k$ are computed by solving finite-dimensional problems for a quadratic integral functional. Such a task is solvable by well-known numerical methods. However, the quantities $\mathcal{M}_\oplus^k$ are related to the majorant $\mathcal{M}_\oplus$, whose second term is represented by the negative norm. In general, computing such a norm may be very expensive. Below we show that under certain assumptions (which hold in the majority of practically interesting cases), this term is estimated by an explicitly computable quantity. This gives a directly computable majorant that preserves all of the main properties of $\mathcal{M}_\oplus$.

Assume that

$$l \in U \qquad (6.3.14)$$

and the variable y^* belongs to the set

$$Q^* := \{y^* \in Y^* \mid \; \Lambda^* y^* \in U\} \quad \text{and} \quad l \subset U,$$

so that for any $w \in V_0$, we have $\langle l + \Lambda^* y^*, w \rangle = (l + \Lambda^* y^*, w)$. In this case,

$$|\, l + \Lambda^* y^* \,| = \sup_{w \in V_0} \frac{\langle l + \Lambda^* y^*, w \rangle}{\|\, \Lambda w \,\|} \le \sup_{w \in V_0} \frac{\|l + \Lambda^* y^*\| \, \|w\|}{\|\, \Lambda w \,\|} \le$$

$$\le \|l + \Lambda^* y^*\| c_1^{-1} \sup_{w \in V_0} \frac{\|w\|}{\|\Lambda w\|} \le c_1^{-1} c_3^{-1} \|l + \Lambda^* y^*\|.$$

Here c_1 and c_3 are the constants in (6.1.2) and (6.1.3). Now, the majorant is represented in the form

$$M_\oplus(v, \beta, y^*) =$$

$$= (1 + \beta) D(\Lambda v, y^*) + \left(1 + \frac{1}{\beta}\right) \frac{c^2}{2} \|l + \Lambda^* y^*\|^2, \quad (6.3.15)$$

where $c^2 = c_1^{-2} c_3^{-2}$.

If $l \in U$, then $p^* \in Q^*$ and, therefore,

$$\inf_{\substack{y^* \in Q^* \\ \beta > 0}} M_\oplus(v, \beta, y^*) \le M_\oplus(v, \varepsilon, p^*) = (1 + \varepsilon) \frac{1}{2} \|\, \Lambda(u - v) \,\|^2,$$

where $\varepsilon > 0$ may be taken arbitrarily small. Hence, we arrive at the following result.

Proposition 6.3.1. *For any* $v \in V_0 + u_0$, $\beta \in \mathbb{R}_+$, *and* $y^* \in Q^*$, *the following relations hold*¿

$$\frac{1}{2} \|\, \Lambda(u - v) \,\|^2 \le \mathcal{M}_\oplus(v, \beta, y^*) \le M_\oplus(v, \beta, y^*), \qquad (6.3.16)$$

$$\frac{1}{2} \|\, \Lambda(u - v) \,\|^2 = \inf_{\substack{y^* \in Q^* \\ \beta > 0}} M_\oplus(v, \beta, y^*). \qquad (6.3.17)$$

These relations mean that the majorant $M_\oplus$ is also *reliable* and *exact*. However, in contrast to $\mathcal{M}_\oplus$, it contains the norm of a Hilbert space U instead of the norm of V_0^*. In particular problems, the former is usually given by a directly computable integral, while the latter one is the norm of a Sobolev space with negative index. Obviously, this makes $M_\oplus$ much more attractive from the viewpoint of practical applications (see Section 10). To justify this conclusion, we prove that a convergent sequence of upper bounds can be constructed by a sequence of finite-dimensional problems associated with the majorant $M_\oplus$.

Assume that $\{Q_k^*\}_{k=1}^\infty \subset Q^*$ is a sequence of finite-dimensional spaces, which are dense in Q^* in the following sense: for any $\varepsilon > 0$ and any $q^* \in Q^*$ one can find a natural number k_ε such that

$$\inf_{q_l^* \in Q_m^*} \|q_m^* - q^*\|_{Q^*} \leq \varepsilon, \quad \forall m > k_\varepsilon, \tag{6.3.18}$$

where $\|q^*\|_{Q^*} := \|q^*\| + \|\Lambda^* q^*\|$.

Proposition 6.3.2. *If the spaces $\{Q_k^*\}_{k=1}^\infty$ are dense in Q^*, then*

$$\lim_{m \to \infty} \inf_{\substack{y_m^* \in Q_m^* \\ \beta \in \mathbb{R}_+}} M_\oplus(v, \beta, y_m^*) = \frac{1}{2} \| \Lambda(v - u) \|^2 . \tag{6.3.19}$$

Proof. Take an arbitrary small $\varepsilon > 0$ and find a respective positive integer k_ε. For $m > k_\varepsilon$, we have

$$\inf_{\substack{y_m^* \in Q_m^* \\ \beta \in \mathbb{R}_+}} M_\oplus(v, \beta, y_m^*) \leq M_\oplus(v, \varepsilon, p_{m\varepsilon}^*) =$$

$$= (1 + \varepsilon)D(\Lambda v, p_{m\varepsilon}^*) + \left(1 + \frac{1}{\beta}\right) \frac{c^2}{2}\|l + \Lambda^* p_{m\varepsilon}^*\|^2, \tag{6.3.20}$$

where $p_{m\varepsilon}^* \in Q_m^*$ is an element such that $\|p_{m\varepsilon}^* - p^*\|_{Q^*} \leq 2\varepsilon$. Since $p^* \in Q^*$, the existence of $p_{m\varepsilon}^*$ follows from (6.3.18). The first term is easily estimated:

$$D(\Lambda v, p_{m\varepsilon}^*) = \frac{1}{2} \| \Lambda v \|^2 + \frac{1}{2} \| p_{m\varepsilon}^* \|_*^2 - (p_{m\varepsilon}^*, \Lambda v) \leq$$

$$\leq D(\Lambda v, p^*) + \frac{1}{2} \left(\mathcal{A}^{-1}(p_{m\varepsilon}^* - p^*), p_{m\varepsilon}^* + p^*\right) + (p^* - p_{m\varepsilon}^*, \Lambda v) \leq$$

$$\leq \frac{1}{2} \| \Lambda(v - u)) \|^2 + \frac{1}{2} \| p_{m\varepsilon}^* - p^* \|_* (\| p_{m\varepsilon}^* + p^* \|_* + 2 \| \Lambda v \|),$$

Hence,

$$D(\Lambda v, p_{m\varepsilon}^*) \leq \frac{1}{2} \| \Lambda(v - u)) \|^2 + \mu_1\varepsilon.$$

For the second term we have

$$\|l + \Lambda^* p_{m\varepsilon}^*\| \leq \|\Lambda^*(p_{m\varepsilon}^* - p^*)\| \leq \|p_{m\varepsilon}^* - p^*\|_{Q^*} < \mu_2\varepsilon,$$

where μ_1 and μ_2 are positive constants. Therefore,

$$M_\oplus(v, \varepsilon, p^*_{m\varepsilon}) \leq (1 + \varepsilon)\frac{1}{2} \parallel \Lambda(v - u) \parallel^2 +$$

$$+ \left(\mu_1 + \frac{c^2 \mu_2^2}{2}\right)(\varepsilon + \varepsilon^2) = \frac{1}{2} \parallel \Lambda(v - u) \parallel^2 + o(\varepsilon). \quad (6.3.21)$$

Now, (6.3.20) and (6.3.21) imply (6.3.19). $\qquad\qquad\square$

In concrete problems, the majorant $M_\oplus$ is a quadratic integral functional. Therefore, the numbers

$$M_\oplus^k = \inf_{\substack{y_k^* \in Q_k^* \\ \beta \in \mathbb{R}_+}} M_\oplus(v, \beta, y_k^*) \quad (6.3.22)$$

can be found by direct minimization procedures. They form a computable sequence of upper bounds such that

$$M_\oplus^k \longrightarrow \frac{1}{2} \parallel \Lambda(v - u) \parallel^2 \quad \text{as} \quad k \to \infty. \quad (6.3.23)$$

Recalling (6.3.12), we conclude that

$$\mathcal{M}_\ominus^k \leq \frac{1}{2} \parallel \Lambda(v - u) \parallel^2 \leq M_\oplus^k. \quad (6.3.24)$$

The quality of these bounds is estimated by the quantities

$$\boxed{\ae_k := \frac{M_\oplus^k}{\mathcal{M}_\ominus^k},} \quad (6.3.25)$$

which are computable and tends to 1 as $k \to +\infty$. We can call them *computable effectivity indexes.*

6.4 Linear elliptic equations of the second order

Let Ω be a bounded connected domain in $\mathbb{R}^n$ with Lipschitz continuous boundary $\partial\Omega$, $V = H^1(\Omega)$, $U = L^2(\Omega, \mathbb{R}^n)$, and Λw be the gradient operator $\nabla w = \left(\frac{\partial w}{\partial x_1}, \ldots, \frac{\partial w}{\partial x_n}\right)$. Define a symmetric matrix $A = \{a_{ij}\}$ such that

$$a_{ij} \in L^\infty \quad (6.4.1)$$

and

$$c_1|\xi|^2 \leq a_{ij}\xi_i\xi_j \leq c_2|\xi|^2, \qquad \forall \xi \in \mathbb{R}^n. \tag{6.4.2}$$

Then, the spaces Y and Y^* are equipped with norms

$$\| y \|^2 = \int_\Omega Ay \cdot y \, dx, \quad \| y^* \|_*^2 = \int_\Omega A^{-1}y^* \cdot y^* \, dx.$$

6.4.1 Dirichlet boundary conditions

Let $V_0 = \overset{\circ}{H}{}^1(\Omega)$. The relation

$$\int_\Omega A\nabla u \cdot \nabla w \, dx + \langle l, w \rangle = 0, \quad \forall w \in V_0, \tag{6.4.3}$$

defines a solution of the boundary-value problem

$$\operatorname{div} A\nabla u = f \quad \text{in } \Omega, \tag{6.4.4}$$
$$u = u_0 \quad \text{on } \partial\Omega. \tag{6.4.5}$$

The relation $(y^*, \Lambda w) = \langle \Lambda^* y^*, w \rangle$ comes in the form

$$\int_\Omega y^* \cdot \nabla w \, dx = \langle -\operatorname{div} y^*, w \rangle, \tag{6.4.6}$$

where $\Lambda^* = -\operatorname{div}$ and the function $\operatorname{div} y^*$ is viewed as an element of $H^{-1}(\Omega)$.

The inequality (6.1.3) is the Friederichs inequality

$$c_\Omega \|\nabla w\| \geq \|w\|, \qquad \forall w \in \overset{\circ}{H}{}^1(\Omega). \tag{6.4.7}$$

Upper estimates of $\| v - u \|$ for an approximation $v \in V_0 + u_0$ follow from (6.2.3). We have

$$\frac{1}{2}\int_\Omega \mathcal{A}\nabla(v - u) \cdot \nabla(v - u) \, dx \leq \mathcal{M}_\oplus(v, \beta, y^*),$$

where

$$\mathcal{M}_\oplus(v, \beta, y^*) =$$
$$\frac{1+\beta}{2}\int_\Omega \left(\nabla v - \mathcal{A}^{-1}y^*\right) \cdot \left(\mathcal{A}\nabla v - y^*\right) \, dx + \frac{1+\beta}{2\beta}|\operatorname{div} y^* - f|^2$$

and

$$| \operatorname{div} y^* - f | = \sup_{w \in V_0} \frac{\langle \operatorname{div} y^* - f, w \rangle}{\| \Lambda w \|} = \sup_{w \in V_0} \frac{\int_\Omega y^* \cdot \nabla w \, dx + \langle f, w \rangle}{\| \Lambda w \|}.$$

Assume that $f \in L^2(\Omega)$ and $y^* \in H(\Omega, \operatorname{div})$, so that

$$\langle f, w \rangle = \int_\Omega f w \, dx, \qquad \int_\Omega y^* \cdot \nabla w \, dx = - \int_\Omega \operatorname{div} y^* w \, dx.$$

Then, the negative norm has an upper estimate

$$| \operatorname{div} y^* - f | \le \frac{c_\Omega}{\sqrt{c_1}} \| \operatorname{div} y^* - f \|.$$

Hence, we arrive at a particular form of the majorant (6.3.15), which yields an estimate

$$\frac{1}{2} \int_\Omega A \nabla (v - u) \cdot \nabla (v - u) \, dx \le$$

$$\le \frac{1+\beta}{2} \int_\Omega (\nabla v - A^{-1} y^*) \cdot (A \nabla v - y^*) \, dx +$$

$$+ \frac{1+\beta}{2\beta} \frac{c_\Omega^2}{c_1} \| \operatorname{div} y^* - f \|_\Omega^2 \quad (6.4.8)$$

valid for any $v \in V_0 + u_0$, $\beta \in \mathbb{R}_+$, and $y^* \in H(\Omega, \operatorname{div})$.

Let $\{ Y_k^* \}$ be finite-dimensional subspaces of Y^* such that

$$Y_k^* \in H(\Omega, \operatorname{div}) \quad \text{for all } k = 1, 2, ...;$$
$$\dim Y_k^* \to +\infty \quad \text{as } k \to \infty.$$

By (6.4.8) we obtain computable upper bounds

$$M_\oplus^k = \inf_{\substack{y_k^* \in Y_k^* \\ \beta \in \mathbb{R}_+}} \left\{ \frac{1+\beta}{2} \int_\Omega (\nabla v - A^{-1} y^*) \cdot (A \nabla v - y^*) \, dx + \right.$$

$$\left. + \frac{1+\beta}{2\beta} \frac{c_\Omega^2}{c_1} \| \operatorname{div} y^* - f \|_\Omega^2 \right\}. \quad (6.4.9)$$

Lower estimates follow from (6.2.4). We have

$$\frac{1}{2} \int_\Omega A \nabla (v - u) \cdot \nabla (v - u) \, dx \ge M_\ominus(v, w), \quad \forall w \in V_0, \quad (6.4.10)$$

where

$$M_\ominus(v,w) = -\frac{1}{2}\int_\Omega A\nabla w \cdot \nabla w\, dx - \int_\Omega A\nabla v \cdot \nabla w dx - \langle f, w\rangle.$$

Let $\{V_{0k}\}$ be finite-dimensional subspaces such that

$$V_{0k}^* \in V_0 \quad \text{for all } k = 1, 2, ...;$$
$$\dim V_{0k} \to +\infty \quad \text{as } k \to \infty.$$

Solving finite-dimensional problems, we find numbers

$$M_\ominus^k = \sup_{w_k \in V_{0k}} M_\ominus(v, w_k). \tag{6.4.11}$$

Both sequences $M_\ominus^k$ and $M_\oplus^k$ tend to $\frac{1}{2}\|v - u\|^2$ as $k \to \infty$, provided that $\{Y_k^*\}$ and $\{V_{0k}\}$ possess necessary approximation properties (see Section 6.3).

Note that if v is a Galerkin approximation computed on V_{0k}, then $M_\ominus(v, w_k) = 0$. This means that to obtain a sensible lower estimate in this case, one must always use a finite-dimensional subspace that is *larger* than V_{0k}.

6.4.2   Neumann boundary condition

If (6.4.5) is replaced by the Neumann boundary condition

$$\nu \cdot A\nabla u + F = 0 \quad \text{on} \quad \partial\Omega, \tag{6.4.12}$$

where ν is the vector of unit outward normal to $\partial\Omega$, then the general scheme can be applied if we set

$$V_0 := \left\{ v \in H^1(\Omega) \;\middle|\; \int_\Omega v\, dx = 0 \right\}$$

and define $\Lambda^* y^* \in V_0^*$ by the relation

$$\langle \Lambda^* y^*, w\rangle = \int_\Omega y^* \cdot \nabla w\, dx, \quad \forall w \in V_0.$$

Assume that

$$f \in L^2(\Omega), \qquad F \in L^2(\partial\Omega) \tag{6.4.13}$$

and that the equilibrium condition

$$\int_{\Omega} f \, dx + \int_{\partial\Omega} F \, dx = 0 \tag{6.4.14}$$

holds. Then the problem (6.4.4), (6.4.12) has a solution defined by the integral identity

$$\int_{\Omega} A\nabla u \cdot \nabla w \, dx + \int_{\Omega} fw \, dx + \int_{\partial\Omega} Fw \, ds = 0, \quad \forall w \in V_0. \tag{6.4.15}$$

In the case under consideration,

$$\langle l, w \rangle = \int_{\Omega} fw \, dx + \int_{\partial\Omega} Fw \, ds$$

and

$$\langle l + \Lambda^* y^*, w \rangle = \int_{\Omega} (y^* \cdot \nabla w + fw) \, dx + \int_{\partial\Omega} Fw \, ds.$$

In general, $|l + \Lambda^* y^*|$ is estimated in terms of the norms

$$\| \operatorname{div} y^* - f \|_{H^{-1}} \quad \text{and} \quad \| y^* \cdot \nu + F \|_{H^{-1/2}}.$$

However, if we assume that y^* possesses a certain regularity, so that

$$y^* \in Q^*(\Omega) := \{ y^* \in Y^* \mid \operatorname{div} y^* \in L^2(\Omega), y^* \cdot \nu \in L^2(\partial\Omega) \},$$

then

$$\langle l + \Lambda^* y^*, w \rangle = \int_{\Omega} (f - \operatorname{div} y^*) w \, dx + \int_{\partial\Omega} (F + y^* \cdot \nu) w \, ds$$

and, therefore,

$$|\langle l + \Lambda^* y^*, w \rangle| \le$$
$$\le \| \operatorname{div} y^* - f \|_{2,\Omega} \|w\|_{2,\Omega} + \| \nu \, y^* + F \|_{2,\partial\Omega} \|w\|_{2,\partial\Omega}. \tag{6.4.16}$$

Let the constant c_Ω be defined as

$$\frac{1}{c_\Omega^2} = \inf_{w \in V_0} \frac{\int_{\Omega} A\nabla w \cdot \nabla w \, dx}{\|w\|_{2,\Omega}^2 + \|w\|_{2,\partial\Omega}^2}. \tag{6.4.17}$$

Since the trace operator is bounded, this constant is finite. Therefore, (6.4.16) implies the estimate

$$|\langle l + \Lambda^* y^*, w \rangle| \leq$$
$$\leq c_\Omega \left(\| \operatorname{div} y^* - f \|_{2,\Omega}^2 + \| \nu\, y^* + F \|_{2,\partial\Omega}^2 \right)^{1/2} \| \Lambda w \|^2$$

and the second term of the majorant is calculated as follows:

$$| l + \Lambda^* y^* | = \sup_{w \in V_0} \frac{\langle l + \Lambda^* y^*, w \rangle}{\| \Lambda w \|} \leq$$
$$\leq c_\Omega \left(\| \operatorname{div} y^* - f \|_{2,\Omega}^2 + \| \nu\, y^* + F \|_{2,\partial\Omega}^2 \right)^{1/2}.$$

The term $D(\Lambda v, y^*)$ is defined in the same way as in the Dirichlet problem.

We see that the distinction in the structure of $M_\oplus$ for the two main boundary-value problems manifests itself in different values of c_Ω and in the fact that the Neumann problem majorant contains an extra term $\| \nu\, y^* + F \|_{2,\partial\Omega}$ that penalizes violations of the boundary condition (6.4.12).

6.4.3   Mixed boundary conditions

Let $\partial\Omega$ consist of two measurable nonintersecting parts $\partial_1\Omega$ and $\partial_2\Omega$, on which different boundary conditions are given:

$$u = u_0 \quad \text{on} \quad \partial_1\Omega, \tag{6.4.18}$$

$$\nu \cdot A\nabla u + F = 0 \quad \text{on} \quad \partial_2\Omega. \tag{6.4.19}$$

Set

$$V_0 := \left\{ v \in H^1(\Omega) \,|\, v = 0 \quad \text{on} \quad \partial_1\Omega \right\}$$

and

$$\langle \Lambda^* y^*, w \rangle = \int_\Omega y^* \cdot \nabla w \, dx, \qquad \forall w \in V_0.$$

Assume that

$$f \in L^2(\Omega), \qquad F \in L^2(\partial_2\Omega). \tag{6.4.20}$$

Then, the functional l has an integral form

$$\langle l, w \rangle = \int_\Omega f w \, dx + \int_{\partial_2\Omega} F w \, ds.$$

For any $w \subset V_0$, we have

$$\langle l + \Lambda^* y^*, w \rangle = \int_\Omega (y^* \cdot \nabla w + fw)\, dx + \int_{\partial_2 \Omega} Fw\, ds =$$

$$= \int_\Omega (\operatorname{div} y^* + f)w\, dx + \int_{\partial_2 \Omega} (y^* \cdot \nu + F)w\, ds,$$

provided that y^* possesses an extra regularity, namely,

$$y^* \in Q^*(\Omega) := \left\{ y^* \in Y^* \mid \operatorname{div} y^* \in L^2(\Omega),\ y^* \cdot \nu \in L^2(\partial_2 \Omega) \right\}.$$

Note that in view of (6.4.20) the exact solution p^* belongs to the set $Q^*(\Omega)$, so that the above-made assumption on the admissible set of y^* is not restrictive. Now, we obtain

$$|\langle l + \Lambda^* y^*, w \rangle| \leq \| \operatorname{div} y^* - f\|_{2,\Omega} \|w\|_{2,\Omega} + \|y^* \cdot \nu + F\|_{2,\partial_2 \Omega} \|w\|_{2,\partial_2 \Omega}.$$

Let γ and γ_* be two numbers such that

$$\gamma > 1, \quad \gamma_* > 1, \quad \frac{1}{\gamma} + \frac{1}{\gamma_*} = 1.$$

Then

$$|\langle l + \Lambda^* y^*, w \rangle| \leq \left(\gamma \| \operatorname{div} y^* - f\|_{2,\Omega}^2 + \gamma_* \|y^* \cdot \nu + F\|_{2,\partial_2 \Omega}^2 \right)^{1/2} \times$$

$$\times \left(\frac{1}{\gamma} \|w\|_{2,\Omega}^2 + \frac{1}{\gamma_*} \|w\|_{2,\partial_2 \Omega}^2 \right)^{1/2}.$$

Since

$$\|w\|_{2,\Omega}^2 \leq C_1^2(\Omega) \|\nabla w\|_{2,\Omega}^2, \quad \forall w \in V_0, \tag{6.4.21}$$

and

$$\|w\|_{2,\partial_2 \Omega}^2 \leq C_2(\Omega, \partial_2 \Omega) \|w\|_{1,2,\Omega}^2, \quad \forall w \in V_0, \tag{6.4.22}$$

we find that

$$\frac{1}{\gamma} \|w\|_{2,\Omega}^2 + \frac{1}{\gamma_*} \|w\|_{2,\partial_2 \Omega}^2 \leq$$

$$\leq \left(C_1^2(\Omega) \frac{1}{\gamma} + C_2^2(\Omega, \partial_2 \Omega) \left(1 + C_1^2(\Omega) \right) \frac{1}{\gamma_*} \right) \|\nabla w\|_{2,\Omega}^2.$$

Therefore, there exist a positive constant C_3 (which depends on Ω, $\partial_2\Omega$, and γ) such that

$$\frac{1}{C_3^2} = \inf_{w \in V_0} \frac{\int_\Omega A\nabla w \cdot \nabla w \, dx}{\frac{1}{\gamma}\|w\|_{2,\Omega}^2 + \frac{1}{\gamma_*}\|w\|_{2,\partial_2\Omega}^2}.$$

Moreover,

$$C_3 \le \left(C_1^2(\Omega)\frac{1}{\gamma} + C_2^2(\Omega,\partial_2\Omega)(1 + C_1^2(\Omega))\frac{1}{\gamma_*} \right) c_1^{-1}, \qquad (6.4.23)$$

so that the constant C_3 has a computable upper bound, provided that the constant $C_1(\Omega)$ in the Friederichs inequality (6.4.21) and the constant $C_2(\Omega,\partial_2\Omega)$ in the trace inequality (6.4.22) are known. It is worthyof note that these constants depend only on Ω and $\partial_2\Omega$. Now, we find that

$$|\langle l + \Lambda^* y^*, w\rangle| \le C_3 \left(\gamma\|\operatorname{div} y^* - f\|_{2,\Omega}^2 + \gamma_*\|y^* \cdot \nu + F\|_{2,\partial_2\Omega}^2 \right)^{1/2} \, \|\!\|\nabla w\|\!\| \, .$$

From this estimate, we obtain

$$\begin{aligned}
|\!| l + \Lambda^* y^* |\!|^2 &\le \\
&\le C_3 \left(\gamma\|\operatorname{div} y^* - f\|_{2,\Omega}^2 + \gamma_*\|y^* \cdot \nu + F\|_{2,\partial_2\Omega}^2 \right) \le \\
&\le \left(C_1^2(\Omega)\frac{1}{\gamma} + C_2^2(\Omega,\partial_2\Omega)(1 + C_1^2(\Omega))\frac{\gamma - 1}{\gamma} \right) \times \\
&\quad \times \left(\gamma\|\operatorname{div} y^* - f\|_{2,\Omega}^2 + \frac{\gamma}{\gamma - 1}\|y^* \cdot \nu + F\|_{2,\partial_2\Omega}^2 \right) c_1^{-2}. \quad (6.4.24)
\end{aligned}$$

A minimum of the right-hand side of (6.4.24) is attained if

$$\gamma = \widehat{\gamma} := 1 + \frac{\|y^* \cdot \nu + F\|_{2,\partial_2\Omega} C_1(\Omega)}{\|\operatorname{div} y^* - f\|_{2,\Omega} C_2(\Omega,\partial_2\Omega)(1 + C_1^2(\Omega))^{1/2}}.$$

In this case,

$$\begin{aligned}
|\!| l + \Lambda^* y^* |\!|^2 &\le \Big(C_1(\Omega)\|\operatorname{div} y^* - f\|_{2,\Omega} + \\
&\quad + C_2(\Omega,\partial_2\Omega)(1 + C_1^2(\Omega))^{1/2}\|y^* \cdot \nu + F\|_{2,\partial_2\Omega} \Big)^2 c_1^{-2}.
\end{aligned}$$
$$(6.4.25)$$

Now, we can present the majorant in two forms. The first one uses (6.4.25), which yields

$$M_\oplus(v,\beta,y^*) = \frac{1+\beta}{2}\int_\Omega (\nabla v - A^{-1}y^*)\cdot(A\nabla v - y^*)\,dx+$$

$$+\frac{1+\beta}{2\beta}\Big(C_1(\Omega)\|\operatorname{div}y^* - f\|_{2,\Omega}+$$

$$+ C_2(\Omega,\partial_2\Omega)(1+C_1^2(\Omega))^{1/2}\|y^*\cdot\nu + F\|_{2,\partial_2\Omega}\Big)^2 c_1^{-2}. \quad (6.4.26)$$

If we take $\gamma = \gamma^* = 2$, then the majorant has a simpler form

$$M_\oplus(v,\beta,y^*) = \frac{1+\beta}{2}\int_\Omega (\nabla v - A^{-1}y^*)\cdot(A\nabla v - y^*)\,dx+$$

$$+\frac{1+\beta}{2\beta}c_1^{-2}\Big(C_1^2(\Omega) + C_2^2(\Omega,\partial_2\Omega)(1+C_1^2(\Omega))\Big)\times$$

$$\times\Big(\|\operatorname{div}y^* - f\|_{2,\Omega}^2 + \|y^*\cdot\nu + F\|_{2,\partial_2\Omega}^2\Big). \quad (6.4.27)$$

This majorant gives an upper bound of the deviation for any $v \in V_0 + u_0$, $y^* \in Q^*$, and $\beta > 0$. It is easy to see that it vanishes if and only if $v = u$ and $y^* = A\nabla u$. Also, it is obvious that the majorant is continuous with respect to the convergence of v in V and y^* in $Q^\cdot$

Remark 6.4.1. The estimate (6.4.26) remains valid for the Neumann problem.

6.4.4 Lower estimates

Lower estimates for the problems considered follow from (6.2.4). They have a common form:

$$\frac{1}{2}\int_\Omega A\nabla(v-u)\cdot\nabla(v-u)\,dx \geq \mathcal{M}_\ominus(v,w), \quad \forall w \in V_0,$$

where

$$\mathcal{M}_\ominus(v,w) = -\frac{1}{2}\int_\Omega A\nabla(w-v)\cdot\nabla w\,dx - \int_\Omega fw\,dx - \int_{\partial_2\Omega} Fw\,ds.$$

Here V_0 depends on the type of boundary conditions, and the integral over $\partial_2\Omega$ must be eliminated in the case of Dirichlét problem.

6.5 The linear elasticity problem

Assume that Ω satisfies the same conditions as in the previous section and $\partial\Omega$ consists of two disjoint parts $\partial_1\Omega$ and $\partial_2\Omega$ and $|\partial_1\Omega| > 0$. The classical statement of a linear elasticity problem is as follows: to find a tensor-valued function σ^* (stress) and a vector-valued function u (displacement) that satisfy the system of equations

$$\sigma^* = \mathbb{L}\varepsilon \quad \text{in} \quad \Omega, \tag{6.5.1}$$

$$\operatorname{div}\sigma^* = f \text{ in} \quad \Omega, \tag{6.5.2}$$

$$u = u_0 \quad \text{on} \quad \partial_1\Omega, \tag{6.5.3}$$

$$\sigma^*\nu + F = 0 \text{ on} \quad \partial_2\Omega. \tag{6.5.4}$$

Here f and F are given forces and $\mathbb{L} = \{L_{ijkm}\}$ is the tensor of elasticity constants, which is subject to the conditions

$$C_1|\varepsilon|^2 \leq \mathbb{L}\varepsilon : \varepsilon \leq C_2|\varepsilon|^2, \quad \forall\varepsilon \in \mathbb{M}_s^{n\times n}, \tag{6.5.5}$$

and

$$\mathbb{L}_{ijkm} = \mathbb{L}_{jikm} = \mathbb{L}_{kmij}, \quad \mathbb{L}_{ijkm} \in L^\infty(\Omega). \tag{6.5.6}$$

Henceforth, we assume that

$$f \in L^2(\Omega, \mathbb{R}^n), \quad F \in L^2(\partial_2\Omega, \mathbb{R}^n). \tag{6.5.7}$$

Then, a generalized solution $u \in V_0 + u_0$ is defined by the identity

$$\int_\Omega \mathbb{L}\varepsilon(u) : \varepsilon(w)\,dx + \langle l, w\rangle = 0, \quad \forall w \in V_0, \tag{6.5.8}$$

where

$$\langle l, w\rangle = \int_\Omega f \cdot w\,dx + \int_{\partial_2\Omega} F \cdot w\,ds.$$

The existence and uniqueness of u follows from (6.5.5) and the Korn's inequality (see, e.g., [64]).

Let v and y^* be some approximations of u and σ^*. Estimates of $v - u$ and $y^* - \sigma^*$ follow from the general scheme if we set

$$U = L^2(\Omega, \mathbb{M}_s^{n\times n}), \qquad V = H^1(\Omega, \mathbb{R}^n),$$
$$V_0 = \{w \in V \,|\, w = 0 \text{ on } \partial_1\Omega\},$$
$$\|\,y\,\|^2 = \int_\Omega \mathbb{L}y : y\,dx, \qquad \|\,y^*\,\|_*^2 = \int_\Omega \mathbb{L}^{-1}y^* : y^*\,dx,$$

and identify Λv with the symmetric part of the gradient tensor

$$\varepsilon(v) = \frac{1}{2}\left(\nabla v + (\nabla v)^T\right).$$

In this case,

$$\langle \Lambda^* y^*, w\rangle = \int_\Omega y^* : \varepsilon(w)\, dx, \quad \forall w \in V_0, \tag{6.5.9}$$

where two dots stand for the scalar product in $\mathbb{M}^{n\times n}$. If

$$y^* \in Q^* := \{y^* \in Y^* \mid \operatorname{div} y^* \in L^2(\Omega, \mathbb{M}^{n\times n}),\ y^*\nu \in L^2(\partial_2\Omega, \mathbb{R}^n)\},$$

then

$$\langle l + \Lambda^* y^*, w\rangle = -\int_\Omega \operatorname{div} y^* \cdot w\, dx + \int_{\partial_2\Omega} (y^*\nu) \cdot w\, d\Gamma. \tag{6.5.10}$$

6.5.1 Upper estimates

By (6.2.3) we obtain the upper estimate of the deviation from the exact solution in terms of "strains":

$$\frac{1}{2}\int_\Omega \mathbb{L}\varepsilon(v-u) : \varepsilon(v-u)\, dx \leq \mathcal{M}_\oplus(v, \beta, y^*),$$

where

$$\mathcal{M}_\oplus(v, \beta, y^*) = \frac{1+\beta}{2} D(\varepsilon v, y^*) + \frac{1+\beta}{2\beta}|\Lambda^* y^* + l|^2$$

and

$$D(\varepsilon(v), y^*) = \frac{1}{2}\int_\Omega \left(\mathbb{L}\varepsilon(v) : \varepsilon(v)\mathbb{L}^{-1}y^* : y^* - 2\varepsilon(v) : y^*\right) dx =$$

$$= \int_\Omega (\varepsilon(u) - \mathbb{L}^{-1}y^*) : (\mathbb{L}\varepsilon(u) - y^*)\, dx.$$

If $y^* \in Q^*$, then

$$|\Lambda^* y^* + l| = \sup_{w \in V_0} \frac{\langle \Lambda^* y^* + l, w \rangle}{\|\Lambda w\|} =$$

$$= \sup_{w \in V_0} \frac{\int\limits_{\Omega} (y^* : \varepsilon(w) + f \cdot w)\, dx + \int\limits_{\partial_2 \Omega} F \cdot w\, ds}{\|\varepsilon(w)\|} =$$

$$= \sup_{w \in V_0} \frac{\int\limits_{\Omega} (f - \operatorname{div} y^*) \cdot w\, dx + \int\limits_{\partial_2 \Omega} (F + y^* \nu) \cdot w\, ds}{\|\varepsilon(w)\|}.$$

Let C_Ω be a constant in the inequality

$$\int\limits_{\Omega} |w|^2\, dx + \int\limits_{\partial_2 \Omega} |w|^2\, ds \leq C_\Omega^2 \|\varepsilon(w)\|_\Omega^2, \quad \forall w \in V_0. \tag{6.5.11}$$

Note that the existence of such a constant follows from the Korn's inequality. By (6.5.11), we obtain

$$|\Lambda^* y^* + l| \leq$$

$$\leq \left(\|\operatorname{div} y^* - f\|_\Omega^2 + \|F + y^* \nu\|_{\partial_2 \Omega}^2 \right)^{1/2} \sup_{w \in V_0} \frac{(\|w\|_\Omega^2 + \|w\|_{\partial_2 \Omega}^2)^{1/2}}{\|\varepsilon(w)\|} \leq$$

$$\leq C_\Omega c_1^{-1/2} \left(\|\operatorname{div} y^* - f\|_\Omega^2 + \|F + y^* \nu\|_{\partial_2 \Omega}^2 \right)^{1/2}$$

and arrive at the majorant $M_\oplus$:

$$M_\oplus(\varepsilon(v), y^*) = \frac{1+\beta}{2} \int\limits_{\Omega} (\varepsilon(u) - \mathbb{L}^{-1} y^*) : (\mathbb{L}\varepsilon(u) - y^*)\, dx +$$

$$+ \frac{1+\beta}{2\beta c_1} C_\Omega^2 \left(\|\operatorname{div} y^* - f\|_\Omega^2 + \|F + y^* \nu\|_{\partial_2 \Omega}^2 \right). \tag{6.5.12}$$

It has a clear physical meaning. The first term of $M_\oplus$ is nonnegative and vanishes if and only if

$$y^* = \mathbb{L}\varepsilon(v).$$

It penalizes violations of the *constitutive relation* (Hooke's law) (6.5.1). The meaning of the second term is obvious: it contains L^2-norms of other two relations (6.5.2) and (6.5.4), which gives errors in the *equilibrium equation* and *boundary condition* for the stress tensor. Thus, the majorant not only gives an idea of the overall value of the error, but also shows its physically sensible parts. The latter information may be of utmost interest, because it suggests a correct way of finding a better approximation.

Let $\{Y_k^*\} \subset H^1(\Omega, \mathbb{M}^{n \times n})$ be a collection of finite-dimensional subspaces that satisfy the condition (6.3.18). Then, (6.5.12) provides a sequence of computable upper bounds

$$M_{\oplus}^k = \inf_{\substack{y^* \in Y_k^* \\ \beta \in \mathbb{R}_+}} \left\{ \frac{1+\beta}{2} \int_{\Omega} \left(\mathbb{L}\varepsilon(v) : \varepsilon(v) + \mathbb{L}^{-1}y^* : y^* - 2\varepsilon(v) : y^* \right) dx \right.$$
$$\left. + \frac{1+\beta}{2\beta c_1} C_{\Omega}^2 \left(\| \operatorname{div} y^* - f \|_{\Omega}^2 + \|F + y^*\nu\|_{\partial_2\Omega}^2 \right) \right\}, \quad (6.5.13)$$

which tends to the exact value of the error.

6.5.2 Lower estimates

Lower estimates follow from (6.2.4). We have

$$\frac{1}{2} \int_{\Omega} \mathbb{L}\varepsilon(v - u) : \varepsilon(v - u) \, dx \geq \mathcal{M}_{\ominus}(v, w), \quad \forall w \in V_0,$$

where

$$\mathcal{M}_{\ominus}(v, w) = -\frac{1}{2} \int_{\Omega} \mathbb{L}\varepsilon(w) : \varepsilon(w) \, dx - \int_{\Omega} \mathbb{L}\varepsilon(v) : \varepsilon(w) \, dx -$$
$$- \int_{\Omega} f \cdot w \, dx - \int_{\partial_2\Omega} F \cdot w \, ds.$$

Maximizing the functional $\mathcal{M}_{\ominus}$ on a certain finite-dimensional space $V_{0k} \subset V_0$, we obtain a computable lower bound

$$\mathcal{M}_{\ominus} = \sup_{w \in V_{0k}} \mathcal{M}_{\ominus}(v, w_k).$$

If the spaces V_{0k} satisfy the conditions stated in Section 3, then the sequence of numbers $\{\mathcal{M}_{\ominus}\}$ tends to $\frac{1}{2} \| \varepsilon(v - u) \|^2$ (cf. (6.3.12)).

6.6　Linear elliptic equations of the fourth order

In this section, we consider the problem

$$\nabla \cdot \nabla \cdot (B\nabla\nabla u) = f \quad \text{in } \Omega, \tag{6.6.1}$$

$$u = \frac{\partial u}{\partial \nu} = 0 \quad \text{on } \partial\Omega. \tag{6.6.2}$$

Here $\Omega \subset \mathbb{R}^2$, ν denotes the outward unit normal to the boundary, and $B = \{b_{ijkl}\} \in \mathcal{L}(\mathbb{M}_s^{2\times2}, \mathbb{M}_s^{2\times2})$. We assume that $b_{ijkl} = b_{jikl} = b_{klij}$,

$$\alpha_1|\eta|^2 \le B\eta : \eta \le \alpha_2|\eta|^2, \quad \forall \eta \in \mathbb{M}_s^{2\times2}, \tag{6.6.3}$$

and

$$f \in L^2(\Omega), \quad b_{ijkl} \in L^\infty(\Omega). \tag{6.6.4}$$

To apply the general scheme, we set

$$U = L^2(\Omega, \mathbb{M}_s^{2\times2}), \qquad V = H^2(\Omega),$$

$$V_0 = \{w \in V \mid w = \frac{\partial w}{\partial \nu} = 0 \text{ on } \partial\Omega\},$$

and define Λ as the Hessian operator. Now, (6.1.1) has the form

$$\int_\Omega B\nabla\nabla u : \nabla\nabla w \, dx = \int_\Omega fw \, dx \quad \forall w \in V_0. \tag{6.6.5}$$

By B^{-1} we denote the inverse tensor, which satisfies the double inequality

$$\alpha_2^{-1}|\eta^*|^2 \le B^{-1}\eta^* : \eta^* \le \alpha_1^{-1}|\eta^*|^2, \quad \forall \eta^* \in \mathbb{M}_s^{2\times2}, \tag{6.6.6}$$

The spaces Y and Y^* are equipped with norms

$$\| y \|^2 = \int_\Omega By : y \, dx; \quad \| y^* \|_*^2 = \int_\Omega B^{-1}y^* : y^* \, dx,$$

$$\langle l, w \rangle = -\int_\Omega fw \, dx,$$

and

$$Q_l^* = \{y^* \in Y^* \mid \int_\Omega y^* : \nabla\nabla w \, dx = \int_\Omega fw \, dx, \quad \forall w \in V_0\}.$$

Since

$$\|\nabla\nabla w\| \geq \alpha_3\|w\|_{2,2,\Omega} \quad \forall w \in V_0,$$

we see that the condition (6.1.3) is satisfied.

Problem (6.6.1) and (6.6.2) is associated with two variational problems.

Problem $\mathcal{P}$. Find $u \in V_0$ such that

$$J(u) = \inf_{v \in V_0} J(v),$$

where

$$J(v) = \frac{1}{2}\int_\Omega B\nabla\nabla v : \nabla\nabla v \, dx - \int_\Omega fw \, dx.$$

Problem $\mathcal{P}^*$. Find $p^* \in Q_l^*$ such that

$$I^*(p^*) = \sup_{\forall q^* \in Q_l^*} I^*(q^*),$$

where

$$I^*(q^*) = -\frac{1}{2}\int_\Omega B^{-1}q^* : q^* \, dx.$$

By Proposition 6.1.1, we obtain two basic relations for deviations:

$$\|\nabla\nabla(v-u)\|^2 + \|q^* - p^*\|_*^2 = 2(J(v) - I^*(q^*)), \tag{6.6.7}$$

and

$$\|\nabla\nabla(v-u)\|^2 + \|q^* - p^*\|_*^2 = 2D(\nabla\nabla v, q^*) =$$
$$= \int_\Omega \left(B\nabla\nabla v : \nabla\nabla v + B^{-1}q^* : q^* - 2\nabla\nabla v : q^*\right) dx, \tag{6.6.8}$$

which hold for any $v \in V_0$ and $q^* \in Q_l^*$.

From (6.1.13) it follows that

$$\frac{1}{2}\|\nabla\nabla(v-u)\|^2 \leq (1+\beta)D(\nabla\nabla v, y^*)+$$
$$+ \left(1 + \frac{1}{\beta}\right)\frac{1}{2}d_l^2(y^*), \tag{6.6.9}$$

where $d_l^2(y^*) = \inf\limits_{q^* \in Q_l^*} \| q^* - y^* \|_*^2$. Note that

$$\int_\Omega y^* : \nabla\nabla w \, dx = \int_\Omega (\operatorname{div}\operatorname{div} y^*) w \, dx, \quad \forall w \in V_0,$$

so that $\Lambda^* : Y^* \to H^{-2}(\Omega)$ is the operator $\operatorname{div}\operatorname{div}$. Next,

$$\langle l + \Lambda^* y^*, w \rangle = \int_\Omega (y^* : \nabla\nabla w - fw) \, dx \qquad (6.6.10)$$

and, therefore,

$$d_l^2(y^*) = |\, l + \Lambda^* y^* \,| = \sup_{w \in V_0} \frac{\int_\Omega (y^* : \nabla\nabla w - fw) \, dx}{\|\, \nabla\nabla w \,\|}.$$

If

$$y^* \in H(\operatorname{div}\operatorname{div}, \Omega) := \left\{ y^* \in L^2(\Omega, \mathbb{M}_s^{n \times n}) \mid \operatorname{div}\operatorname{div} y^* \in L^2(\Omega) \right\},$$

then this quantity is estimated by the relation

$$|\, l + \Lambda^* y^* \,| \leq \sup_{w \in V_0} \frac{\| \operatorname{div}\operatorname{div} y^* - f \|_\Omega \|w\|_\Omega}{\|\, \nabla\nabla w \,\|} \leq$$

$$\leq \sup_{w \in V_0} \frac{\| \operatorname{div}\operatorname{div} y^* - f \|_\Omega \|w\|_\Omega}{\alpha_1 \|\nabla\nabla w\|} \leq \frac{C_{1\Omega}}{\alpha_1} \| \operatorname{div}\operatorname{div} y^* - f \|_\Omega, \quad (6.6.11)$$

in which $C_{1\Omega}$ is a constant in the inequality

$$\|w\|_\Omega \leq C_{1\Omega} \|\nabla\nabla w\|_\Omega \qquad \forall w \in V_0. \qquad (6.6.12)$$

By (6.6.9), we obtain

$$\frac{1}{2} \|\, \nabla\nabla(v - u) \,\|^2 \leq (1 + \beta) D(\nabla\nabla v, y^*) +$$

$$+ \left(1 + \frac{1}{\beta}\right) \frac{C_{1\Omega}^2}{2\alpha_1^2} \| \operatorname{div}\operatorname{div} y^* - f \|_\Omega^2, \quad (6.6.13)$$

In this estimate, y^* is an arbitrary tensor-valued function from $H(\operatorname{div}\operatorname{div}, \Omega)$ and β is a positive real number. However, this is rather demanding in relation to the dual variable y^* (which must have square summable $\operatorname{div}\operatorname{div}$).

To avoid technical difficulties that rises from this condition, we estimate the negative norm as follows:

$$|\,l + \Lambda^* y^*\,| \leq \frac{C_{2\Omega}}{\alpha_1}\|\operatorname{div} y^* - \eta^*\|_\Omega + \frac{C_{1\Omega}}{\alpha_1}\|\operatorname{div}\eta^* - f\|_\Omega. \qquad (6.6.14)$$

Here, η^* is an arbitrary vector-valued function from $H(\operatorname{div},\Omega)$ and $C_{1\Omega}$ is a constant in the inequality

$$\|\nabla w\|_\Omega \leq C_{1\Omega}\|\nabla\nabla w\|_\Omega \qquad \forall w \in V_0. \qquad (6.6.15)$$

Then, we arrive at the estimate

$$\frac{1}{2}\,\|\,\nabla\nabla(v-u)\,\|^2 \leq (1+\beta)D(\nabla\nabla v, y^*)+$$
$$+\left(1+\frac{1}{\beta}\right)\frac{1}{2\alpha_1^2}\left(C_{2\Omega}\|\operatorname{div} y^* - \eta^*\|_\Omega + C_{1\Omega}\|\operatorname{div}\eta^* - f\|_\Omega\right)^2, \quad (6.6.16)$$

in which $y^* \in L^2(\Omega, \mathbb{M}_s^{2\times 2})$ must have square summable divergence and $\eta^* \in H(\operatorname{div},\Omega)$. Note that $C_{1\Omega} \leq C_\Omega C_{2\Omega}$, where C_Ω is a constant in the Friederichs inequality. In view of this, we obtain a slightly different form of the deviation estimate (cf. [149]):

$$\frac{1}{2}\,\|\,\nabla\nabla(v-u)\,\|^2 \leq (1+\beta)D(\nabla\nabla v, y^*)+$$
$$+\left(1+\frac{1}{\beta}\right)\frac{C_{2\Omega}^2}{2\alpha_1^2}\left(\|\operatorname{div} y^* - \eta^*\|_\Omega + C_\Omega\|\operatorname{div}\eta^* - f\|_\Omega\right)^2, \quad (6.6.17)$$

For boundary conditions of other types, the deviation majorants can be derived by arguments similar to those used in Section 4.

Lower estimates follow from (6.2.4). We have

$$\frac{1}{2}\,\|\,\nabla\nabla(v-w)\,\|^2 \geq \mathcal{M}_\ominus(v,w) \quad w \in V_0, \qquad (6.6.18)$$

where

$$\mathcal{M}_\ominus(v,w) := -\frac{1}{2}\,\|\,\nabla\nabla w\,\|^2 - \int_\Omega (B\nabla\nabla v : \nabla\nabla w - fw)dx.$$

6.7 Relationship with other methods

The majorant $\mathcal{M}_\oplus(v, \beta, y^*)$ involves an arbitrary function y^*. One way of finding a suitable y^* has been discussed. It leads to an auxiliary finite-dimensional problem (6.3.3). However, one can also define y^* by attracting the information encompassed in the approximate solution v. In this section, we are focused on this way. Our aim is to show that such *modus operandi* leads to a wide spectrum of a posteriori error estimates. For finite element approximations, some of them coincide with a posteriori estimates earlier discussed in Chapter 4.

Consider the abstract problem presented in §1 of this chapter. Assume that

$$\langle l, w \rangle = (g, w), \quad \forall w \in V, \tag{6.7.1}$$

where $g \in U$. In this case,

$$p^* \in Q^* := \{ y^* \in Y^* \mid \Lambda^* y^* \in U \} \tag{6.7.2}$$

and, moreover,

$$p^* \in Q_l^* := \{ y^* \in Q^* \mid (\Lambda^* y^* + g, w) = 0, \quad \forall w \in V_0 \}. \tag{6.7.3}$$

Let

$$y_1^* = \mathcal{A}\Lambda v. \tag{6.7.4}$$

Setting

$$y^* = \Pi y_1^*, \tag{6.7.5}$$

where Π is a certain continuous mapping, we obtain a function y^* that can be substituted into the majorant $\mathcal{M}_\oplus$.

6.7.1   Residual based estimates

The simplest way is to set

$$y^* = y_1^*, \tag{6.7.6}$$

which means that Π is the identity mapping of Y^* to itself. In this case,

$$D(\Lambda v, y_1^*) = 0$$

and $\mathcal{M}_\oplus(v, \beta, y_1^*)$ contains only the second term, which upon minimization with respect to β gives the first form of the majorant

$$\mathcal{M}_\oplus^{(1)}(v) = \frac{1}{2} |l + \Lambda^* \mathcal{A}\Lambda v|, \tag{6.7.7}$$

which implies the estimate

$$\| \Lambda(v - u) \| \leq | l + \Lambda^* \mathcal{A}\Lambda v | = \sup_{w \in V_0} \frac{(g, w) + (\mathcal{A}\Lambda v, \Lambda w)}{\| \Lambda w \|}. \tag{6.7.8}$$

If v is obtained by the finite element method (so that v coincides with a Galerkin approximation $u_h \in V_h := V_{0h} + u_0$), then the right-hand side of (6.7.8) is estimated by the method presented in Chapter 4. As a result, we obtain residual type a posteriori error estimates that involve integral terms associated with finite elements and interelement jumps.

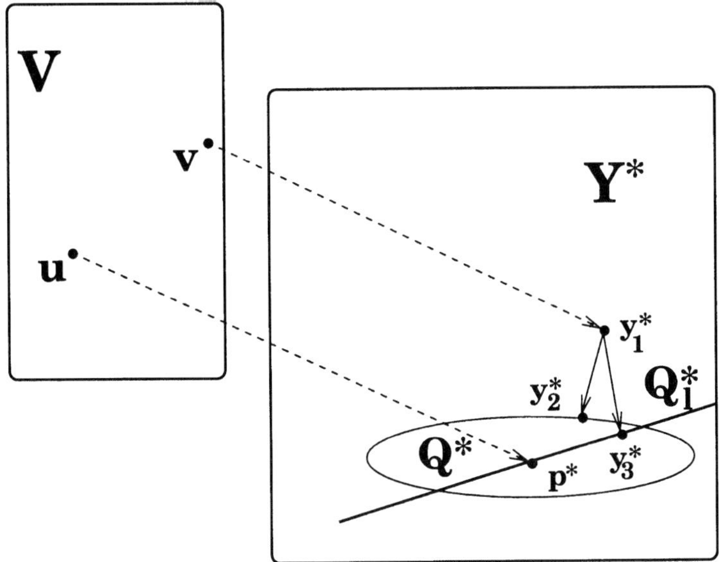

Figure 6.7.1: Various choices of the dual variable.

6.7.2 Estimates based on the "regularization" of the dual variable

Obviously, the best choice of y^* in $\mathcal{M}_\oplus(v, \beta, y^*)$ is $y^* = p^* \in Q^*$ (cf. 6.7.3). Therefore, if $y_1^* \notin Q^*$ and we know an inexpensive continuous mapping from Y^* to Q^*, then the image of y_1^* obtained by such a mapping could be a better approximation of p^*. We call such a post-processing procedure the *regularization* of the dual variable and denote the respective operator by Π_{reg}. By Π_{reg}, we define another element. This situation is presented in Fig. 6.7.1, which is taken from [172].

$$y_2^* = \Pi_{\text{reg}} y_1^* \in Q^* \tag{6.7.9}$$

and the quantity $\mathcal{M}_\oplus(v, \beta, y_1^*)$, which gives the second form of the majorant

$$\mathcal{M}_\oplus^{(2)}(v) = \inf_{\beta \in \mathbb{R}_+} \left\{ (1+\beta) D(\Lambda v, \Pi_{\text{reg}}(\mathcal{A}\Lambda v)) + \right.$$
$$\left. + \left(1 + \frac{1}{\beta}\right) \frac{1}{2} |l + \Lambda^* \Pi_{\text{reg}}(\mathcal{A}\Lambda v)|^2 \right\}. \tag{6.7.10}$$

Now

$$\left| l + \Lambda^* \Pi_{\text{reg}}(\mathcal{A}\Lambda v) \right| = \sup_{w \in V_0} \frac{(g + \Lambda^* y_2^*, w)}{\| \Lambda w \|} \leq C_3^{-1} \| g + \Lambda^* y_2^* \|.$$

Assume that $v = u_h$, where u_h is a Galerkin approximation computed for a finite element space $V_{0h} \subset V_0$ and $\Pi_{\text{reg}} = G$, where G is a "Λ–averaging" operator (cf. IV.4). Then

$$D(\Lambda v, \Pi_{\text{reg}}(\mathcal{A}\Lambda v)) = \frac{1}{2} \| \mathcal{A}\Lambda u_h - G(\mathcal{A}\Lambda u_h) \|_*^2 \qquad (6.7.11)$$

and by (6.7.10) we derive the estimate

$$\| \Lambda(u_h - u) \|^2 \leq (1 + \beta) \| \mathcal{A}\Lambda u_h - G(\mathcal{A}\Lambda u_h) \|_*^2 +$$

$$+ \left(1 + \frac{1}{\beta} \right) C_3^{-2} \| g + \Lambda^* G(\mathcal{A}\Lambda u_h) \|^2, \quad (6.7.12)$$

whose right-hand side is easily computable. If we set $\mathrm{T} = \mathcal{A}\Lambda$, then the first part of (6.7.12) is precisely the error indicator (4.4.3). Of interest is to discuss conditions that make this term dominant.

Let the regularity of u and properties of approximation spaces provide that

$$\| \Lambda(u_h - u) \| \sim Ch^{\alpha}, \qquad (6.7.13)$$

where a constant C does not depend on h.

Assume that the operator G possesses a superconvergence property (see IV.4), i.e

$$\| \mathcal{A}\Lambda u - G\mathcal{A}\Lambda u_h \|_* \leq Ch^{\alpha+\varepsilon}, \quad \varepsilon > 0. \qquad (6.7.14)$$

In this case, the majorant (6.7.10) consists of two terms that have different asymptotic rates. Indeed,

$$(g + \Lambda^* G\mathcal{A}\Lambda u_h, w) = (G\mathcal{A}\Lambda u_h - \mathcal{A}\Lambda u, \Lambda w)$$

and we find that

$$\left| l + \Lambda^* G\mathcal{A}\Lambda u_h \right| \leq \| \mathcal{A}\Lambda u - G\mathcal{A}\Lambda u_h \|_* \leq Ch^{\alpha+\varepsilon}.$$

We see that $\mathcal{M}_\oplus^{(2)}(u_h)$ tends to zero as h^α, but its second term has a higher convergence rate. Thus, the first term given by (6.7.11) dominates, provided that h is sufficiently small and we conclude that

$$\mathcal{M}_\oplus^{(2)}(u_h) \cong D(\Lambda u_h, G\mathcal{A}\Lambda u_h). \qquad (6.7.15)$$

The quantity $D(\Lambda u_h, G\mathcal{A}\Lambda u_h)$ is easily computable. Therefore, (6.7.15) gives rise to various practically attractive error indicators.

6.7.3 Estimates based on the "equilibration" of the dual variable

Finally, consider the case in which the operator Π maps Y^* to the set $Q_l^* \subset Q^*$. In this case, we define

$$y_3^* = \Pi y_1^* \in Q_l^*. \tag{6.7.16}$$

Hence,

$$\langle \Lambda^* y_3^* + l, w \rangle = 0, \quad \forall w \in V_0,$$

so that $|\Lambda^* y_3^* + l| = 0$. Now the majorant contains only one term:

$$\mathcal{M}_\oplus^{(3)}(v) = D(\Lambda v, y_3^*). \tag{6.7.17}$$

The meaning of this type error majorants is easy to explain by the paradigm of the linear elasticity problem (see VI.5). In this problem, y^* presents a stress field, and the set Q_l^* consists of the tensor-valued functions satisfying the equilibrium equation

$$\int\limits_{\Omega} (y^* : \varepsilon(w) + f \cdot w) \, dx + \int\limits_{\partial_2 \Omega} F \cdot w \, ds = 0, \quad \forall w \in V_0. \tag{6.7.18}$$

For this reason, it is natural to call Π an *equilibration operator* and denote it by Π_{eq}. The respective majorant has the form

$$\mathcal{M}_\oplus^{(3)}(v) = \frac{1}{2} \int\limits_{\Omega} \Big(\varepsilon(v) - \mathbb{L}^{-1} \Pi_{eq}(\mathbb{L}\varepsilon(v)) \Big) : \Big(\mathbb{L}\varepsilon(v) - \Pi_{eq}(\mathbb{L}\varepsilon(v)) \Big) \, dx. \tag{6.7.19}$$

Let $\tau^*(v) = \mathbb{L}\varepsilon(v)$ be the stress function associated with $\varepsilon(v)$. Then, we rewrite $\mathcal{M}_\oplus^{(3)}$ as follows:

$$\mathcal{M}_\oplus^{(3)}(v) = \frac{1}{2} \int\limits_{\Omega} \mathbb{L}^{-1}(\tau^*(v) - \Pi_{eq}\tau^*(v)) : (\tau^*(v) - \Pi_{eq}\tau^*(v)) \, dx =$$

$$= \frac{1}{2} \| \tau^*(v) - \Pi_{eq}\tau^*(v) \|_*^2 . \tag{6.7.20}$$

The relations (6.7.19) and (6.7.20) show that if a computed stress field τ^* is post-processed in such a way that it satisfies (6.7.18), then the majorant is given by the error in the Hooke's law, or, what is equivalent, by the $\| \cdot \|_*$-norm of the difference between τ^* and its post-processed image $\Pi_{eq}\tau^*$. We recall that this type error estimators for finite element approximations have been discussed in Chapter 4. In general, it is rather difficult to construct an operator Π_{eq} with the afore-mentioned projection properties. Instead, one may try to use an operator $\widetilde{\Pi}_{eq}$, which provides an approximate equilibration. In thise case, the second term of the majorant $\mathcal{M}_{\oplus}(v, \beta, \widetilde{\Pi}_{eq}\tau^*)$ does not vanish and should be taken into account.

6.7.4   A priori projection type error estimates

Now, our goal is to show that classical a priori projection type error estimates also follow from the general deviation estimate (6.2.3). Let $u_h \in V_h$ be a Galerkin approximation of u. We have

$$\| \Lambda(u - u_h) \| \le 2(1 + \beta)D(\Lambda u_h, y^*) + \left(1 + \frac{1}{\beta}\right) |\Lambda^* y^* + l|^2 \quad (6.7.21)$$

Set here $y^* = \mathcal{A}\Lambda v_h$, where v_h is an arbitrary element of V_h. Then,

$$|\Lambda^* y^* + l| = \sup_{w \in V_0} \frac{(y^* - p^*, \Lambda w)}{\| \Lambda w \|} =$$

$$= \sup_{w \in V_0} \frac{(\mathcal{A}\Lambda(v_h - u), \Lambda w)}{\| \Lambda w \|} \le \| \Lambda(v_h - u) \| .$$

It is easy to see that

$$D(\Lambda u_h, \mathcal{A}\Lambda v_h) = J(v_h) - J(u_h). \quad (6.7.22)$$

Indeed,

$$D(\Lambda u_h, \mathcal{A}\Lambda v_h) = \frac{1}{2}(\mathcal{A}\Lambda v_h, \Lambda v_h) + \langle l, v_h \rangle - \frac{1}{2}(\mathcal{A}\Lambda u_h, \Lambda u_h) - \langle l, u_h \rangle +$$

$$+ (\mathcal{A}\Lambda u_h, \Lambda(u_h - v_h)) + \langle l, u_h - v_h \rangle.$$

Since $u_h \in V_h$ is a Galerkin approximation, the last two terms vanish and we arrive at (6.7.22).

Recalling (6.1.10), we find that

$$\| \Lambda(u_h - u) \|^2 = 2(J(u_h) - J(u)),$$
$$\| \Lambda(v_h - u) \|^2 = 2(J(v_h) - J(u)).$$

Therefore,

$$2D(\Lambda u_h, \mathcal{A}\Lambda v_h) = 2(J(v_h) - J(u)) - 2(J(u_h) - J(u)) =$$

$$= \| \Lambda(v_h - u) \|^2 - \| \Lambda(u_h - u) \|^2 .$$

Now, (6.7.21) leads to the estimate

$$(2+\beta) \parallel \Lambda(u - u_h) \parallel^2 \leq$$

$$\leq (1+\beta) \parallel \Lambda(v_h - u) \parallel^2 + \left(1 + \frac{1}{\beta}\right) \parallel \Lambda(v_h - u) \parallel^2,$$

which shows that

$$\parallel \Lambda(u - u_h) \parallel^2 \leq \left(1 + \frac{1}{\beta(2+\beta)}\right) \parallel \Lambda(u - v_h) \parallel^2 .$$

Since β is an arbitrary positive number, we arrive at the projection type error estimate (cf. 2.3.14)

$$\parallel \Lambda(u - u_h) \parallel \leq \inf_{v_h \in V_h} \parallel \Lambda(u - v_h) \parallel . \tag{6.7.23}$$

Finally, we note that the general deviation estimate (6.2.3) also implies another projection type error estimate. Let us set $v = u_h$, $y^* = y_h^* := \mathcal{A}\nabla u_h$ and use (6.1.13). Since

$$D(\Lambda u_h, y_h^*) = 0,$$

we have

$$\parallel \Lambda(u_h - u) \parallel^2 \leq \parallel y_h^* - q^* \parallel_*^2 \quad \forall q^* \in Q_l^* .$$

From here, it follows the estimate

$$\parallel \Lambda(u - u_h) \parallel \leq \inf_{q^* \in Q_l^*} \parallel y_h^* - q^* \parallel_*, \tag{6.7.24}$$

which is in a sense dual to (6.7.23). It shows that an upper bound of the error is given by the distance (in the space Y^*) between y_h^* (which is a dual counterpart of the Galerkin approximation u_h) and the set Q_l^* containing the exact solution p^* of Problem $\mathcal{P}^*$.

6.8 Error estimates taking account of the indeterminacy in the data

6.8.1 General concept

In the majority of applied problems, exact values of the data are unknown. For example, in linear elasticity problems material constants may be known only approximately, so that instead of exact numbers we have some intervals. This indeterminacy must be taken into account in the error analysis especially if the influence of the arising additional errors is significant. Below we consider this question by the paradigm of the problem

$$\Lambda^* \mathcal{A} \Lambda u = l \ \text{ in } \Omega, \qquad u = u_0 \ \text{ on } \partial\Omega. \tag{6.8.1}$$

where the operator $\mathcal{A}$ and the functional l are defined with some indeterminacy. Their actual values are unknown. Instead, we know the sets of admissible data. Formally, this means that

$$\mathcal{A} \in \mathcal{U}_{\mathcal{A}} \subset \mathcal{L}(U, U), \tag{6.8.2}$$

$$l \in \mathcal{U}_l \subset V_0^*, \tag{6.8.3}$$

where $\mathcal{U}_{\mathcal{A}}$ and $\mathcal{U}_l$ are bounded and connected sets that take into account possible variations of the data.

It is easily observed that the majorant $\mathcal{M}_\oplus$ and the minorant $\mathcal{M}_\ominus$ derived in Section 6.2 explicitly depend on $\mathcal{A}$ and l. Because of this fact, functional type majorants and minorants can serve as numerical tools naturally adopted to error estimation if data of a boundary-value problem are not exactly determined. Below, we show how such an estimate can be obtained. To make the exposition clearer, we demonstrate this with the paradigm of the problem (6.4.4)–(6.4.5). Possible generalizations of these results to other linear elliptic equations are rather obvious. In essence, they require choosing proper spaces V and Y and finding the constant $C(\Omega, A)$, which also depends on properties of the operator Λ.

First note that the solutions of problems with above data form the set

$$\Upsilon(\mathcal{U}_{\mathcal{A}}, \mathcal{U}_l) := \left\{ \widetilde{u} \in V_0 + u_0 \, \middle| \, \begin{array}{c} \widetilde{u} \text{ satisfies (6.8.1) for some} \\ \mathcal{A} \in \mathcal{U}_{\mathcal{A}} \text{ and } l \in \mathcal{U}_l \end{array} \right\}.$$

Let $v \in V_0 + u_0$ be an approximation of an unknown exact solution. Since the data are indeterminate, the error estimation problem comes in two different forms. The first problem is to find the quantity

$$e_{\min}^2(v, \Upsilon) = \inf_{\widetilde{u} \in \Upsilon} \frac{1}{2} \, \|\Lambda(v - \widetilde{u})\|^2 , \tag{6.8.4}$$

where the multiplier $1/2$ is present for convenience. The quantity $e_{\min}$ measures the distance between v and the set Υ. It is equal to zero if v satisfies (6.8.1) for some pair $(\mathcal{A}, l) \in \mathcal{U}_{\mathcal{A}} \times \mathcal{U}_l$. This quantity provides a lower bound of the true error or *the error in the best-case situation* when the exact solution is the function of Υ closest to v.

Another problem is to find the quantity

$$e_{\max}^2(v, \Upsilon) = \sup_{\tilde{u} \in \Upsilon} \frac{1}{2} \, \|\Lambda(v - \tilde{u})\|^2 , \qquad (6.8.5)$$

which shows an *upper bound* of the error. It takes into account computational errors and errors caused by indeterminacy and shows *the error in the worst-case situation* when the exact solution is an element of Υ that is most distant of v. This quantity is always positive and its value gives an idea of the accuracy limit dictated by the effect of indeterminacy in the data.

Thus,

$$e_{\min}(v, \Upsilon) \leq e(v) \leq e_{\max}(v, \Upsilon), \qquad (6.8.6)$$

where the actual error $e(v)$ is principally unknown and we may only hope to find its bounds. In general, the exact values of $e_{\min}$ and $e_{\max}$ could hardly be found. However, using functional type a posteriori estimates, one can find their computable bounds. Indeed, the majorant $\mathcal{M}_\oplus$ and the minorant $\mathcal{M}_\ominus$ explicitly depend on $\mathcal{A}$ and l, which opens a way for computing errors caused by the indeterminacy in values of the problem data. Below we show how such an account can be performed.

Assume that the set Υ is known. Our aim is to find practically computable numbers $e_\ominus(v, \Upsilon)$ and $e_\oplus(v, \Upsilon)$ such that for any $v \in V_0 + u_0$ the following relations hold:

$$e_\ominus(v, \Upsilon) \leq e_{\min}(v, \Upsilon) \leq e_{\max}(v, \Upsilon) \leq e_\oplus(v, \Upsilon). \qquad (6.8.7)$$

These numbers give bounds of the total error that is caused by the computational inaccuracy and data indeterminacy. It is worth noting that such type estimates can also be applied to differential equations with random coefficients.

6.8.2 Upper bound of the error

In the subsequent text, we take the problem (6.4.4)–(6.4.5) as a basic example. In this case,

$$V = H^1(\Omega), \quad Y = L^2(\Omega, \mathbb{R}^n), \quad V_0 := \overset{\circ}{H}{}^1(\Omega),$$
$$V_0^* = H^{-1}(\Omega), \quad \Lambda v = \nabla v, \quad \Lambda^* y^* = -\operatorname{div} y^*,$$

and $\mathcal{A}$ is a mapping given by the relation $y^*(x) \to A(x)y^*(x)$, where $A(x)$ is a symmetric positive definite matrix.

Assume that the indeterminacy in the coefficients of the differential equation (6.8.1) is described by fixing some "mean" elements

$$A_0 \in L^\infty(\Omega; \mathbb{M}_s^{n \times n}) \quad \text{and} \quad l_0 \in L^2(\Omega)$$

and defining bounds of possible variations. In this case, the elements of the sets $\mathcal{U}_A$ and $\mathcal{U}_l$ are represented by elements of the following two functional sets:

$$\mathcal{U}_A := \left\{ A \in L^\infty(\Omega; \mathbb{M}_s^{n \times n}) \mid A = A_0 + \varepsilon E, \ E \in \mathcal{E} \right\},$$
$$\mathcal{U}_f := \left\{ f \in L^2(\Omega) \mid f = f_0 + \delta\varphi, \ \varphi \in \mathcal{F} \right\},$$

where

$$\mathcal{E} := \left\{ E \in L^\infty(\Omega; \mathbb{M}_s^{n \times n}) \mid \||E|\|_{\infty,\Omega} \le 1 \right\},$$
$$\mathcal{F} := \left\{ \varphi \in L^2(\Omega) \mid \|\varphi\|_{2,\Omega} \le 1 \right\}.$$

Here, ε and δ are small parameters characterizing the range of indeterminacy. Henceforth, we assume that the parameter ε is small enough, so that the problems remain uniformly elliptic for all possible data. Formally, this means that the condition

$$c_1|\xi|^2 \le A_0\xi \cdot \xi \le c_2|\xi|^2, \qquad \forall \xi \in \mathbb{R}^n, \tag{6.8.8}$$

implies similar estimates for all $A \subset \mathcal{U}_A$. Since $|E\xi \cdot \xi| \le |E| |\xi|^2$, we find that

$$A\xi \cdot \xi \ge A_0\xi \cdot \xi - \varepsilon|E| |\xi|^2 \ge (c_1 - \varepsilon)|\xi|^2, \tag{6.8.9}$$
$$A\xi \cdot \xi \le A_0\xi \cdot \xi + \varepsilon|E| |\xi|^2 \le (c_2 + \varepsilon)|\xi|^2. \tag{6.8.10}$$

From here, it is clear that ε must be subject to the condition

$$\varepsilon < c_1. \tag{6.8.11}$$

For the inverse matrix, we have

$$c_2^{-1}|\xi|^2 \le A_0^{-1}\xi \cdot \xi \le c_1^{-1}|\xi|^2, \tag{6.8.12}$$
$$(c_2 + \varepsilon)^{-1}|\xi|^2 \le A^{-1}\xi \cdot \xi \le (c_1 - \varepsilon)^{-1}|\xi|^2, \tag{6.8.13}$$

where $A \in \mathcal{U}_A$. To use (6.4.8), we must estimate the functional

$$D(\Lambda v, y^*) := \frac{1}{2}(A\Lambda v, \Lambda v) + \frac{1}{2}(A^{-1}\Lambda^* y^*, \Lambda^* y^*) - (\Lambda v, y^*)$$

for any $A \in \mathcal{U}_{\mathcal{A}}$. Denote the unit matrix by $\mathbb{I}$. Then

$$A^{-1} = \left(A_0(\mathbb{I} + \varepsilon A_0^{-1} E)\right)^{-1} = \left(\mathbb{I} + \varepsilon A_0^{-1} E\right)^{-1} A_0^{-1}.$$

Note that

$$\varepsilon |A_0^{-1} E| \leq \varepsilon \|A_0^{-1}\| \, |E| \leq \varepsilon c_1^{-1} < 1,$$

and, therefore, $(\mathbb{I} + \varepsilon B)^{-1} = \mathbb{I} + \sum_{j=1}^{\infty} (-1)^j \varepsilon^j B^j$, where $B = A_0^{-1} E$. Hence,

$$A^{-1} y^* \cdot y^* = (I + \varepsilon B)^{-1} A_0^{-1} y^* \cdot y^* =$$

$$= A_0^{-1} y^* \cdot y^* - \varepsilon B A_0^{-1} y^* \cdot y^* + \sum_{j=2}^{\infty} (-1)^j \varepsilon^j B^j A_0^{-1} y^* \cdot y^*.$$

Since $E \in \mathcal{E}$, we have

$$\int_{\Omega} B^j A_0^{-1} y^* \cdot y^* \, dx \leq \int_{\Omega} |A_0^{-1}|^{j+1} \, |E|^j \, |y^*|^2 \, dx \leq c_1^{-(j+1)} \|y^*\|^2$$

and

$$\int_{\Omega} A^{-1} y^* \cdot y^* \, dx \leq \int_{\Omega} \left(A_0^{-1} y^* \cdot y^* - \varepsilon B A_0^{-1} y^* \cdot y^*\right) dx +$$

$$+ \left(\sum_{j=2}^{\infty} (-1)^j \varepsilon^j c_1^{-(j+1)}\right) \|y^*\|^2.$$

Moreover,

$$\int_{\Omega} A \nabla v \cdot \nabla v \, dx = \int_{\Omega} \left(A_0 \nabla v \cdot \nabla v + \varepsilon E \nabla v \cdot \nabla v\right) dx.$$

From these relations, we find that the first term of the duality error majorant explicitly depends on ε. It is estimated as follows:

$$D(\nabla v, y^*) \leq D_0(\nabla v, y^*) + \frac{\varepsilon}{2} \int_{\Omega} \left(E \nabla v \cdot \nabla v - B A_0^{-1} y^* \cdot y^*\right) dx +$$

$$+ \frac{1}{2c_1} \|y^*\|^2 \sum_{j=2}^{\infty} \left(-\frac{\varepsilon}{c_1}\right)^j, \quad (6.8.14)$$

where

$$D_0(\nabla v, y^*) = \int_\Omega \left(\frac{1}{2} A_0 \nabla v \cdot \nabla v + \frac{1}{2} A_0^{-1} y^* \cdot y^* - \nabla v \cdot y^* \right) dx.$$

Since all the matrices are symmetric, we have

$$E \nabla v \cdot \nabla v - A_0^{-1} E A_0^{-1} y^* \cdot y^* = E(\nabla v - A_0^{-1} y^*) \cdot (\nabla v + A_0^{-1} y^*).$$

Now, (6.8.14) implies the estimate

$$D(\nabla v, y^*) \le D_0(\nabla v, y^*) + \frac{\varepsilon}{2} \int_\Omega E(\nabla v - A_0^{-1} y^*) \cdot (\nabla v + A_0^{-1} y^*)\, dx +$$

$$+ \left(\frac{\varepsilon}{c_1} \right)^2 \frac{1}{2(\varepsilon + c_1)} \| y^* \|^2. \quad (6.8.15)$$

Recall that (see (6.3.17)

$$\frac{1}{2} \| \nabla(v - u) \|^2 = \inf_{\substack{y^* \in Q^* \\ \beta > 0}} \mathcal{M}_\oplus(v, \beta, y^*). \quad (6.8.16)$$

For the problem considered, we have

$$\| \nabla(v - u) \|^2 = \int_\Omega (A_0 + \varepsilon E) \nabla(v - u) \cdot \nabla(v - u)\, dx \ge$$

$$\ge (c_1 - \varepsilon) \| \nabla(v - u) \|^2. \quad (6.8.17)$$

Therefore,

$$e_{\max}^2(v, \Upsilon) = \sup_{\tilde{u} \in \Upsilon} \frac{1}{2} \| \nabla(v - \tilde{u}) \|^2 \le \frac{1}{2(c_1 - \varepsilon)} \sup_{\tilde{u} \in \Upsilon} \| \nabla(v - \tilde{u}) \|^2 =$$

$$= \frac{1}{(c_1 - \varepsilon)} \sup_{\tilde{u} \in \Upsilon} \inf_{\substack{y^* \in Q^* \\ \beta > 0}} \mathcal{M}_\oplus(v, \beta, y^*).$$

By definition, $\Upsilon(\mathcal{U}_A, \mathcal{U}_f)$ is a subset of $V_0 + u_0$ that contains solutions of the boundary-value problems (6.8.1) with

$$A = A_0 + \varepsilon E \quad \text{and} \quad l = f_0 + \delta \varphi.$$

Since any such problem has a unique solution, we can replace $\sup\limits_{\tilde{u}}$ by $\sup\limits_{(A,f)}$. Hence, we obtain

$$(c_1 - \varepsilon)e_{\max}^2(v, \Upsilon) \leq \sup_{\substack{\tilde{u} \in \Upsilon}} \inf_{\substack{y^* \in Q^* \\ \beta > 0}} \mathcal{M}_{\oplus}(v, \beta, y^*) \leq$$

$$\leq \inf_{\substack{y^* \in Q^* \\ \beta > 0}} \sup_{\tilde{u} \in \Upsilon} \mathcal{M}_{\oplus}(v, \beta, y^*) =$$

$$= \inf_{\substack{y^* \in Q^* \\ \beta > 0 \\ f \in \mathcal{U}_f}} \sup_{A \in \mathcal{U}_A} \left\{ (1+\beta)D(\nabla v, y^*) + \frac{(1+\beta)C^2(\Omega, A)}{2\beta} \| \operatorname{div} y^* - f \|^2 \right\}.$$

From here, we deduce the main estimate

$$e_{\max}^2(v, \Upsilon) \leq \frac{1}{c_1} \left(1 + \frac{\varepsilon}{c_1 - \varepsilon} \right) \left\{ (1+\beta) \sup_{A \in \mathcal{U}_A} D(\nabla v, y^*) + \right.$$

$$\left. + \frac{(1+\beta)}{2\beta} \sup_{A \in \mathcal{U}_A} C^2(\Omega, A) \sup_{f \in \mathcal{U}_f} \| \operatorname{div} y^* - f \|^2 \right\}, \quad (6.8.18)$$

which is valid for any $y^* \in Q^*$ and $\beta > 0$. The first term on the right-hand side in (6.8.18) is estimated by (6.8.15). Let us consider the second one. We have

$$\frac{1}{C^2(\Omega, A)} = \inf_{w \in V_0} \frac{\int_{\Omega} A \nabla w \cdot \nabla w \, dx}{\|w\|^2},$$

where

$$\int_{\Omega} A \nabla w \cdot \nabla w \, dx \geq \int_{\Omega} A_0 \nabla w \cdot \nabla w \, dx - \varepsilon \|\nabla w\|^2.$$

Hence,

$$\frac{1}{C^2(\Omega, A)} \geq (1 - \varepsilon c_1^{-1}) \inf_{w \in V_0} \frac{\int_{\Omega} A_0 \nabla w \cdot \nabla w \, dx}{\|w\|^2} = \frac{c_1 - \varepsilon}{c_1} \frac{1}{C^2(\Omega, A_0)}$$

and

$$C^2(\Omega, A) \leq \left(1 + \frac{\varepsilon}{c_1 - \varepsilon} \right) C^2(\Omega, A_0). \quad (6.8.19)$$

Now, we see that

$$\sup_{\substack{A\in\mathcal{U}_A \\ f\in\mathcal{U}_f}} \left\{ (1+\beta)D(\nabla v, y^*) + \left(1+\frac{1}{\beta}\right) \frac{C^2(\Omega, A)}{2} \|\operatorname{div} y^* - f\|^2 \right\} \leq$$

$$\leq (1+\beta)\left(D_0(\nabla v, y^*) + \left(\frac{\varepsilon}{c_1}\right)^2 \frac{1}{2(\varepsilon + c_1)} \|y^*\|^2 + \right.$$

$$\left. + \frac{\varepsilon}{2}\sup_{E\in\mathcal{E}}\int_\Omega E(\nabla v - A_0^{-1}y^*)\cdot(\nabla v + A_0^{-1}y^*)\,dx \right) +$$

$$+ \left(1 + \frac{\varepsilon}{c_1 - \varepsilon}\right)\frac{(1+\beta)C^2(\Omega, A_0)}{2\beta}\sup_{\varphi\in\mathcal{F}}\int_\Omega |\operatorname{div} y^* - f_0 - \delta\varphi|^2 dx. \quad (6.8.20)$$

Note that for any $g \in L^2(\Omega)$

$$\sup_{\varphi\in\mathcal{F}}\int_\Omega (g - \phi)^2 dx = \|g\|^2 + 2\|g\| + 1.$$

By this relation, we find the value of the very last term of (6.8.20). Recalling (6.8.19), we arrive at the estimate

$$e_{\max}^2(v, \Upsilon) \leq$$

$$\leq \frac{1}{c_1}\left(1 + \frac{\varepsilon}{c_1 - \varepsilon}\right)\inf_{\substack{y^*\in Q^* \\ \beta>0}}\left\{ (1+\beta)\left(D_0(\nabla v, y^*) + \left(\frac{\varepsilon}{c_1}\right)^2 \frac{\|y^*\|^2}{2(\varepsilon + c_1)} + \right.\right.$$

$$\left. + \frac{\varepsilon}{2}\int_\Omega |(\nabla v - A_0^{-1}y^*)\cdot(\nabla v + A_0^{-1}y^*)|\,dx\right) +$$

$$+ \left(1 + \frac{\varepsilon}{c_1 - \varepsilon}\right)\frac{(1+\beta)C^2(\Omega, A_0)}{2\beta}\left(\|\operatorname{div} y^* - f_0\|^2 + 2\delta\|\operatorname{div} y^* - f_0\| + \delta^2 \right)\right\}.$$

$$(6.8.21)$$

To represent the right-hand side of this estimate in a more transparent form, we introduce the following quantities:

$$
\begin{aligned}
M_{00}(v,\beta,y^*) \;:=\;& (1+\beta)D_0(\nabla v, y^*) + \\
&+ \left(1+\frac{1}{\beta}\right)\frac{C^2(\Omega, A_0)}{2}\|\operatorname{div} y^* - f_0\|^2,
\end{aligned}
$$

$$
\begin{aligned}
M_{10}(v,\beta,y^*) \;:=\;& \frac{\varepsilon}{2}\left(\int_\Omega \left|(\nabla v - A_0^{-1}y^*)\cdot(\nabla v + A_0^{-1}y^*)\right|\,dx + \right. \\
&\left. + \left(1+\frac{1}{\beta}\right)\frac{C^2(\Omega,A_0)}{c_1-\varepsilon}\int_\Omega |\operatorname{div} y^* - f_0|^2\,dx\right),
\end{aligned}
$$

$$
M_{01}(v,\beta,y^*) \;:=\; \delta\left(1+\frac{1}{\beta}\right)C^2(\Omega,A_0)\|\operatorname{div} y^* - f_0\|,
$$

$$
M_{11}(v,\beta,y^*) \;:=\; \varepsilon\delta\left(1+\frac{1}{\beta}\right)\frac{C^2(\Omega,A_0)}{c_1-\varepsilon}\|\operatorname{div} y^* - f_0\|,
$$

$$
\begin{aligned}
M_{22}(v,\beta,y^*) \;:=\;& (1+\beta)\left(\frac{\varepsilon}{c_1}\right)^2\frac{\|y^*\|^2}{2(\varepsilon+c_1)} + \\
&+ \left(1+\frac{1}{\beta}\right)\frac{c_1 C^2(\Omega,A_0)}{2(c_1-\varepsilon)}\delta^2.
\end{aligned}
$$

Now, we see the structure of the upper bound $e_\oplus(v,\Upsilon)$ in (6.8.7):

$$
\begin{aligned}
e_\oplus^2(v,\Upsilon) =& \\
=& \frac{1}{c_1}\left(1+\frac{\varepsilon}{c_1-\varepsilon}\right)\inf_{\substack{y^*\in Q^* \\ \beta>0}}\left(\sum_{s,t=1}^{2} M_{st}(v,\beta,y^*) + M_{22}(v,\beta,y^*)\right).
\end{aligned}
\tag{6.8.22}
$$

The term $M_{00}(v,\beta,y^*)$ does not contain the small parameters δ and ε. It coincides with the majorant constructed for the "mean" problem ($A = A_0$, $l = f_0$) and represents the major part of the approximation error. The terms M_{10}, M_{01}, and M_{11} are given by some combinations of the weighted residual and small parameters ε and δ. In principle, all these terms can be made arbitrarily small by taking v close enough to the exact solution u of the problem with $A = A_0$ and $l = f_0$ and y^* close enough to $A_0\nabla u$. In contrast, the term $M_{22}(v,\beta;y^*)$ is always positive. This term contains the *inherent* part of the error, which does not depend on the accuracy of numerical approximations. Indeed, in all cases we have

$$
M_{22}(v,\beta,y^*) \geq \theta_0 := C^2(\Omega,A_0)\frac{c_1\delta^2}{2(c_1-\varepsilon)}.
$$

This quantity does not depend on the choice of v, β, and y^*. It gives an idea of the accuracy limit that could be achieved within the framework of the worst-case scenario. For example, if A_0 is a unit matrix and $\Omega = (0.1) \times (0.1)$, then $c_1 = 1$ and $C^2(\Omega, A_0) = 1/2\pi^2$. Assume that the coefficients are defined with the ambiguity rate $\varepsilon = 1\%$ and l with $\delta = 2\%$. In this case, $\theta_0 \approx 10^{-5}$.

A series of computable upper bounds is obtained if we replace Q^* by a sequence of finite-dimensional subspaces $\{Q_k^*\} \subset Q^*$. Then, we have

$$
e_\oplus^2(v, \Upsilon) \le e_{k\oplus}^2(v, \Upsilon) =
$$

$$
= \frac{1}{c_1}\left(1 + \frac{\varepsilon}{c_1 - \varepsilon}\right) \inf_{\substack{y^* \in Q_k^* \\ \beta > 0}} \left\{\sum_{s,t=1}^2 M_{st}(v, \beta, y^*) + M_{22}(v, \beta, y^*)\right\}. \tag{6.8.23}
$$

If $Q_k^* \subset Q_{k+1}^*$, then the sequence $\{e_{k\oplus}^2(v, \Upsilon)\}$ monotonically decreases but may not tend to zero.

6.8.3 Lower bound of the error

To find a lower bound, we use the relation

$$
\frac{1}{2} \int_\Omega A\nabla(v - u) \cdot \nabla(v - u) \, dx = \sup_{w \in V_0} \mathcal{M}_\ominus(v, w),
$$

where

$$
\mathcal{M}_\ominus(v, w) = -\int_\Omega \left(\frac{1}{2}A\nabla w \cdot \nabla w + A\nabla v \cdot \nabla w + fw\right) dx.
$$

Recall that $A\xi \cdot \xi \le (\varepsilon + c_2)|\xi|^2$. Then, by (6.8.4), we see that for any $(A, f) \in \mathcal{U}_A \times \mathcal{U}_f$ and the respective solution $\tilde{u}$, the following relation holds:

$$
\frac{1}{2}\|\nabla(v - \tilde{u})\|^2 \ge \frac{1}{2(\varepsilon + c_2)} \int_\Omega A\nabla(v - \tilde{u}) \cdot \nabla(v - \tilde{u}) \, dx =
$$

$$
= \frac{1}{\varepsilon + c_2} \sup_{w \in V_0} \mathcal{M}_\ominus(v, w).
$$

Therefore,

$$
e_{\min}(v, \Upsilon) = \inf_{\tilde{u} \in \Upsilon} \frac{1}{2} \|\nabla(v - \tilde{u})\|^2 \geq
$$

$$
\geq \frac{1}{\varepsilon + c_2} \inf_{(A,f) \in \mathcal{U}_A \times \mathcal{U}_f} \sup_{w \in V_0} \mathcal{M}_\ominus(v, w) \geq
$$

$$
\geq \frac{1}{\varepsilon + c_2} \sup_{w \in V_0} \inf_{(A,f) \in \mathcal{U}_A \times \mathcal{U}_f} \mathcal{M}_\ominus(v, w) =
$$

$$
= \frac{1}{\varepsilon + c_2} \sup_{w \in V_0} \left\{ - \int_\Omega \left(\frac{1}{2} A_0 \nabla w \cdot \nabla w + A_0 \nabla v \cdot \nabla w + f_0 w \right) dx + \right.
$$

$$
\left. + \inf_{\substack{E \in \mathcal{E} \\ \varphi \in \mathcal{F}}} \left(-\varepsilon \int_\Omega \left(\frac{1}{2} E \nabla w \cdot \nabla w + E \nabla v \cdot \nabla w \right) dx - \delta \int_\Omega w\varphi \, dx \right) \right\}.
$$

$$(6.8.24)$$

Here,

$$
\inf_{E \in \mathcal{E}} \left\{ - \int_\Omega \left(\frac{1}{2} E \nabla w \cdot \nabla w + E \nabla v \cdot \nabla w \right) dx \right\} =
$$

$$
= - \int_\Omega \left| \left(\frac{1}{2} \nabla w + \nabla v \right) \otimes \nabla w \right| dx
$$

and

$$
\inf_{\varphi \in \mathcal{F}} \left\{ - \int_\Omega w\varphi \, dx \right\} = -\|w\|.
$$

Now, we obtain

$$
e_{\min}^2(v, \Upsilon) \geq
$$

$$
\geq \frac{1}{\varepsilon + c_2} \sup_{w \in V_0} \left\{ - \int_\Omega \left(\frac{1}{2} A_0 \nabla w \cdot \nabla w + A_0 \nabla v \cdot \nabla w + f_0 w \right) dx - \right.
$$

$$
\left. - \varepsilon \int_\Omega \left| \left(\frac{1}{2} \nabla w + \nabla v \right) \otimes \nabla w \right| dx - \delta \|w\| \right\}. \quad (6.8.25)
$$

Introduce the quantities

$$m_{00}(v, w) \; = \; -\int_\Omega \left(\frac{1}{2} A_0 \nabla w \cdot \nabla w + A_0 \nabla v \cdot \nabla w + f_0 w \right) \, dx,$$

$$m_{10}(v, w) \; = \; -\varepsilon \int_\Omega \left| \left(\frac{1}{2} \nabla w \otimes \nabla w + \nabla v \otimes \nabla w \right) \right| \, dx,$$

$$m_{01}(w) \; = \; -\delta \|w\|.$$

Then, we represent the lower bound in the form

$$e_\ominus^2(v, \Upsilon) := \frac{1}{\varepsilon + c_2} \sup_{w \in V_0} \{m_{00}(v, w) + m_{01}(v, w) + m_{10}(w)\} \ge 0. \quad (6.8.26)$$

In this estimate, the term $m_{00}(v, w)$ contains the major part of the approximation error. It vanishes if v is a solution of the "mean" problem with $A = A_0$ and $l = f_0$. Two other terms reflect the influence of the small parameters δ and ε.

To obtain a computable lower bound, we replace V_0 by a finite-dimensional subspace V_{0k} and solve a finite-dimensional problem

$$e_{\min}^2(v, \Upsilon) \ge e_{k\ominus}^2(v, \Upsilon) = \frac{1}{\varepsilon + c_2} \sup_{w \in V_{0k}} \{m_{00}(v, w) + m_{01}(v, w) + m_{10}(w)\}.$$

The quantities $e_{k\ominus}^2(v, \Upsilon)$ show the efficiency of further computational efforts within the framework of the best-case scenario. If they are large, then approximation errors are significant, and using a more precise method of approximation is sensible. On the contrary, if all these quantities are small, then an approximate solution found is probably close to a function $u \in \Upsilon$, so that further mesh refinements do not give much additional information.

6.9   Error estimation in terms of linear functionals

In this section, we apply two-sided error estimates given by $\mathcal{M}_\oplus$ (or $M_\oplus$) and $\mathcal{M}_\ominus$ to deriving sharp bounds of errors evaluated in terms of a linear functional ℓ.

Consider the problem (6.1.1) with $l = f$ and another problem that is to find $v \in V_0 + u_0$ such that

$$(\mathcal{A}\Lambda v, \Lambda w) + \langle \ell, w \rangle = 0, \quad \forall w \in V_0, \tag{6.9.1}$$

where ℓ is a given functional in V_0^*. By Proposition 4.5.1, we have

$$\langle \ell, u - \widetilde{u} \rangle := E(\widetilde{u}, \widetilde{v}) = E_0(\widetilde{u}, \widetilde{v}) + E_1(\widetilde{u}, \widetilde{v}), \tag{6.9.2}$$

where $\widetilde{u}$ is an approximation of u, $\widetilde{v}$ is an approximation of v, and

$$E_0(\widetilde{u}, \widetilde{v}) = \langle f, \widetilde{v} \rangle - (\mathcal{A}\Lambda\widetilde{u}, \Lambda\widetilde{v}), \tag{6.9.3}$$

$$E_1(\widetilde{u}, \widetilde{v}) = (\mathcal{A}\Lambda\delta_f, \Lambda\delta_\ell), \quad \delta_f = (u - \widetilde{u}), \ \delta_\ell = (v - \widetilde{v}). \tag{6.9.4}$$

Two-sided bounds of the term $(\mathcal{A}\Lambda\delta_\ell, \Lambda\delta_1)$ can be found by two-sided energy estimates as follows. Note that for any positive α we have

$$2(\mathcal{A}\Lambda\delta_\ell, \Lambda\delta_f) =$$
$$= \| \Lambda(\alpha\delta_f + \frac{1}{\alpha}\delta_\ell) \|^2 - \alpha^2 \| \Lambda\delta_f \|^2 - \frac{1}{\alpha^2} \| \Lambda\delta_\ell \|^2 . \tag{6.9.5}$$

By the majorant $\mathcal{M}_\oplus$ (or $M_\oplus$) and the minorant $\mathcal{M}_\ominus$, we find two-sided estimates of the energy norms of δ_1 and δ_ℓ. Let y_f^*, y_ℓ^*, β_f, β_ℓ, w_f, and w_ℓ be the functions and parameters that correspond to these estimates, i.e,

$$m_f := \mathcal{M}_\ominus(\widetilde{u}, w_f) \le \frac{1}{2} \| \Lambda\delta_f \|^2 \le \mathcal{M}_\oplus(\widetilde{u}, \beta_f, y_f^*) := M_f,$$

$$m_\ell := \mathcal{M}_\ominus(\widetilde{v}, w_\ell) \le \frac{1}{2} \| \Lambda\delta_\ell \|^2 \le \mathcal{M}_\oplus(\widetilde{v}, \beta_\ell, y_\ell^*) := M_\ell.$$

In this case, the quantity

$$\| \Lambda(\alpha\delta_f + \frac{1}{\alpha}\delta_\ell) \|^2 = \| \Lambda(\alpha u + \frac{1}{\alpha}v - \alpha\widetilde{u} - \frac{1}{\alpha}\widetilde{v}) \|^2 \tag{6.9.6}$$

can be viewed as the norm of the difference between the exact solution $u_{f\ell}^\alpha \in V_0 + (\alpha + \frac{1}{\alpha})u_0$ of the problem

$$(\mathcal{A}\Lambda u_{f\ell}, \Lambda w) + \langle \alpha f + \frac{1}{\alpha}\ell, w \rangle = 0, \quad \forall w \in V_0 \tag{6.9.7}$$

and the function $v_{f\ell}^\alpha = \alpha\widetilde{u} + \frac{1}{\alpha}\widetilde{v} \in V_0 + (\alpha + \frac{1}{\alpha})u_0$, which is an approximation of $u_{f\ell}^\alpha$. To obtain two-sided bounds of the norm (6.9.6) we use already known functions y_f^*, y_ℓ^*, w_f, and w_ℓ and compute the numbers

$$m_{f\ell} = \mathcal{M}_\ominus(v_{f\ell}^\alpha, \alpha w_f + \frac{1}{\alpha}w_\ell),$$

$$M_{f\ell} = \mathcal{M}_\oplus(v_{f\ell}^\alpha, \beta, \alpha y_f^* + \frac{1}{\alpha}y_\ell^*).$$

We note that $m_{f\ell} = m_{f\ell}(\alpha)$, $M_{f\ell} = M_{f\ell}(\alpha, \beta)$ and positive numbers α and β can be taken arbitrary. By (6.9.5), we obtain

$$(\mathcal{A}\Lambda\delta_f, \Lambda\delta_\ell) \leq \inf_{\alpha,\beta \in \mathbb{R}_+} \{M_{f\ell} - \alpha^2 m_f - \frac{m_\ell}{\alpha^2}\} := \mathfrak{M}_\oplus,$$

$$(\mathcal{A}\Lambda\delta_f, \Lambda\delta_\ell) \geq \sup_{\alpha \in \mathbb{R}_+} \{m_{f\ell} - \alpha^2 M_f - \frac{M_\ell}{\alpha^2}\} := \mathfrak{M}_\ominus.$$

Recalling (6.9.2), we now deduce a two-sided estimate

$$\mathfrak{M}_\ominus + E_0(\widetilde{u}, \widetilde{v}) \leq \langle \ell, u - \widetilde{u} \rangle \leq \mathfrak{M}_\oplus + E_0(\widetilde{u}, \widetilde{v}), \qquad (6.9.8)$$

Note that if y_f^*, y_ℓ^*, w_f, and w_ℓ provide accurate two-sided estimates of the energy error norms in the problems (6.1.1) and (6.9.1), then $M_{f\ell}$ and $m_{f\ell}$ furnish accurate two-sided estimates in the problem (6.9.7) and, consequently, (6.9.8) yields sharp upper and lower bounds of $\langle \ell, u - \widetilde{u} \rangle$.

6.10   Practical implementation

In this section, we analyze the effect of duality error majorants (DEM) and compare them with the results obtained by other methods.

6.10.1   General scheme

Two-sided estimates obtained in Section 6.2 suggest a general algorithm for solving Problem $\mathcal{P}$ with any a priori given accuracy Δ. First we choose a set of finite-dimensional spaces $\{V_k\} \subset V$ ($\dim V_k = N_k \to +\infty$), which is dense in V. If the original problem and its discrete analogs are well-posed, then the respective sequence of approximate solutions $\{u_k\}$ tends to u as k tends to infinity. Next, we choose a set of finite-dimensional spaces $\{Y_s^*\} \subset Y^*$ ($\dim Y_s^* = M_s \to +\infty$) that is dense in Y^*. In principal, the spaces V_k and Y_s^* are quite independent. However, it is useful to correlate them in such a way that there exists a simple mapping of $\Lambda v_k \in V_k$ to Y_s^*. In the finite element method, such a mapping is easy to construct if V_k and Y_s^* are based on coherent meshes.

Then, we proceed as follows:

Step 1 Choose a desired tolerance Δ, set $k = 1$, $s = 1$, and define V_k and Y_s^*.

Step 2 Find u_k by solving Problem $\mathcal{P}$ on the space V_k.

Step 3 Find $M_\oplus^{ks} = \inf\limits_{\substack{y^* \in Y_s^* \\ \beta \in \mathbb{R}_+}} M_\oplus(u_k, \beta, y^*)$.

Step 4 If $M_\oplus^{ks} < \Delta$, then go to Step 8.

Step 5 Find $\mathcal{M}_\ominus^k = \sup\limits_{w \in V_{k+1}} \mathcal{M}_\ominus(u_k, w)$.

Step 6 If $\mathcal{M}_\ominus^k > \Delta$, then replace k by $k + 1$, V_k by a refined space V_{k+1} and go to Step 2.

Step 7 Replace s by $s + 1$, Y_s^* by a refined space Y_{s+1}^* and go to Step 3.

Step 8 Print u_k and stop.

The first step defines spaces originally used for obtaining an approximate solution (Step 2) and finding the first (rough) upper estimate of the error (Step 3). If Δ exceeds the majorant (Step 4), then the desired accuracy is achieved and the process passes to the final Step 8. If it is stated that the minorant exceeds Δ, then u_k does not have the required accuracy and it is necessary to solve Problem $\mathcal{P}$ on a refined subspace V_{k+1} (Steps 6 and 2).

The situation that occurs if

$$\mathcal{M}_\ominus^k < \Delta < M_\oplus^{ks},$$

deserves a special comment. In this case, it remains unclear whether or not u_k is close enough to u. To answer this question additional efforts are required. One possibility is to refine V_k and another is to refine Y_s^*, trying either to find a better approximation or to compute a better estimate for the existing approximation. Since $u_k \to u$ and our two-sided estimates are exact (see Section 3) , this theoretical algorithm will end with finding a proper approximation u_k. However, any particular computer has a certain limited power, so that any problem can be solved only if

$$\Delta \geq \Delta_0,$$

where Δ_0 depends both on the problem considered and the computer used.

Certainly, the above algorithm is rather schematic and can be viewed only as a skeleton of reliable numerical algorithms to be used in practice. Such algorithms should include numerous improvements focused first of all on accelerating the process. It could happen that these computations are terminated by time limitations. In this case, the very last value of $M_\oplus^{kl}$ shows the best accuracy achieved, which gives an idea of the required power of a computer to be used for finding an approximate solution with the desired tolerance Δ.

Finally, it seems worthwhile to add one more remark. Steps 3 and 5 are related to variational problems, so that one may ask about the sensitivity of the algorithm with respect to inaccuracy of their solutions. To clarify this point, we recall that $M_\oplus(v_k, \beta, y^*)$ provides an upper bound of the error *for any* $y^* \in Y^*$ and $M_\ominus(v_k, w)$ provides a lower bound *for any* $w \in V$. For this reason, exact solutions of these variational are, in general, not required. For example, it may occur that on some stage of the minimization procedure the value of $M_\oplus$ becomes less than Δ. Then computations may be terminated regardless of that how close "current" function y^* is to the minimizer y_s^*.

6.10.2   Minimization of the majorant

Consider the majorant $M_\oplus(v, \beta, y^*)$ defined by the relation (6.3.15). A coarse upper bound of the error is easy to compute by setting $y^* = G\Lambda v$, where G is an operator that maps Λv to Q^* (e.g., if v is a finite element approximation, then G is the standard gradient averaging operator). Then, we obtain the error majorant

$$M_\oplus(v, \beta, G\Lambda v) = (1 + \beta)D(\Lambda v, G\Lambda v) + \left(1 + \frac{1}{\beta}\right) \frac{c^2}{2}\|l + \Lambda^* G\Lambda v\|^2.$$

A sharper estimate requires a minimization of $M_\oplus$ with respect to the variables y^* and β, which can be done by a direct minimization of $M_\oplus$ or by finding a minimiser as a solution of the respective the system of linear simultaneous equations. The latter way is considered below.

Assume that

$$y^* \in Y_s^* := \mathrm{Span}\{y_1^*, y_2^*, ..., y_s^*\},$$

where $y_i^* \in Y^*$ are given trial functions. Then any y^* can be represented in the form

$$y^* = \sum_{j=1}^{k} \gamma_j y_j^*. \tag{6.10.1}$$

Define the matrix $\mathbb{M}^\beta$ with entries

$$m_{st}^\beta = (1 + \beta)\left((\mathcal{A}^{-1}y_s^*, y_t^*) + \frac{c^2}{\beta}(\Lambda^* y_s^*, \Lambda^* y_t^*)\right), \quad s, t = 1, l$$

and the vector $\mathbb{F}^\beta$ with components

$$F_s^\beta = (1 + \beta)\left((\Lambda v, y_s^*) - \frac{c^2}{\beta}(\Lambda^* y_s^*, l)\right).$$

Then

$$M_\oplus(v, \beta, y^*) = \Phi(\beta, \boldsymbol{\gamma}) := \frac{1}{2}\mathbb{M}^\beta \boldsymbol{\gamma} : \boldsymbol{\gamma} - \mathbb{F}^\beta \cdot \boldsymbol{\gamma} + \mu^\beta,$$

where

$$\mu^\beta = \frac{(1+\beta)}{2}\left((\mathcal{A}\Lambda v, \Lambda v) + \frac{c^2}{\beta}\|l\|^2\right)$$

and $\boldsymbol{\gamma} = \{\gamma_s\}$.

In view of the property of $M_\oplus$, we have

$$\frac{1}{2}\,\|\,\Lambda(v-u)\,\|^2 \leq \inf_{\substack{\beta\in\mathbb{R}_+,\\ \boldsymbol{\gamma}\in Y_s^*}}\ \Phi(\beta, \boldsymbol{\gamma}). \tag{6.10.2}$$

The problem on the right-hand side in (6.10.2) can be solved with help of a direct minimization method that creates a sequence $\{\beta_j, \boldsymbol{\gamma}_j\}$ as follows:

(A) Find $\beta_j \in \mathbb{R}_+$ such that

$$\Phi(\beta_j, \boldsymbol{\gamma}_j) = \inf_{\beta\in\mathbb{R}_+} \Phi(\beta, \boldsymbol{\gamma}_j) := M_\oplus^j$$

and

(B) Find $\boldsymbol{\gamma}_{j+1} \in Y_s^*$ such that

$$\Phi(\beta_j, \boldsymbol{\gamma}_{j+1}) = \inf_{\boldsymbol{\gamma}\in Y_s^*} \Phi(\beta_j, \boldsymbol{\gamma}).$$

This algorithm creates a decreasing sequence of numbers $M_\oplus^j$ tending to an upper bound of the error. Each number is a reliable upper bound of the error, so that if $M_\oplus^j \leq \Delta$, then the process may be terminated.

Remark 6.10.1. Minimization of the majorant can be performed by the least-square methods developed for some classes of elliptic problems (see, e.g., [33]). Also, we note that the minimization problem can be partitioned into several subproblems associated with some blocks of $\boldsymbol{\gamma}$. Therefore, one can accelerate the minimization process by various parallelization procedures.

Remark 6.10.2. Let the functions y_s^* be properly chosen and $\mathbb{M}^\beta$ has an inverse matrix for any positive β. Then, $\Phi(\beta, \boldsymbol{\gamma})$ is minimized by the vector

$$\overline{\boldsymbol{\gamma}} = \mathbb{M}_\beta^{-1}\mathbb{F}^\beta$$

and the respective minimal value of this functional is easy to find. In this case, for a given β we obtain the estimate

$$\frac{1}{2}\,\|\,\Lambda(v-u)\,\|^2 \leq \mu^\beta - \frac{1}{2}\mathbb{M}_\beta^{-1}\mathbb{F}^\beta \cdot \mathbb{F}^\beta. \tag{6.10.3}$$

6.10.3 Effectivity index and shape index

In the examples considered below, we evaluate the quality of error estimates by means of two numbers. The first number is the so-called *effectivity index*

$$i_{\text{eff}} := \frac{M_\oplus}{e} = 1 + \frac{M_\oplus - e}{e}, \qquad (6.10.4)$$

where

$$e = \frac{1}{2} \parallel \Lambda(v - u) \parallel$$

and $M_\oplus$ is an estimate of this quantity obtained by an error majorant. If i_{eff} is close to 1, then $M_\oplus$ gives a sharp upper bound.

The second number characterizes the distribution of subdomain errors. Let $\mathcal{T}$ be a partition of Ω into m nonoverlapping subdomains Ω_i, $i = 1, 2, ...m$. For obvious reasons, it is desirable to have an error majorant that not only shows the total value of the error, but also indicates errors in various subdomains. We suggest to evaluate the latter capacity by a special number called the *error shape index*

$$i_{\text{esh}} = 1 + \frac{\sum\limits_{i=1}^{m} |\epsilon_i - \mu_i|}{e} \geq 1. \qquad (6.10.5)$$

In the above relation, ϵ_i is an error associated with Ω_i and μ_i denotes an approximate value of this local error computed by the majorant. If i_{esh} is close to 1 then the majorant provides a sharp error estimate and, in addition, the corresponding numbers μ_i serve as good indicators of subdomain errors.

Now we turn to examples.

6.10.4 Non–Galerkin approximations

Below, we present results of several numerical tests that demonstrate the capability of the DEM error estimation method to correctly estimate errors of approximate solutions regardless of their closeness to the exact one.

We begin with a simple 1-D problem

$$(\alpha(x)u')' = f(x), \qquad (6.10.6)$$

$$u(a) = 0, \quad u(b) = u_b. \qquad (6.10.7)$$

It is easy to see that (6.10.6) is the Euler equation for the functional

$$J(v) = \int\limits_a^b \left(\frac{1}{2}\alpha(x) \mid v' \mid^2 + f(x)v \right) dx.$$

Assume that

$$\alpha \in L^\infty(a, b), \quad \alpha \geq c_1 > 0, \quad f \in L^2(a, b). \tag{6.10.8}$$

Then, the problem (6.10.6)–(6.10.7) is uniquely solvable. To estimate the errors, we apply the general scheme with

$$U = L^2(a, b), \quad Y^* = L^2(a, b),$$
$$V_0 = \overset{\circ}{H}{}^1(a, b), \quad Q^* = H^1(a, b),$$

and

$$V_0 + u_0 = \{v \in H^1(a, b) \mid v(a) = 0, \ v(b) = u_b\}.$$

In this simple case, Λ is the differentiation operator $\Lambda^* y^* = v^* \in H^{-1}$, where

$$\int_a^b y^* \phi' \, dx = -\int_a^b v^* \phi \, dx, \quad \forall \phi \in \overset{\circ}{H}{}^1(a, b).$$

Now, the duality error majorant $M_\oplus$ has the form (cf. (6.3.15))

$$M_\oplus(v, \beta, y^*) = \frac{1 + \beta}{2} \int_a^b |\, \alpha v' - y^* \,|^2 \, dx +$$

$$+ \frac{1 + \beta}{2\beta} c^2_{(a,b)} \int_a^b |y^{*\prime} - f|^2 \, dx. \tag{6.10.9}$$

and the minorant $M_\ominus$ has the form

$$M_\ominus(v, w) = -\int_a^b \left(\frac{1}{2}|w'|^2 + \alpha v' w' + f w \right). \tag{6.10.10}$$

The exact solution admits the integral representation

$$u(x) = \int_a^x \frac{1}{\alpha(t)} \int_a^t f(z) \, dz \, dt + \frac{x}{b} \left(u_b - \int_a^b \frac{1}{\alpha(t)} \int_a^t f(z) \, dz \, dt \right). \tag{6.10.11}$$

In the tests presented below, we consider piecewise affine approximations v computed on regular meshes with N intervals. Our aim is to show that $M_\oplus$ and $M_\ominus$ provide good error estimates for exact solutions of the respective finite-dimensional problems as well as for solutions that do not satisfy the Galerkin orthogonality condition. Therefore, we compute approximations by a direct minimization of J and analyze errors arising on different stages of this process. To find a sharp upper bound, we minimize $M_\oplus$ with respect to y^* and β starting from the function $y_0^* = G(v')$. Here, $G : L^\infty(a, b) \to H^1(a, b)$ is a simple averaging operator. For example, G may be defined by the relations

$$G(v')(x_i) = \frac{1}{2}(v'(x_i - 0) + v'(x_i + 0)),$$
$$G(v')(x_0) = 2G(v')(x_1) - G(v')(x_2),$$
$$G(v')(x_N) = 2G(v')(x_{N-1}) - G(v')(x_{N-2}).$$

Computing

$$\inf_{\beta > 0} M_\oplus(v, \beta, y_0^*),$$

we find the first (rough) upper bound of the error. It is further improved by minimizing $M_\oplus$ with respect to y^*.

To compute a lower bound, we define a recovery operator R that maps Y^* to $V_0 + u_0$:

$$Ry^*(x) := \int_a^x \frac{y^*(t)}{\alpha(t)}\, dt + \frac{x}{b}\left(u_b - \int_a^b \frac{y^*(t)}{\alpha(t)}\, dt\right).$$

It is easy to see that $Ry^* = u$ if $y^* = \alpha u'$. The quantity

$$\mathcal{M}_\ominus(v, Ry^* - v)$$

gives a lower bound of the deviation.

Let us begin with an elementary example in which $\alpha(x) = 1$, $f(x) = 2$, $a = 0$, $b = 1$, $u_b = 1$, and $C_{(a,b)} = 1/\pi^2$. In this case, it is easy to find the exact solution

$$u = \frac{c}{2} x^2 + \left(1 - \frac{c}{2}\right) x$$

and compare the actual errors with their estimates. Below we do this for various approximations v.

First we take a "very rough" approximation $v = x$ (for $c = 0.5$ the functions u and v are depicted in Fig. 6.10.1(a). In this simplest case,

$$\frac{1}{2}\|(v - u)'\|^2 = \int_0^1 (1 - 2x)^2 dx = c^2/48 \approx 0.006\, c^2.$$

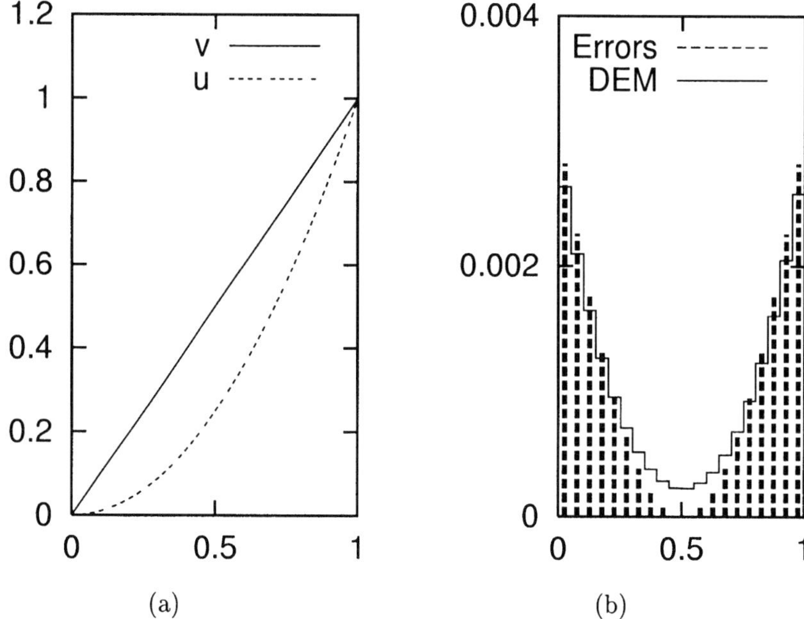

Figure 6.10.1: (a) Exact solution and an approximation, (b) distribution of local errors.

For any partition of the interval $(0,1)$, the function $v = x$ can be viewed as a piecewise affine approximation of u. The corresponding subinterval errors ϵ_i^2 for the uniform partition with 20 subintervals are depicted in Fig. 6.10.1b. We compare them with the errors computed by the majorant (6.10.9) with different y^*.

If we set $y^* = y_1^* \equiv 1$ (cf. (6.7.4)), then the first term vanishes and

$$M_{\oplus}(v, \beta, y_1^*) \to c^2/2\pi^2 \approx 0.05 \, c^2 \quad \text{as } \beta \to +\infty.$$

Hence, we see that the quantity obtained by the only one (residual) term of $M_{\oplus}$ gives a rough upper bound.

However, if we take $y^* =\equiv c\,x + 1 - c/2$, then the second term vanishes and we obtain another upper estimate:

$$M_{\oplus}(v, \beta, y_2^*) \to \frac{c^2}{2} \int_0^1 \left(x - \frac{1}{2} \right)^2 \, dx = \frac{c^2}{48} \quad \text{as } \beta \to 0,$$

which is exact. Thus, we see that the value of a computed upper bound strongly depends on the function y^*.

Now let us show that this dependence also takes place when the majorant is used as an indicator of local errors. Namely, if y^* is defined in the process of minimization of $M_\oplus$, then the local errors given by the majorant practically coincide with the true distribution (typical picture is presented in Fig. 6.10.1(b). However, attempts to use simplified forms of the majorant that eliminate either the first or the second term may lead to quite different results. In our simple case, this fact is easy to demonstrate.

Indeed, the choice $y^* = y_1^* \equiv 1$ zeroises the first term, so that

$$M_\oplus(v, \beta, y_1^*) = \frac{1+\beta}{2\beta} c_{(a,b)}^2 \int\limits_a^b |f|^2 \, dx.$$

In the integral sense, this majorant gives a corect result, but cannot serve as an error indicator, because its integrand given by the residual term is constant ($f = 2$), while the real error distribution demonstrates strong heterogeneity.

It is easy to present even a more striking example showing that the use of averaged derivatives without taking into account the second (residual) term of $M_\oplus$ may lead to a confusion. In our case, all sensible averagings of $v' = 1$ give exactly the same function, so that $G(v') = G(1) = 1$. Therefore,

$$G(v') - v' \equiv 0$$

and the estimator $\|G(v') - v'\|$ (which is, in fact, the first term of $M_\oplus$ for $y^* = G(v')$) leads to a wrong conclusion that the error is equal to zero. However, taking into account the second term changes the situation and gives a correct upper bound of the error. This observation shows that error indicators based on comparing the gradients of approximate solutions with their averaged values are not applicable to arbitrary conforming approximations without the second (residual) term.

To give further illustrations, we consider the functions

$$u_\delta = u + \delta\phi,$$

where δ is a number and ϕ is a certain function (perturbation). Approximate solutions (whose errors are measured) are piecewise affine continuous interpolants of u_δ defined on a uniform mesh with 20 subintervals. If $\delta = 0$, then v coincides with the interpolant of u. In the first series of tests, we take $\phi = x\sin(\pi x)$ and $\delta = 0.1, 0.01, 0.001,$ and 0. The results are presented in Table 6.10.1. For $\phi = x\sin(2\pi x)$, the results are exposed in Table 6.10.2.

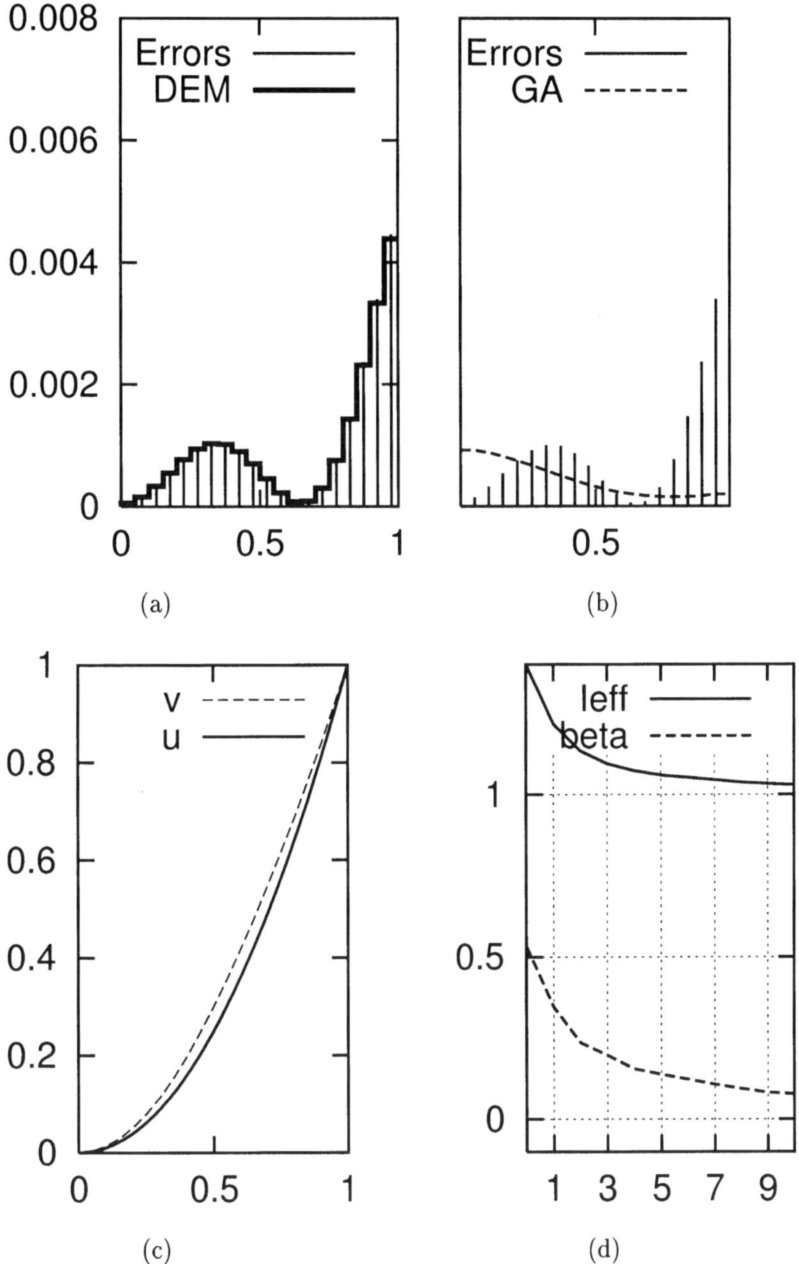

Figure 6.10.2: Error estimation for $\delta = 0.1$.

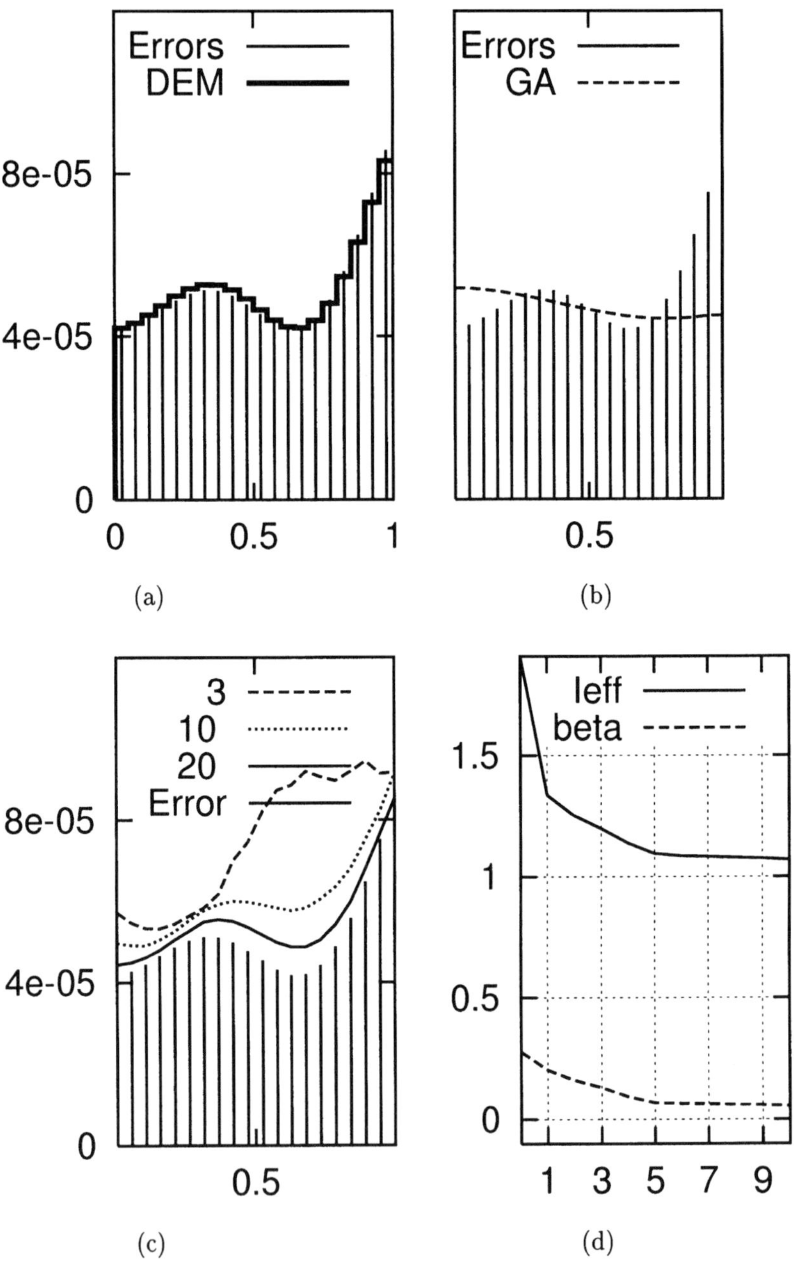

Figure 6.10.3: Error estimation for $\delta = 0.01$.

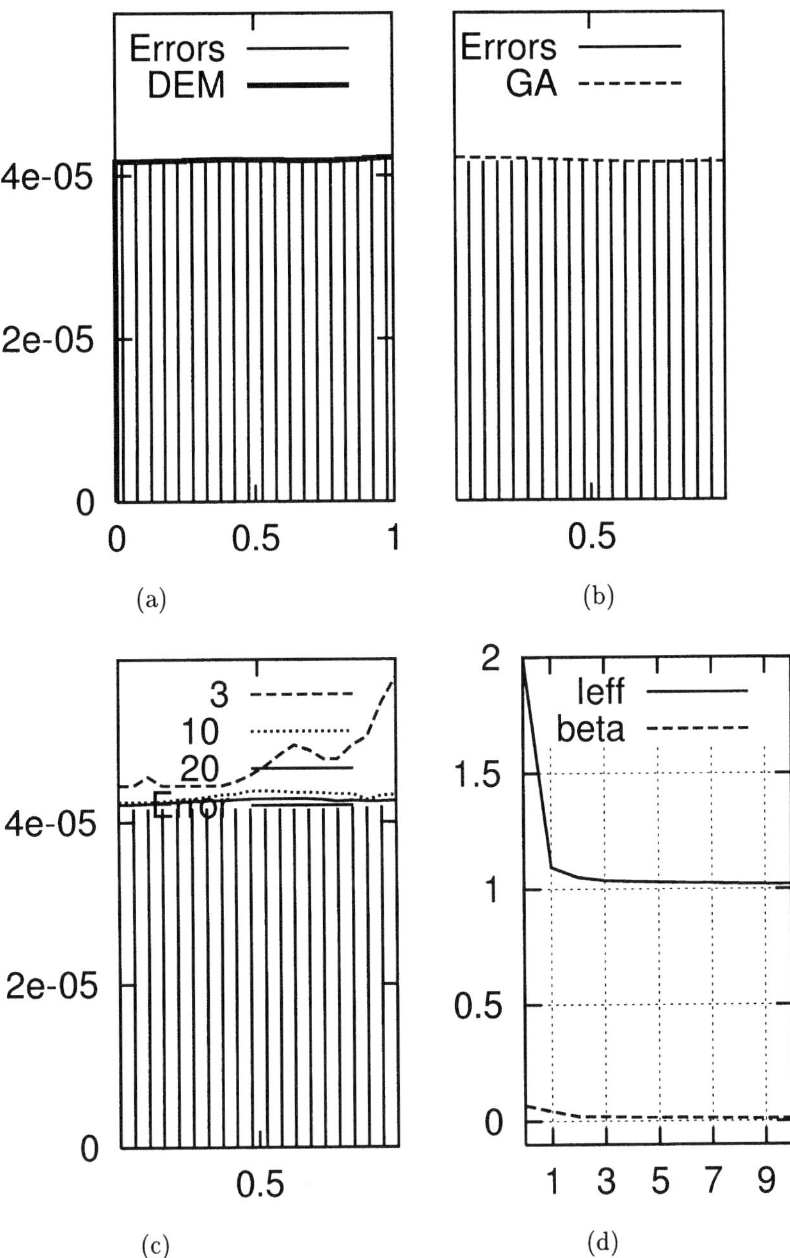

Figure 6.10.4: Error estimation for $\delta = 0.001$.

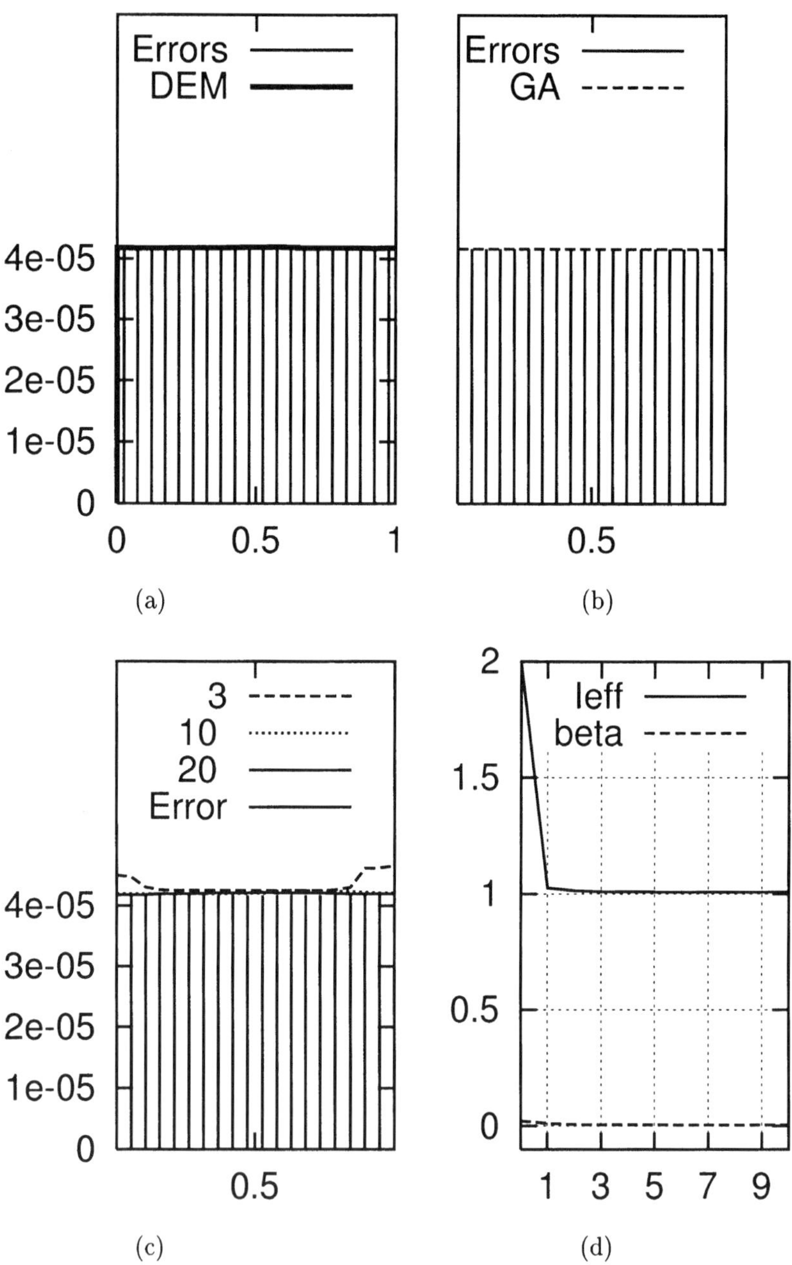

Figure 6.10.5: Error estimation for $\delta = 0$.

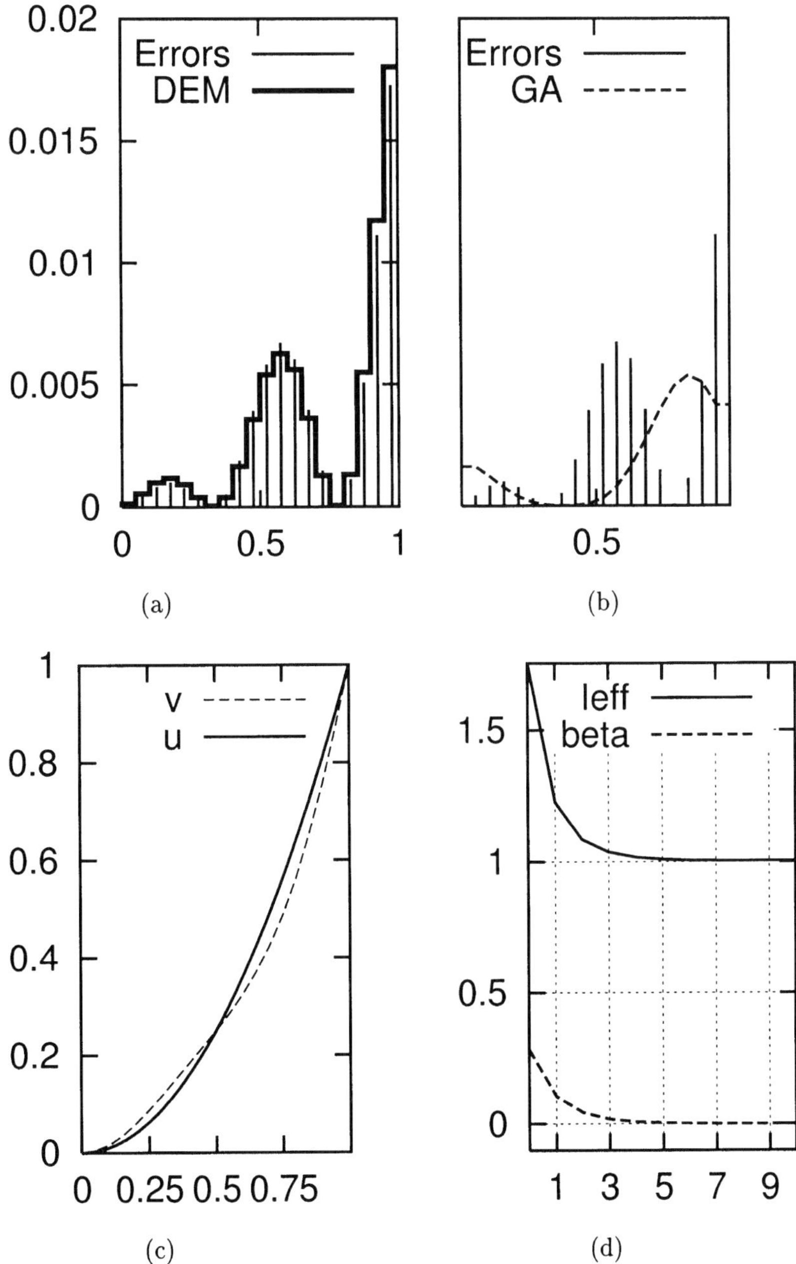

Figure 6.10.6: Error estimation for $\delta = 0.1$.

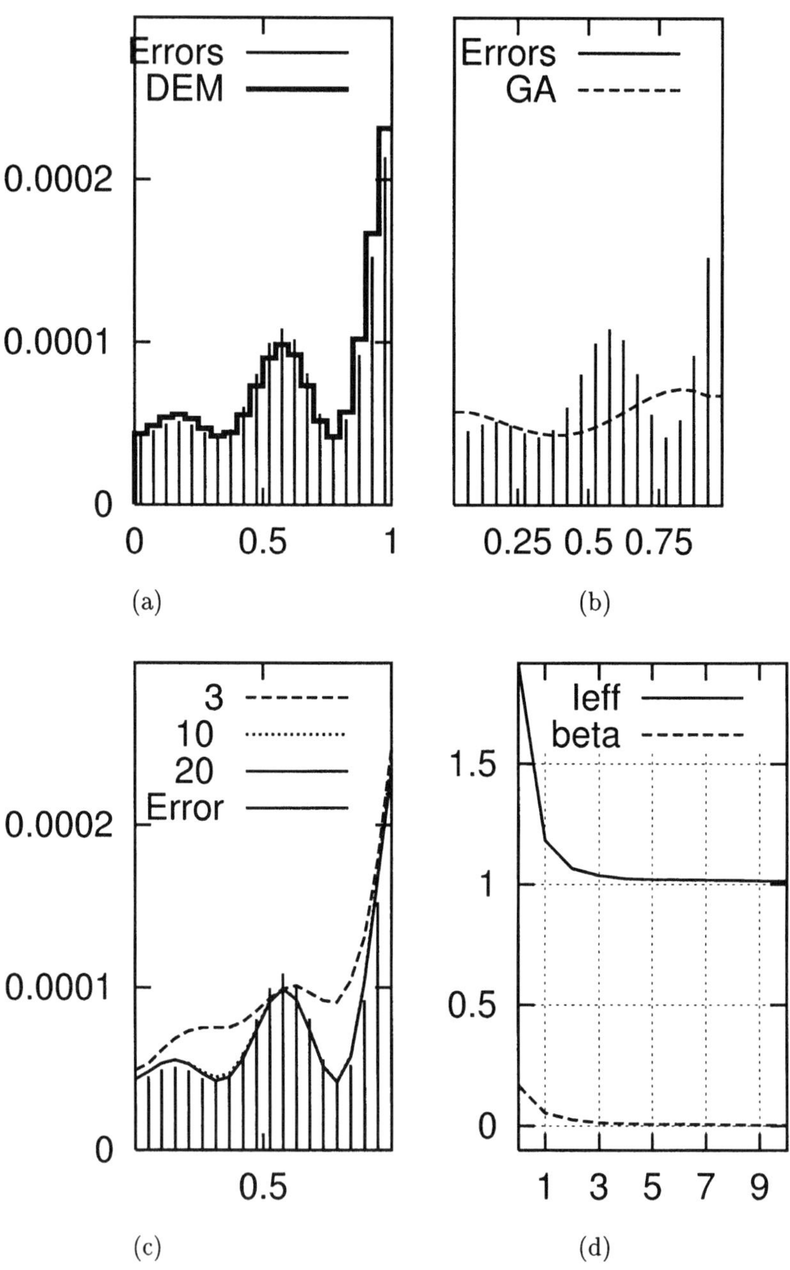

Figure 6.10.7: Error estimation for $\delta = 0.01$.

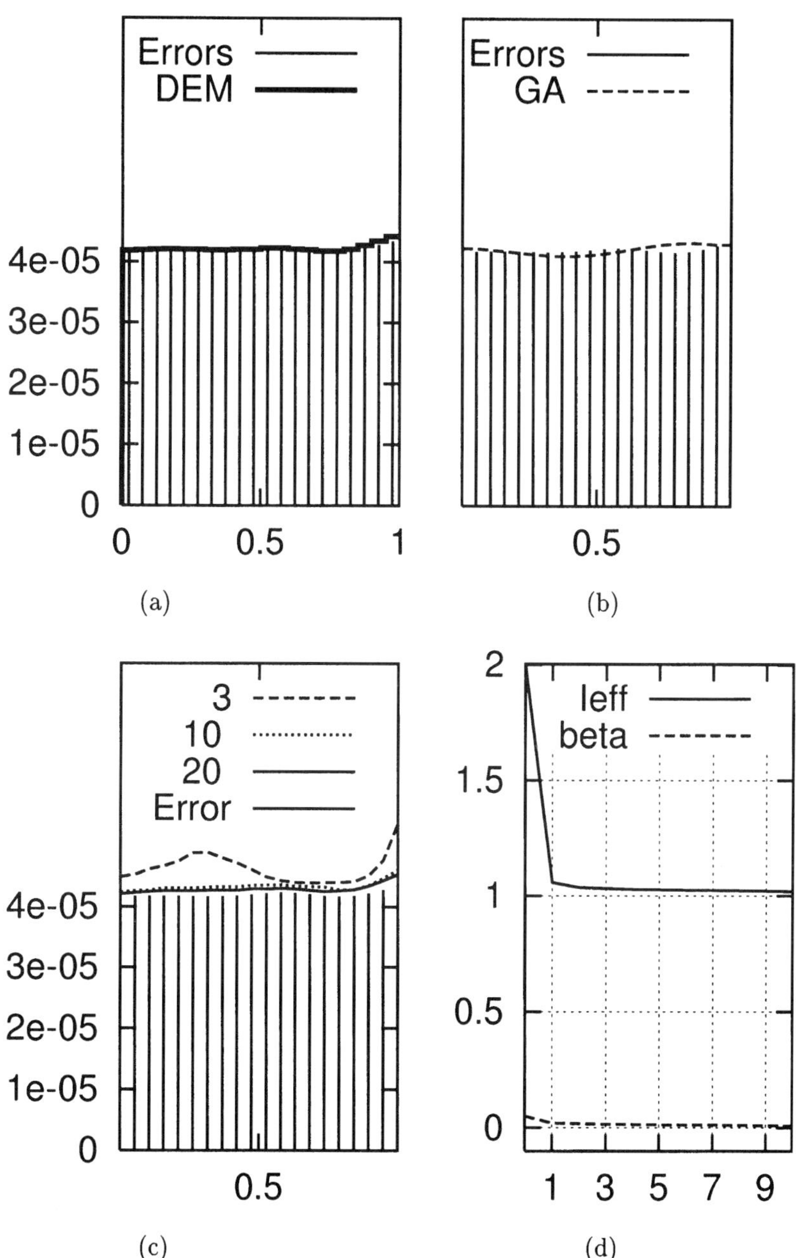

Figure 6.10.8: Error estimation for $\delta = 0.0001$.

Table 6.10.1:

δ	e	$2M_\oplus$	$2\mathcal{M}_\ominus$	i_{eff}	i_{esh}
0.1	0.019692	0.019743	0.019683	1.003	1.018
0.01	0.001022	0.001025	0.001013	1.003	1.011
0.001	0.000835	0.000839	0.000827	1.005	1.002
0	0.000833	0.000836	0.000825	1.004	1.002

Table 6.10.2:

δ	e	$2M_\oplus$	$2\mathcal{M}_\ominus$	i_{eff}	i_{esh}
0.1	0.068365	0.068543	0.068357	1.003	1.045
0.01	0.001509	0.001520	0.001500	1.003	1.042
0.001	0.000840	0.000844	0.000832	1.005	1.004
0	0.000833	0.000836	0.000825	1.004	1.002

Moreover, we compare the actual values of subinterval errors with the indicators given by the gradient averaging (GA) and duality error majorant (DEM) methods. In the first case, our indicator is a function proportional to $(v' - G(v'))^2$, where G is an averaging operator that defines nodal values as a simple average of two neighbor values. In the pictures, the values of this indicator are properly scaled in order to fit the actual value of the error. The second indicator is simply the integrand of the majorant $M_\oplus$.

Four pictures presented in Fig. 6.10.2 refer to the first line of Table 6.10.1. Picture (c) depicts u and v. True distributions of local errors are shown in (a) and (b) by impulses and compared with the DEM and GA indicators. Picture (d) gives an idea of how the values of the majorant and the parameter β change in the process of minimization. Here, the values of the x-axis are the iteration numbers (or time units). In Fig. 6.10.3 these numbers also appear in the picture (c), where they show how the integrand of $M_\oplus$ converges to the true error in the process of minimization. Subsequent pictures are associated with other lines of the above tables. We see that, irrespective of the actual appearance of the approximate solution v, the majorant $M_\oplus$ yields an accurate estimate of the global norm of the error and also rather accurately reproduces the distribution of local errors. Also, we observe that GA indicator provides a good error indication for small values of δ.

Also, we tested the DEM error estimation method for various two-dimensional problems. One typical example is presented below. Let u be a

Table 6.10.3:

N	Estimate	i_{eff}	i_{esh}
1	1.0812	1.477	2.184
3	0.9791	1.334	1.790
6	0.8069	1.102	1.215
10	0.7418	1.013	1.027

solution of the problem

$$\Delta u = 4x^2 \quad \text{in} \ \ \Omega \in (0,1) \times (0,1),$$

$$u|_{x=0} = y, \quad u|_{x=1} = y + 1\frac{1}{3},$$

$$u|_{y=0} = \frac{1}{3}x^4 + x, \quad u|_{y=1} = \frac{1}{3}x^4 + x + 1.$$

The exact solution of this problem is $u = \frac{1}{3}x^4 + x + y$. We compare it with the approximate one $v = u + xy\sin(\pi x)\sin(\pi y^*)$. To find the best upper bound for the error e, we minimize the majorant on the spaces Y_s^* constructed by means of the polynomial trial functions $x^i y^j$, $i, j = 0, 1, 2....$. Fig. 6.10.9(a) shows the distribution of local errors and the errors computed by minimization of the majorant on Y_s^* with 3, 6, and 10 trial functions are depicted in Figs. 6.10.9(b), 6.10.10(a), and 6.10.10(b), respectively. The numbers are given in Table 6.10.3. We see that the DEM furnishes a good estimate of the energy norm of the error and of its shape even for $N = 6$. If $N = 10$, then the norm and the shape of the error are found with a high accuracy. Certainly, in more complicated examples it may be necessary to invoke more trial functions in order to accurately estimate an error and its distribution. Moreover, if Ω has a complicated structure, then "global type" trial functions that we used may be inconvenient. In such a case, it is better to construct Y_s^* by local functions usually used in finite element methods.

6.10.5 Duality error majorants for finite element approximations

Now, we focus our attention on finite element approximations of 2-D boundary-value problems. We consider problems with homogeneous boundary conditions in domains with polygonal boundaries. Results obtained by the DEM method are compared with other three error indicators widely used for finite element approximations. They are as follows:

- the indicator

$$\eta_1^2 = \|A\nabla u_h - G(A\nabla u_h)\|_\Omega^2,$$

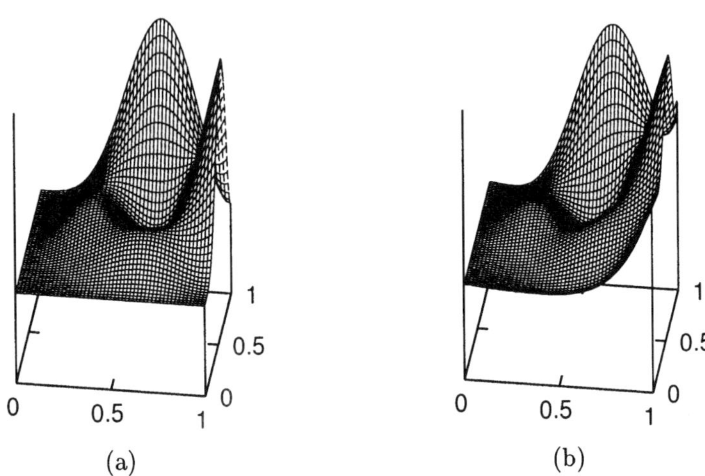

Figure 6.10.9: (a) Error and (b) integrand of the majorant for $N = 3$.

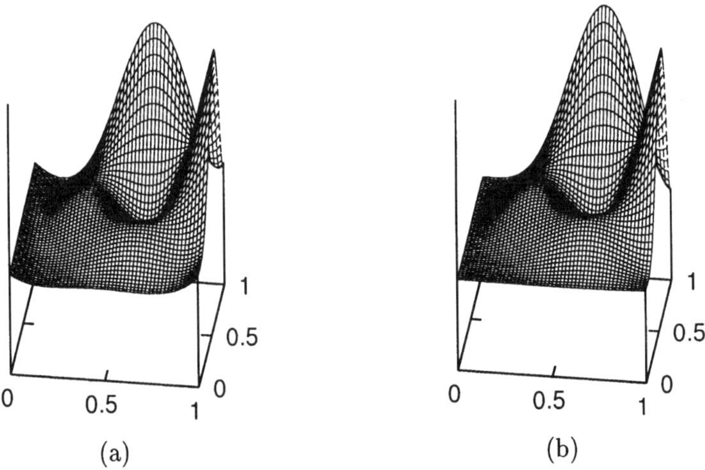

Figure 6.10.10: Integrands of the majorant for (a) $N = 6$ and (b) $N = 10$.

where $G(A\nabla u_h)$ is a piecewise affine function defined at the central point x_i of a patch Ω_i by the relation (cf. (6.4.6))

$$G(A\nabla u_h)|_{x_i} = \sum_{T_{ij} \in \Omega_i} \frac{|T_{ij}|}{|\Omega_i|} \, (A\nabla u_h)|_{T_{ij}}.$$

- the indicator η_2 based on the L^2-projection of $A\nabla u_h$ onto the space V_h of piecewise affine vector-valued functions (cf. 6.4.10):

$$\eta_2^2 = \inf_{y^* \in V_h} \|A\nabla u_h - y^*\|_\Omega^2 \,.$$

- the indicator

$$\eta_3^2 = \sum_{E \in \mathfrak{E}_h} |E| \, \|j(A\nabla u_h \cdot \nu)\|_E^2 \,,$$

where $\mathfrak{E}_h$ is the set of interior edges and ν is the unit vector normal to the edge E.

The error indicators η_1 and η_2 were considered in Chapter 4. It is easy to see that η_3 coincides with the part of the explicit residual error estimator that contains jumps. It is known (see, e.g., [46]) that for low order finite element approximations usually edge terms dominate in the residual type estimates and, therefore, can be used as an error indicator. We compare the results obtained by these indicators with those computed by DEM, paying attention to (a) the accuracy of the energy error estimate and (b) the quality of the local error indication. It should be noted that in many cases (especially for problems with smooth data), all indicators demonstrate close results. However, in other cases results are different. Below, we present several examples taken from [77], which show that various situations may occur.

In the first pair of examples, we consider two cases that differ only by coefficients of the matrix A. As we will see, the "standard" indicators give good results in the first case and are much less reliable in the second one. The error estimates computed by the DEM technology are quite reliable in both cases.

Example 6.10.1. Let us consider the equation

$$\operatorname{div} A\nabla u + f = 0$$

in a square domain $\Omega \in \mathbb{R}^2$ and take

$$\begin{cases} f = 1 & \text{for } x \in (0.5, 1.5), \\ f = 0 & \text{otherwise}, \end{cases}$$

$$\begin{cases} a_{11} = 10 & \text{for } x \in (0.5, 1.5), \\ a_{11} = 1 & \text{otherwise}, \end{cases}$$

$$a_{12} = a_{21} = 0, \ a_{22} = 1.$$

In this example, all the tests are computed for a certain fixed partition of the domain. To visualize the error indication, we mark elements by two colors (see Fig. 6.10.11): *dark* (if the error on a selected element is less than the mean value computed for the whole domain) and *light* (in the opposite case). This information can be further used in the simplest mesh-adaptation procedure that refines elements of the second group. The quantity p shows the number of correctly classified elements (in percent). The results of error estimation are compared with the actual values of errors obtained by using exact solutions. If in an example considered the exact solution is unknown, then the so-called "reference" solution is used instead. Such a solution is rigorously computed on a mesh that is much finer than those used in the error estimation tests. In all the examples, the mesh used for reference solutions has eight times more degrees of freedom than a primal mesh.

Numerical results are collected in Table 6.10.4. In this example,

$$\| e_h \| := \| u_h - u \| = 0.02578$$

and the normalized value of the error is

$$\frac{\| e_h \|}{\| u \|} = 5.3\%.$$

The respective results for the effectivity index are depicted in Fig. 6.10.12. From the values of p in Table 6.10.4 and the diagrams in Figs. 6.10.11 and

Table 6.10.4:

	η_1	η_2	η_3	$\sqrt{M}$
estimate	0.03060	0.02952	–	0.02656
I	1.19	1.15	–	1.03
p	90.2	90.3	93.0	98.9

6.10.12, we conclude that the DEM gives an accurate error indication and a sharp upper bound of the error norm ($p = 98.9\%$ and $I = 1.03$). The behavior of the effectivity index depicted in Fig. 6.10.12 shows that one time unit is enough for obtaining a sharp error estimate. It is easy to see that the majorant can also be used as an effective indicator of local errors. We present the true distribution of these errors and their estimates computed by the majorant as 3D graphs that present their averaged values on each element (see 6.10.12). By comparing these two pictures, we observe that the majorant accurately shows the distribution of local errors.

The indicators η_1–η_3 are also quite good. Among them, η_3 ($p = 93.0\%$) is the best (but this indicator does not give a guaranteed upper bound of the total error). However, the quality of these estimates is lower than of those made by the DEM method.

Example 6.10.2. In the second example, we take the same data as in Example 1, but define the coefficient a_{11} as follows:

$$\begin{cases} a_{11} = 1 & \text{for } x \in (0.5, 1.5), \\ a_{11} = 10 & \text{otherwise.} \end{cases}$$

In this case, the energy norm of the error is 0.02608. The respective results are collected in Table 6.10.5 and Figs. 6.10.13 and 6.10.14. It is clear that here, as in the previous example, the DEM technology provides a high accuracy of the error estimation. It was observed that, in the minimization

Table 6.10.5:

	η_1	η_2	η_3	$\sqrt{M}$
estimate	0.03346	0.03211	–	0.02680
I	1.28	1.23	–	1.03
p	61.1	62.5	63.3	97.9

process, the second term of the majorant diminishes with respect to the first one. As in the previous example, finding a sharp energy estimate takes about two time units.

In contrast to the DEM, all indicators give an inaccurate error distribution (see Fig. 6.10.13), but η_1 and η_2 provide quite good estimates of the global norm.

The above results show that, in some situations, the indicators η_1, η_2, and η_3 may work quite good, but in another one their quality may be not good enough to guarantee the most effective mesh-refinement. Since it is difficult to predict a priori which case we are dealing with, the quality of such error indication may be not optimal. The DEM provides quite reliable error estimates in all cases, but the determinition of a sharp distribution of local errors may require an extra CPU-time. We also remark that the

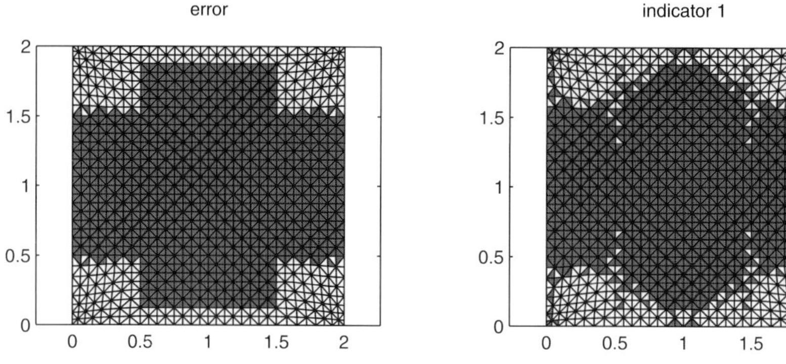

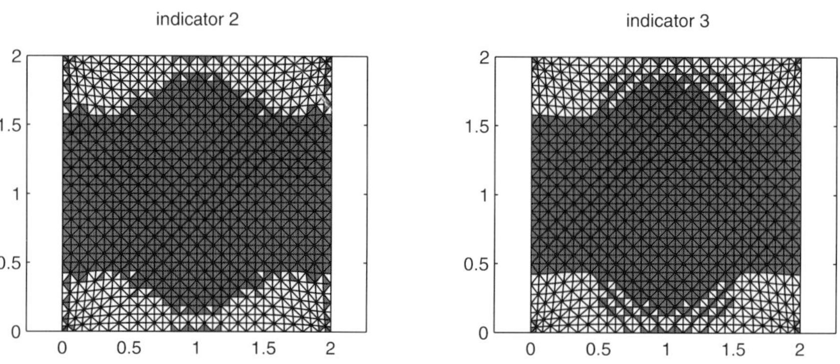

Figure 6.10.11:

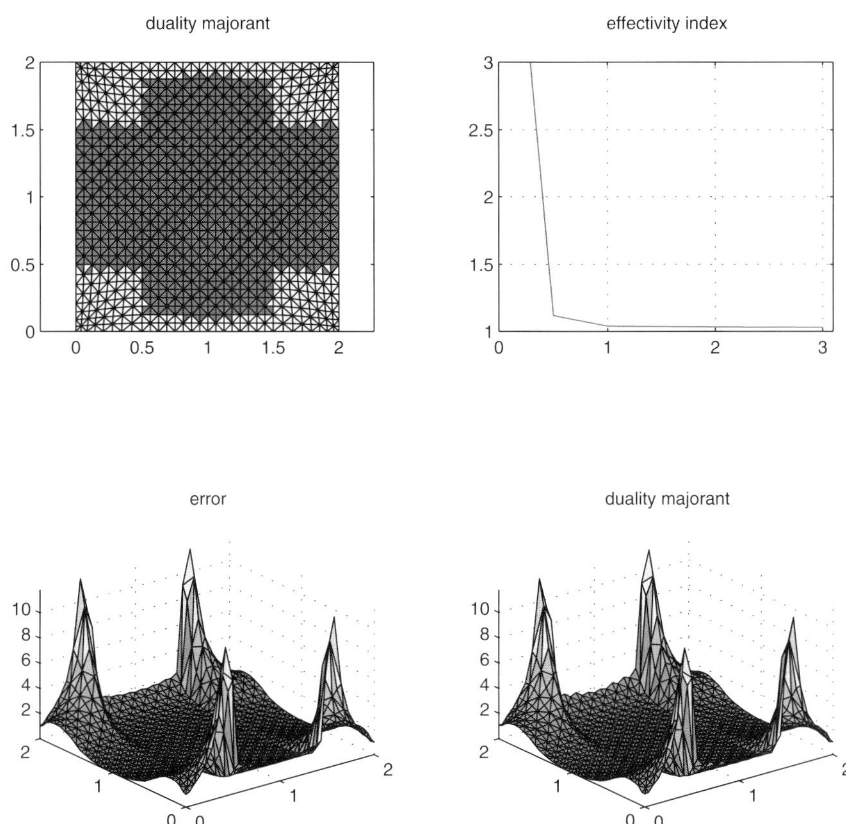

Figure 6.10.12:

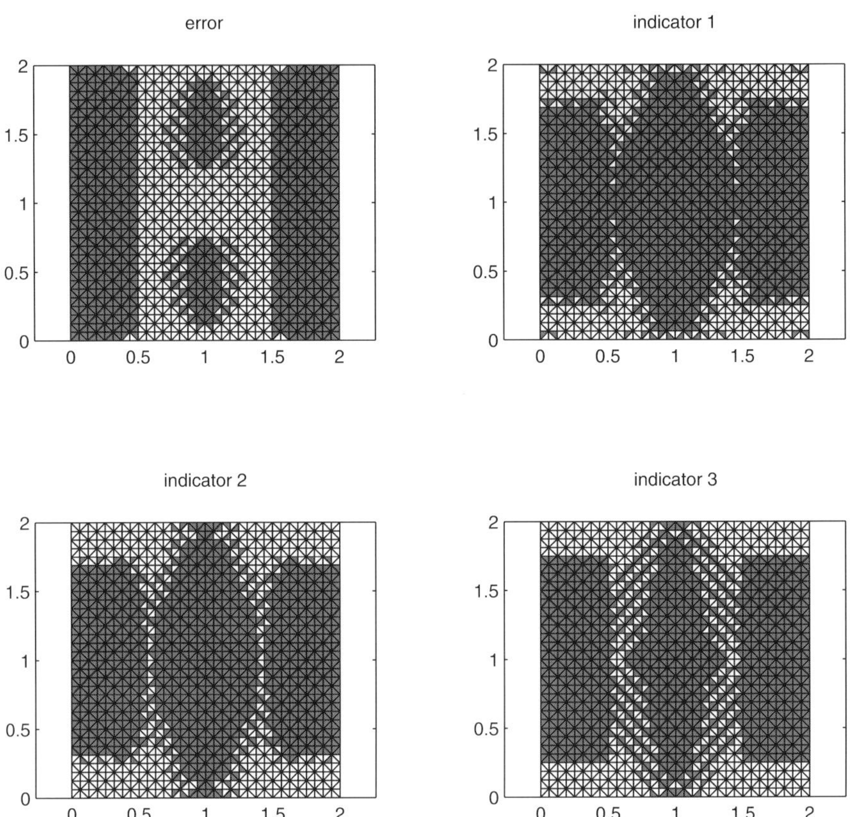

Figure 6.10.13:

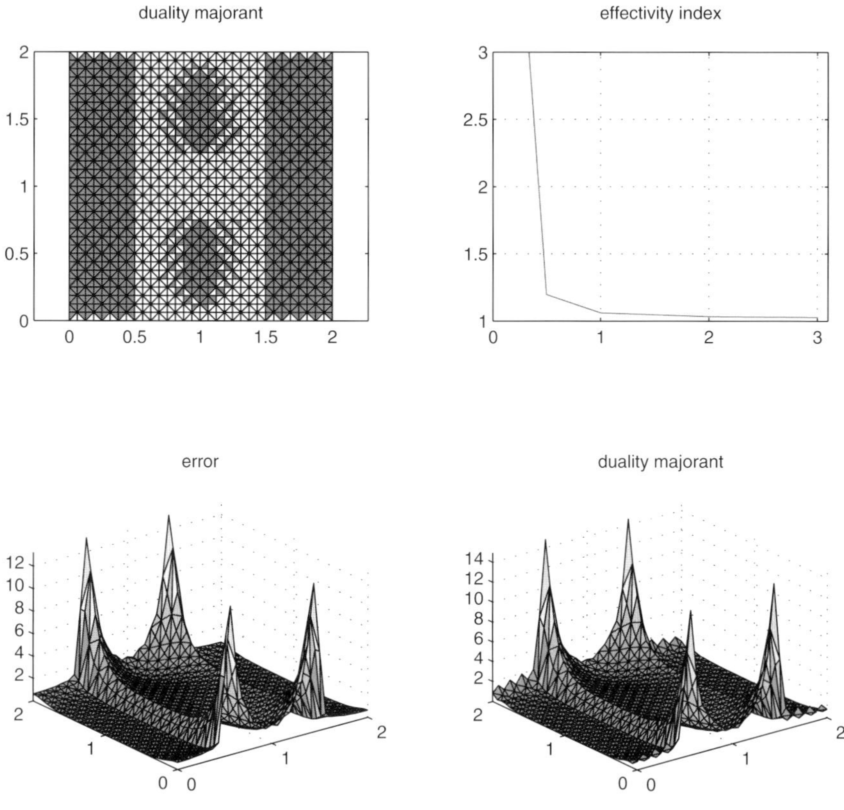

Figure 6.10.14:

indicators give some values that are difficult to improve. In contrast, the quality of the DEM can be increased up to any required level (it is only the matter of time).

In [77, 181], the reader can find a more detailed discussion of these questions and results of various numerical tests, in which the known a posteriori error estimation methods for finite element approximations were compared with those computed by the duality error majorants.

At the end of this part, we make the following note. Since majorants contain constants depending on the respective domain Ω, we must either find them analytically or compute their upper bounds by a certain numerical procedure. For problems with Dirichlét boundary conditions such a procedure is easy to construct. In other cases, this task my require a more complicated analysis (see [145]). Finding such global constants is an important task that requires a deeper investigation.

6.10.6 Computational costs

Let us estimate computational expenditures required to compute an upper error bound by means of the duality error majorant.

Assume that Ω is a square domain and a finite element sampling is given by a regular $(n \times n)$ mesh with 6 elements in a patch. For large n we can estimate the total amount of elements N by the quantity $2n^2$ and the number of edges M by $3n^2$. In this case, the computation of the majorant

$$(1 + \beta)\|\nabla u_h - y^*\|_{2,\Omega}^2 + \left(1 + \frac{1}{\beta}\right) C_\Omega^2 \|\operatorname{div} y^* + f\|_{2,\Omega}^2$$

requires the following steps:

(a) compute ∇u_h on each simplex T (which amounts to approximately N simple computations);

(b) find the averaged gradient $y^* = G(\nabla u_h)$ (the respective expenditures are proportional to the number of nodes and require only simple summation operations);

(c) find $y^* - \nabla u_h$ for any simplex T (approximately N simple computations);

(d) find $\operatorname{div} \nabla u_h$ for any simplex T (approximately N simple computations);

(e) find one global constant C_Ω. This step may not lead to extra computations provided that C_Ω (or an upper estimate of it) is defined analytically. Otherwise, it leads to a single complicated computation of a constant. However, it is important that this constant does not depend on the mesh, so that in a process of mesh refinement no further computations of this constant are necessary;

(f) compute $N + N$ volume integrals and find $M_\oplus$ ($2N$ simple computations).

Certainly, the upper bound of the error computed in such a simple way may be coarse. However, changing y^* by some minimization procedure we can rapidly decrease the value of the index up to the numbers about 1.5 or 2. Further improvements of the effectivity index usually requires more time.

We may compare these expenditures with the costs required to compute the explicit residual type estimate

$$\sum_{i=1}^{N} c_{1i}\| \operatorname{div} \nabla u_h + f\|^2_{2,T_i} + \sum_{i,j=1 \ (i>j)}^{M} c_{2ij}\|\mathrm{j}(\nu_{ij} \cdot \nabla u_h)\|^2_{2,E_{ij}}.$$

In this case, we need to perform the following steps:

(a′) compute ∇u_h on each T_i (N simple computations);

(b′) compute normal vectors ν_{ij} on each edge E_{ij} (M simple computations, which, however, must be repeated upon changing a mesh);

(c′) find $\nu_{ij} \cdot \nabla u_h$ for any simplex T (approximately $3N$ simple computations);

(d′) find $\operatorname{div} \nabla u_h$ for each simplex T (N simple computations);

(e′) find the constants c_{1i} and c_{2ij} ($N + M$ complicated computations, which must be repeated upon changing a mesh);

(f′) compute N volume integrals and M line integrals ($N + M$ simple computations).

It is not difficult to see that (a) and (a′) are identical, while (b), (c), (b′), and (c′) are not expensive and lead to comparable expenditures. The steps (d) and (d′) are identical. However, (e) is much simpler than (e′), because we need to estimate only one global constant. In (e′), we define local constants that depend on the mesh and must be recomputed after each mesh refinement. Even if such a recomputation is based upon some a priori computations made for a certain referenced element (or elements), the determination of all constants must take into account new geometric data (angles, areas, etc) and would not be cheap. Also, a change in the degrees of polynomials forming the approximation space may lead to an expensive recomputation of all interpolation constants. Finally, we observe that (f) and (f') lead to approximately the same expenditures.

Thus, we conclude that expenditures required for computing the majorant for a certain y^* (e.g., for $y^* = G(A\nabla u_h)$) does not exceed those that must be spend for the standard residual based error estimator.

6.10.7   Conclusion

Functional type a posteriori error estimates are universal error estimation tools that can adequately deal with conforming approximations of all types. Its combination with an effective numerical method for solving Problem $\mathcal{P}$ can give a reliable numerical strategy able to find approximate solutions with any a priori given accuracy.

If an approximation v is close to the exact solution, then the DEM method often gives a good upper estimate even on the first step (e.g., without any minimization of $M_\oplus$ with respect to y^*). In this case, we define y^* by a simple averaging procedure and find an optimal parameter β in $M_\oplus(v, y^*, \beta)$, which is a simple and easily solved task.

In general, finding a sharp error estimate requires solving an additional minimization problem whose difficulty is comparable with the difficulty of the original problem. This effectively means that real expenditures for finding approximate solutions with a guaranteed accuracy can noticeably exceed those required for finding a solution without verifying its accuracy. We believe that, in general, such a "reliability fee" is unavoidable and one cannot obtain a posteriori error estimates that are simultaneously universal, exact, and computationally cheap.

6.11   Comments

At the end of this section, we prove one useful result that justifies the asymptotic behavior of the majorant $M_\oplus$ usually observed in numerical experiments.

Proposition 6.11.1. *Let* $(\beta_k, y_k^*) \in \mathbb{R}_+ \times Q^*$ *be a minimizing sequence for the majorant* $M_\oplus(v, \beta, y^*)$ *and*

$$\beta_k \to 0. \tag{6.11.1}$$

Then the first term of the majorant tends to the energy norm of the error and the second one to zero.

Proof. Since (β_k, y_k^*) is a minimizing sequence, we have

$$M_\oplus(v, \beta_k, y_k^*) = (1 + \beta_k)D(\Lambda v, y_k^*) + \left(1 + \frac{1}{\beta_k}\right)\frac{c^2}{2}\|l + \Lambda^* y_k^*\|^2 \leq C,$$

where C is a positive constant independent of k. The sequence β_k is bounded, so that for all k we have $\beta_k \leq \bar{\beta}$. Hence,

$$\| y_k^* \|_*^2 \leq C \qquad \text{and} \qquad \frac{1}{\beta_k}\|\Lambda^*(y_k^* - p^*)\|^2 \leq C. \tag{6.11.2}$$

From this fact, we conclude that there exists a subsequence (for the sake of simplicity we keep for it the same notation y_k^*) such that

$$y_k^* \rightharpoonup y^* \qquad \text{in } Y^*, \tag{6.11.3}$$
$$\Lambda^* y_k^* \to \Lambda^* p^* \quad \text{in } U. \tag{6.11.4}$$

Therefore,

$$\lim_{m\to\infty} M_\oplus(v, \beta_k, y_k^*) \geq$$
$$\geq \lim_{m\to\infty} D(v, y_k^*) \geq \frac{1}{2}\| \Lambda v \|^2 + \frac{1}{2}\| y^* \|_*^2 - (y^*, \Lambda v). \tag{6.11.5}$$

Recall that (β_k, y_k^*) is a minimizing sequence for $M_\oplus$, so that

$$\lim_{k\to\infty} M_\oplus(v, \beta_k, y_k^*) = \frac{1}{2}\| \Lambda(v - u) \|^2$$

and, therefore,

$$\frac{1}{2}\| \Lambda(v - u) \|^2 \geq \frac{1}{2}\| \Lambda v \|^2 + \frac{1}{2}\| y^* \|_*^2 - (y^*, \Lambda v).$$

This inequality leads to another one

$$-(p^*, \Lambda v) + \frac{1}{2}\| p^* \|_*^2 \geq -(y^*, \Lambda v) + \frac{1}{2}\| y^* \|_*^2 . \tag{6.11.6}$$

Note that

$$\lim_{k\to+\infty} (y_k^* - p^*, \Lambda w) + (p^* - y^*, \Lambda w) = 0, \quad \forall w \in V_0.$$

In view of (6.11.4), the limit on the left-hand side is zero, and we find that

$$(y^*, \Lambda w) = (p^*, \Lambda w), \quad \forall w \in V_0.$$

Take $w = v - u_0 \in V_0$ and use the relation

$$(y^* - p^*, \Lambda v) = (y^* - p^*, \Lambda u_0) + (y^* - p^*, \Lambda(v - u_0)) = (y^* - p^*, \Lambda u_0).$$

Then, (6.11.6) yields

$$(p^*, \Lambda u_0) - \frac{1}{2} \parallel p^* \parallel_*^2 \leq (y^*, \Lambda u_0) - \frac{1}{2} \parallel y^* \parallel_*^2 . \qquad (6.11.7)$$

Since p^* is a maximizer of Problem $\mathcal{P}^*$, we have

$$I^*(p^*) \geq I^*(y^*)$$

which is equivalent to

$$(p^*, \Lambda u_0) - \frac{1}{2} \parallel p^* \parallel_*^2 \geq (y^*, \Lambda u_0) - \frac{1}{2} \parallel y^* \parallel_*^2 . \qquad (6.11.8)$$

Now (6.11.7) and (6.11.8) leads to the equality

$$(p^*, \Lambda v) - \frac{1}{2} \parallel p^* \parallel_*^2 = (y^*, \Lambda v) - \frac{1}{2} \parallel y^* \parallel_*^2 . \qquad (6.11.9)$$

However, Problem $\mathcal{P}^*$, has only one maximizer, so that (6.11.9) means that $y^* = p^*$.

Now, we obtain

$$\frac{1}{2} \parallel \Lambda(v - u) \parallel^2 = \lim_{m \to \infty} M_\oplus(v, \beta_k, y_k^*) = \mu_1 + \mu_2, \qquad (6.11.10)$$

where

$$\mu_1 = \lim_{k \to \infty} (1 + \beta_k) D(\Lambda v, y_k^*),$$

$$\mu_2 = \lim_{k \to \infty} \left(1 + \frac{1}{\beta_k} \right) \frac{c^2}{2} \| l + \Lambda^* p_k^* \| \geq 0.$$

We know that y_k^* weakly converges to p^* in Y^*. Therefore (we use lower semicontinuity of the norm),

$$\lim_{k \to \infty} \left(\frac{1}{2} \parallel \Lambda v \parallel^2 + \frac{1}{2} \parallel y_k^* \parallel_*^2 - (\Lambda v, y_k^*) \right) \geq$$

$$\geq \frac{1}{2} \parallel \Lambda v \parallel^2 + \frac{1}{2} \parallel p^* \parallel_*^2 - (\Lambda v, p^*) = \frac{1}{2} \parallel \Lambda(v - u) \parallel^2,$$

and from (6.11.10) we conclude that $\mu_1 = \frac{1}{2} \| \Lambda(v-u) \|^2$ and $\mu_2 = 0$. Now, we have

$$
\mu_1 = \frac{1}{2} \| \Lambda v \|^2 + \frac{1}{2} \| p^* \|_*^2 - (\Lambda v, p^*) =
$$
$$
= \lim_{k\to\infty} \left(\frac{1}{2} \| \Lambda v \|^2 + \frac{1}{2} \| y_k^* \|_*^2 - (\Lambda v, y_k^*) + \beta_k D(\Lambda v, y_k^*) \right) =
$$
$$
= \frac{1}{2} \| \Lambda v \|^2 - (\Lambda v, p^*) + \lim_{k\to\infty} \frac{1}{2} \| y_k^* \|_*^2 .
$$

Thus, $\| y_k^* \|_* \to \| p^* \|_k$ and, consequently, y_k^* strongly convergence to p^*. In view of (6.11.4), we conclude that y_k^* strongly converges to p^* in Q^*. $\qquad\square$

Remark 6.11.1. Such a behavior of the majorant (when the first term tends to the energy norm of the error and the second one rapidly decreases) was really observed in the vast majority of numerical tests. Some of them are presented in [77, 78, 173, 181].

Remark 6.11.2. Consider linear elliptic equations of the second order. In this case, $Q^* = H(\mathrm{div}, \Omega)$. Assume that y_k^* converges to p^* in Q^*, then it is not difficult to show that the integrand of $M_\oplus$ (denote it by $\mu(x)$) approximates the integrand of the error (denote it by $e(x)$) in the following sense. Let $\epsilon > 0$ be an arbitrary small number and

$$
\omega := \{ x \in \Omega \mid |\mu(x) - e(x)| \geq \epsilon \}.
$$

It is proved that $|\omega|$ tends to zero as k tends to infinity (see [182]). This fact justifies *local properties* of the majorant that can also be used as an efficient indicator of local errors.

6.12 Notes for the Chapter

The history of the variational approach to deriving a posteriori error estimates dates back to the works of W. Prager and J. L. Synge [163] and S. Mikhlin [143]. In [143], it was shown that if a problem is stated as a minimization problem for a quadratic functional J, then the difference between $J(v)$ and the exact lower bound provides an upper bound of the error. Since the latter is bounded from below by the values of the so-called dual (or complementary) functional, it is natural to use the respective dual problem for the computation of a posteriori error estimates (see, e.g., H. Gaevskiĭ, H. K. Gröger and K. Zacharias [82], D. W. Kelly [118] M. Mosolov and V. Myasnikov [146]). Such bounds can also be computed by the orthogonal projection method (see S. Zaremba [221], H. Weil [218], and M. I. Vishik [215]), which creates a sequence of trial functions for the dual problem. However, in the majority of cases admissible functions of the dual problem belong to a linear manifold defined by some differential relations (we denote it Q_l^*). In general, the determination of such functions poses a difficult task. Probably, this fact hindered the development of a posteriori error estimation methods based upon duality theory. For a class of variational problems with convex integrands dependent on the gradient of the minimizing function, a way of overcoming this difficulty was found and justified in S. Repin [168, 170, 171]. In S. Repin [174] and some other papers, such a justification is presented for a wider class of uniformly convex functionals. Functional type a posteriori error estimates for boundary-value problems with the biharmonic operator were derived in P. Neittaanmäki and S. Repin [149]. Two-sided a posteriori estimates of the functional type were derived in S. Repin [175] for linear and some nonlinear elliptic problems. The variational approach described in Chapter 6 is based on these results. Also, [175] presents another (nonvariational) method of deriving a posteriori estimates. With the help of this method, estimates of the deviation from the exact solution were obtained for a problems that have no variational statement (see [176]).

For linear elliptic equations of the second order, a numerical verification of a posteriori error estimates of the functional type was carried out in the papers by S. Repin [173], M. Frolov, P. Neittaanmäki and S. Repin [77, 78, 181]. For equations of the fourth order, several tests were performed in M. Frolov, P. Neittaanmäki and S. Repin [78]. For linear elasticity problems, such tests can be found in A. Muzalevskii and S. Repin [147].

Finally, we mention several results related to the subject considered, which has not been exposed in Chapter 6. Functional type a posteriori error estimates for approximations that do not satisfy Dirichlét boundary conditions precisely were obtained in S. Repin, S. Sauter and A. Smolianski [182, 183]. Such estimates can be useful if it is difficult to precisely approximate the boundary or if a numerical method takes into account boundary conditions in a generalized sense (see, e.g., R. Glowinski, T.-W. Pan and J. Periaux [92]). In S. Repin [180] and S. Repin, S. Sauter and A. Smolianski [184], functional type a posteriori error estimates were used for estimation of dimensional reduction errors arising when a 3D model is approximated by a simplified 2D model.

Chapter 7

A posteriori estimates for nonlinear variational problems

7.1   Nonlinear variational problems

Let Y be a reflexive Banach space, Y^* be the space of linear continuous functionals on Y, and (y^*, y) denote the value of a functional $y^* \in Y^*$ on $y \in Y$. The norms of Y and Y^* are denoted by $\|\cdot\|$ and $\|\cdot\|_*$, respectively. In contrast to the case considered in the previous chapter, these two spaces are now essentially different. Henceforth, we use another Banach space $\mathcal{V}$, which has a topologically dual counterpart $\mathcal{V}^*$. The product of $v \in \mathcal{V}$ and $v^* \in \mathcal{V}^*$ is denoted by $\langle v^*, v \rangle$. Introduce a bounded linear operator $\Lambda : \mathcal{V} \to Y$ and the corresponding conjugate operator $\Lambda^* : \mathcal{V}^* \to Y^*$ satisfying the relation

$$(y^*, \Lambda w) = \langle \Lambda^* y^*, w \rangle, \quad \forall w \in \mathcal{V}. \tag{7.1.1}$$

Also, it is assumed that

$$\|\Lambda w\| \geq c_1 \|w\|_{\mathcal{V}}, \quad \forall v \in \mathcal{V}. \tag{7.1.2}$$

Consider the functional

$$J(v) = G(\Lambda v) + F(v), \tag{7.1.3}$$

where $G \in \Gamma_0(Y)$ and $F \in \Gamma_0(\mathcal{V})$. We assume that this functional is coercive on $\mathcal{V}$, i.e.,

$$J(v) \to +\infty \quad \text{as } \|v\|_{\mathcal{V}} \to +\infty. \tag{7.1.4}$$

The basic variational problem that we analyze is as follows.

Problem $\mathcal{P}$. Find $u \in \mathcal{V}$ such that

$$J(u) = \inf \mathcal{P} := \inf_{v \in \mathcal{V}} J(v). \tag{7.1.5}$$

Applying Theorem 5.4.1, we establish the existence of a minimizer u.

Theorem 7.1.1. *Let $G : Y \to \mathbb{R}$ and $F : \mathcal{V} \to \mathbb{R}$ be two convex lower semicontinuous functionals satisfying the condition (7.1.4). Then Problem $\mathcal{P}$ has a solution.*

Note that

$$G(\Lambda v) + F(v) = \sup_{y^* \in Y^*} \left\{ (y^*, \Lambda v) - G^*(y^*) + F(v) \right\},$$

where $G^* : Y^* \to \mathbb{R}$ is the functional conjugate to G. Therefore,

$$\inf \mathcal{P} = \inf_{v \in \mathcal{V}} \sup_{y^* \in Y^*} L(v, y^*), \tag{7.1.6}$$

where

$$L(v, y^*) := F(v) + (y^*, \Lambda v) - G^*(y^*).$$

Let $0_\mathcal{V}$ denote the zero element of $\mathcal{V}$. Assume that

$$F(0_\mathcal{V}) \quad \text{is finite}, \tag{7.1.7}$$
$$G^*(y^*) \quad \text{is coercive on } Y^*. \tag{7.1.8}$$

The Lagrangian L is convex and lower semicontinuous with respect to the variable v and concave and upper semicontinuous with respect to the variable y^*. By (7.1.7) and (7.1.8), we see that

$$L(0_\mathcal{V}, y^*) = F(0_\mathcal{V}) - G^*(y^*) \to -\infty \quad \text{as } \|y^*\| \to +\infty.$$

Hence (see Theorem 5.4.4 and the comments)

$$\inf \mathcal{P} = \sup \mathcal{P}^* = \sup_{y^* \in Y^*} \inf_{v \in \mathcal{V}} L(v, y^*). \tag{7.1.9}$$

Define the functional

$$I^*(y^*) = \inf_{v \in \mathcal{V}} L(v, y^*) = -G^*(y^*) + \inf_{v \in \mathcal{V}} ((y^*, \Lambda v) + F(v)) =$$
$$= -G^*(y^*) - \sup_{v \in \mathcal{V}} (\langle -\Lambda^* y^*, v \rangle - F(v)) = -G^*(y^*) - F^*(-\Lambda^* y^*).$$

Now, (7.1.9) leads to a new variational problem.

Problem $\mathcal{P}^*$. Find $p^* \in Y^*$ such that

$$I^*(p^*) = \sup \mathcal{P}^* := \sup_{y^* \in Y^*} I^*(y^*). \qquad (7.1.10)$$

Theorem 7.1.2. *Let the conditions of Theorem 7.1.1 and the assumptions (7.1.7) and (7.1.8) hold. Then Problem $\mathcal{P}^*$ has a solution.*

Proof. We have

$$F^*(-\Lambda^* y^*) \geq \langle -\Lambda^* y^*, 0_\nu \rangle - F(0_\nu).$$

In view of (7.1.7), this relation means that $F^*(-\Lambda^* y^*)$ is uniformly bounded from below. The functional $G^*(y^*)$ is coercive (cf. (7.1.8)). Thus, the functional $-I^*(y^*)$ is also coercive on the reflexive space Y^* and the existence of a maximizer follows from Theorem 5.4.1. $\qquad \square$

The pair (u, p^*) forms a saddle point of L on $\mathcal{V} \times Y^*$ (see Theorem 5.4.2), i.e.

$$L(u, y^*) \leq L(u, p^*) \leq L(v, p^*), \quad \forall v \in \mathcal{V}, \, y^* \in Y^*.$$

The left-hand side of the inequality yields the relation

$$(y^* - p^*, \Lambda u) \leq G^*(y^*) - G^*(p^*), \quad \forall y^* \in Y^*,$$

which means that

$$\Lambda u \in \partial G^*(p^*) \quad \text{or} \quad p^* \in \partial G(\Lambda u). \qquad (7.1.11)$$

Similarly, the right-hand side of the inequality implies

$$F(v) - F(u) \geq (p^*, \Lambda(u - v)) = \langle -\Lambda^* p^*, v - u \rangle$$

or

$$-\Lambda^* p^* \in \partial F(u) \quad \text{or} \quad u \in \partial F^*(-\Lambda^* p^*). \qquad (7.1.12)$$

By recalling Proposition 5.3.2, we may represent (7.1.11) and (7.1.12) in terms of the respective compound functionals:

$$D_G(\Lambda u, p^*) := G(\Lambda u) + G^*(p^*) - (p^*, \Lambda u) = 0, \qquad (7.1.13)$$

$$D_F(u, -\Lambda^* p^*) := F(u) + F^*(-\Lambda^* p^*) + \langle \Lambda^* p^*, u \rangle = 0. \qquad (7.1.14)$$

7.2   General form of the error majorant for the functional $J(v) = G(\Lambda v) + F(v)$

Assume that G and G^* are uniformly convex in the balls $\mathcal{B}_\delta := B(0_Y, \delta) \in Y$ and $\mathcal{B}^*_{\delta^*} := B(0_{Y^*}, \delta^*) \in Y^*$, respectively (see (5.5.3)). Then, they satisfy the relations

$$G(\frac{y_1 + y_2}{2}) + \Upsilon_\delta(\frac{y_1 - y_2}{2}) \leq \frac{1}{2}\left(G(y_1) + G(y_2)\right), \qquad (7.2.1)$$

$$G^*(\frac{y_1^* + y_2^*}{2}) + \Upsilon^*_{\delta^*}(\frac{y_1^* - y_2^*}{2}) \leq \frac{1}{2}\left(G^*(y_1) + G^*(y_2)\right), \qquad (7.2.2)$$

where $\Upsilon_\delta : \mathcal{V} \to \mathbb{R}_+$ and $\Upsilon^*_{\delta^*} : Y^* \to \mathbb{R}_+$ are nonnegative functionals vanishing at zero. Henceforth, we divide the arguments of the forcing functionals by 2 to make the relations more symmetric.

The general form of a functional type a posteriori error majorant is given by the theorem stated below, which was originally established in [168] (see also [174, 175]). It gives an upper bound of the differences $v - u$ (for the primal problem) and $y^* - p^*$ (for the dual one) evaluated in terms of the functionals Υ and Υ^*.

Theorem 7.2.1. *Let the functionals F and G satisfy (7.2.1), (7.2.2), and the conditions of Theorems 7.1.1 and 7.1.2. If, in addition,*

$$\Lambda u \in \mathcal{B}_\delta \quad \text{and} \quad p^* \in \mathcal{B}^*_{\delta^*},$$

then for any $v \in \mathcal{V}$ and $y^* \in Y^*$ such that $\Lambda v \in \mathcal{B}_\delta$ and $y^* \in \mathcal{B}^*_{\delta^*}$, the estimate

$$\boxed{\Upsilon_\delta\left(\tfrac{\Lambda(v-u)}{2}\right) + \Upsilon^*_{\delta^*}\left(\tfrac{y^*-p^*}{2}\right) \leq \mathcal{M}_\oplus(v, y^*),} \qquad (7.2.3)$$

holds, where

$$\mathcal{M}_\oplus(v, y^*) := \frac{1}{2}\Big[D_F(v, -\Lambda^* y^*) + D_G(\Lambda v, y^*)\Big].$$

Proof. From (7.2.1), we conclude that

$$\Upsilon_\delta\left(\tfrac{\Lambda(v-u)}{2}\right) \leq \frac{1}{2}\left(G(\Lambda v) + G(\Lambda u) - 2G(\Lambda(\tfrac{u+v}{2}))\right).$$

Since $F \in \mathbb{CF}(V)$, this estimate can be rewritten as follows

$$\Upsilon_\delta \left(\tfrac{\Lambda(v-u)}{2} \right) \leq \frac{1}{2} \left(J(v) + J(u) - 2J \left(\tfrac{u+v}{2}, \Lambda \left(\tfrac{u+v}{2} \right) \right) \right) \leq$$

$$\leq \frac{1}{2} \left(J(v) - J(u) \right). \quad (7.2.4)$$

Set $y_1^* = y^*$ and $y_2^* = p^*$ in (7.2.2). Then, we obtain

$$G^* \left(\tfrac{p^*+y^*}{2} \right) + \Upsilon_{\delta^*}^* \left(\tfrac{y^*-p^*}{2} \right) \leq \frac{1}{2} (G^*(p^*) + G^*(y^*)).$$

Since $F^* \in \mathbb{CF}(V^*)$, we have

$$F^* \left(-\Lambda^* \left(\tfrac{p^*+y^*}{2} \right) \right) \leq \frac{1}{2} \left(F^*(-\Lambda^* p^*) + F^*(-\Lambda^* y^*) \right).$$

The latter two inequalities imply

$$\Upsilon_{\delta^*}^* \left(\frac{y^* - p^*}{2} \right) \; \leq \; \frac{1}{2} (G^*(p^*) + F^*(-\Lambda^* p^*)) + \frac{1}{2} (G^*(y^*) + F^*(-\Lambda^* y^*)) -$$

$$-G^* \left(\tfrac{p^*+y^*}{2} \right) - F^* \left(-\Lambda^* \left(\tfrac{p^*+y^*}{2} \right) \right).$$

Note that

$$-G^* \left(\tfrac{p^*+y^*}{2} \right) - F^* \left(-\Lambda^* \left(\tfrac{p^*+y^*}{2} \right) \right) \leq -G^*(p^*) - F^*(-\Lambda^* p^*).$$

Therefore,

$$\Upsilon_{\delta^*}^* \left(\tfrac{y^*-p^*}{2} \right) \leq \frac{1}{2} \left(G^*(y^*) + F^*(-\Lambda^* y^*) - G^*(p^*) - F^*(-\Lambda^* p^*) \right). \quad (7.2.5)$$

Since

$$J(u) = -G^*(p^*) - F^*(-\Lambda^* p^*), \quad (7.2.6)$$

the relations (7.2.4) and (7.2.5) give the estimate

$$\Upsilon_\delta \left(\frac{\Lambda(v-u)}{2} \right) + \Upsilon_{\delta^*}^* \left(\frac{y^* - p^*}{2} \right) \leq$$

$$\leq \frac{1}{2} \Big(G(\Lambda v) + G^*(y^*) + F(v) + F^*(-\Lambda^* y^*) \Big),$$

which coincides with (7.2.3). $\qquad \square$

It is worthwhile to add some comments on the meaning of the estimate (7.2.3). We see that the left-hand side of this estimate is formed by two compound functionals, which are nonnegative and vanish if and only if v and y^* satisfy the relations (7.1.1) and (7.1.2). Obviously, both relations are satisfied in the unique case: $v = u$ and $y^* = p^*$. The left-hand side of (7.2.3) is given by two functionals that are measures of the deviations $\Lambda(v - u)$ and $y^* - p^*$ in the spaces Y and Y^*, respectively. We believe that these measures are in a sense natural for the problems associated with the functional (7.1.3). If only one of the functionals is uniformly convex (say, the functional G) and the other one (G^*) is only convex, then the estimate (7.2.3) remains valid with only one functional (Υ_δ).

From the estimate (7.2.3) it follows that

$$
\boxed{\Upsilon_\delta \left(\tfrac{\Lambda(v-u)}{2} \right) \leq \mathcal{M}_\oplus(v, y^*), \quad \forall y^* \in Y^*}
\tag{7.2.7}
$$

and

$$
\boxed{\Upsilon_{\delta*}^* \left(\tfrac{y^*-p^*}{2} \right) \leq \mathcal{M}_\oplus(v, y^*), \qquad \forall v \in \mathcal{V}.}
\tag{7.2.8}
$$

From (7.2.7), we conclude that the functional Υ_δ serves as a natural measure of the deviation $\Lambda(v - u)$. Certainly, to compute a sharp estimate it is necessary to minimize $\mathcal{M}_\oplus(v, y^*)$ with respect to $y^* \in Y^*$. For any $v \in \mathcal{V}$ such a minimum is attained for $y^* = p^*$. Indeed,

$$
\begin{aligned}
2(\mathcal{M}_\oplus(v, y^*) - \mathcal{M}_\oplus(v, p^*)) = {} & F(v) + F^*(-\Lambda^* y^*) + \langle \Lambda^* y^*, v \rangle \\
& + G(\Lambda v) + G^*(y^*) - (y^*, \Lambda v) - F(v) - F^*(-\Lambda^* p^*) \\
& \langle -\Lambda^* p^*, v \rangle - G(\Lambda v) - G^*(p^*) + (p^*, \Lambda v) \\
& = F^*(-\Lambda^* y^*) + G^*(y^*) - F^*(-\Lambda^* p^*) - G^*(p^*).
\end{aligned}
$$

Hence,

$$
\mathcal{M}_\oplus(v, y^*) - \mathcal{M}_\oplus(v, p^*) = \frac{1}{2}\left(I^*(p^*) - I^*(y^*)\right) \geq 0,
\tag{7.2.9}
$$

and, therefore,

$$
\inf_{y^* \in Y^*} \mathcal{M}_\oplus(v, y^*) = \mathcal{M}_\oplus(v, p^*).
\tag{7.2.10}
$$

Similarly, the estimate (7.2.8) shows that $\Upsilon_{\delta*}^*$ serves as a measure of the difference between p^* and its approximation y^*. In this estimate, the function

v is "free" and the estimate is sharp if v is properly chosen. Since

$$
\begin{aligned}
2(\mathcal{M}_\oplus(v,y^*) - \mathcal{M}_\oplus(u,y^*)) &= F(v) + F^*(-\Lambda^*y^*) + \langle \Lambda^*y^*, v\rangle \\
&\quad + G(\Lambda v) + G^*(y^*) - (y^*, \Lambda v) - F(u) - F^*(-\Lambda^*y^*) - \langle \Lambda^*y^*, u\rangle \\
&\quad - G(\Lambda u) - G^*(y^*) + (y^*, \Lambda u) \\
&= F(v) + G(\Lambda v) - F(u) - G(\Lambda u),
\end{aligned}
$$

we conclude that

$$
\mathcal{M}_\oplus(v,y^*) - \mathcal{M}_\oplus(u,y^*) = \frac{1}{2}\left(J(v) - J(u)\right). \tag{7.2.11}
$$

Hence,

$$
\inf_{v\in\mathcal{V}} \mathcal{M}_\oplus(v,y^*) = \mathcal{M}_\oplus(u,y^*). \tag{7.2.12}
$$

Therefore, the quality of an upper bound computed by $\mathcal{M}_\oplus(v,y^*)$ is the better the closer v to u.

Finally, we establish two important relations that hold for the majorant $\mathcal{M}_\oplus$. We have

$$
\begin{aligned}
2(\mathcal{M}_\oplus(v,p^*) + \mathcal{M}_\oplus(u,y^*)) \\
= G(\Lambda v) + G^*(p^*) + F^*(-\Lambda^*p^*) + F(v) \\
+ G(\Lambda u) + G^*(y^*) + F^*(-\Lambda^*y^*) + F(u).
\end{aligned}
$$

Since $\inf \mathcal{P} = \sup \mathcal{P}^*$, we arrive at the first relation:

$$
\boxed{\mathcal{M}_\oplus(v,p^*) + \mathcal{M}_\oplus(u,y^*) = \tfrac{1}{2}\left(J(v) - I^*(y^*)\right),} \tag{7.2.13}
$$

which is valid for any $v \in \mathcal{V}$ and $y^* \in Y^*$. Set here $v = u$ and use (7.2.9). Then, we deduce another relation

$$
\boxed{\mathcal{M}_\oplus(v,p^*) + \mathcal{M}_\oplus(u,y^*) = \mathcal{M}_\oplus(v,y^*).} \tag{7.2.14}
$$

Remark 7.2.1. It could happen that F is also uniformly convex in a ball $\mathcal{B}'_\delta \in \mathcal{V}$ with a forcing functional $\gamma_\delta : \mathcal{V} \to \mathbb{R}_+$. Then, for any $v \in \mathcal{B}'_\delta$ such that $\Lambda v \in \mathcal{B}_\delta$, the estimate (7.2.7) can be enforced as follows:

$$
\Upsilon_\delta\left(\frac{\Lambda(v-u)}{2}\right) + \gamma_\delta\left(\frac{v-u}{2}\right) \le \mathcal{M}_\oplus(v,y^*), \qquad \forall y^* \in Y^*. \tag{7.2.15}
$$

Similarly, if F^* is uniformly convex in a ball $\mathcal{B}^{*\prime}_{\delta*} \in \mathcal{V}^*$ with a forcing functional $\gamma_{\delta*} : \mathcal{V}^* \to \mathbb{R}_+$, $y^* \in \mathcal{B}^{*\prime}_{\delta*}$, and $\Lambda^* y^* \in \mathcal{B}^*_{\delta*}$, then (7.2.8) is replaced by the estimate

$$\Upsilon^*_{\delta*}\left(\frac{y^* - p^*}{2}\right) + \gamma^*_{\delta*}\left(\frac{\Lambda^*(y^* - p^*)}{2}\right) \leq \mathcal{M}_\oplus(v, y^*), \qquad \forall v \in \mathcal{V}. \quad (7.2.16)$$

7.3   Variational problems with nonhomogeneous boundary conditions

Boundary-value problems often contain nonhomogeneous boundary conditions, so that the respective variational problem is defined on an affine set $V_0 + u_0$, where V_0 is a subspace of a Banach space V and u_0 is a given element of V used to specify these conditions. This case can be reduced to the one considered in the previous section. Note that

$$\inf_{w \in V_0 + u_0} J(w) = \inf_{w \in V_0} \widehat{J}(w),$$

$$\widehat{J}(w) = \widehat{G}(\Lambda w) + \widehat{F}(w), \quad \widehat{G}(y) = G(y + y_0),$$

$$\widehat{F}(w) = F(w + u_0), \qquad\qquad y_0 = \Lambda u_0 \in Y.$$

Thus, we may consider the following problem: find $\widehat{u} \in V_0$ such that

$$\widehat{J}(\widehat{u}, \Lambda\widehat{u}) = \inf_{w \in V_0} \widehat{J}(w). \tag{7.3.1}$$

Assume that the functional F and the operator Λ are defined on the whole space V and the relation (7.1.2) holds if we set $V = V_0$. It is easy to see that the functionals $\widehat{G}$ and $\widehat{F}$ are uniformly convex, provided that G and F are uniformly convex. Indeed, if $G : Y \to \mathbb{R}$ is uniformly convex in a ball $\mathcal{B}_\delta$ with a forcing functional Υ_δ, then for $(y_i + y_0) \in \mathcal{B}_\delta$, $i = 1, 2$, we have

$$G\left(\tfrac{y_1 + y_2}{2} + y_0\right) + \Upsilon_\delta\left(\tfrac{y_2 - y_1}{2}\right) \leq \tfrac{1}{2}\left(G(y_1 + y_0) + G(y_2 + y_0)\right).$$

This means that

$$\widehat{G}\left(\tfrac{y_1 + y_2}{2}\right) + \Upsilon_\delta\left(\tfrac{y_2 - y_1}{2}\right) \leq \tfrac{1}{2}\left(\widehat{G}(y_1) + \widehat{G}(y_2)\right)$$

or, in other words, $\widehat{G}$ is uniformly convex on the set

$$\{\mathcal{B}_\delta - y_0\} := \{y \in Y \mid y + y_0 \in \mathcal{B}_\delta\}.$$

Thus, we can apply (7.2.7) to the problem (7.3.1), which implies the estimate

$$\Upsilon_\delta\left(\tfrac{\Lambda(w-\widehat{u})}{2}\right) \le \tfrac{1}{2}\left(D_{\widehat{G}}(\Lambda w, y^*) + D_{\widehat{F}}(w, -\Lambda^* y^*)\right)$$

that holds for $\Lambda\widehat{u}$, $\Lambda w \in \{\mathcal{B}_\delta - y_0\}$, and $y^* \in Y^*$.

Let $v = w + u_0$. Then,

$$\Lambda v = \Lambda w + y_0 \in \mathcal{B}_\delta, \quad \Lambda u = \Lambda\widehat{u} + y_0 \in \mathcal{B}_\delta$$

and we rewrite the estimate as follows:

$$\Upsilon_\delta\left(\tfrac{\Lambda(v-u)}{2}\right)\le \tfrac{1}{2}\left(D_{\widehat{G}}(\Lambda v - y_0, y^*) + D_{\widehat{F}}(v - u_0, -\Lambda^* y^*)\right). \tag{7.3.2}$$

Note that

$$\widehat{G}^*(y^*) = \sup_{y\in Y}\{(y^*,y) - G(y + y_0)\} = G^*(y^*) - (y^*, y_0).$$

Therefore,

$$\begin{aligned}
D_{\widehat{G}}(\Lambda w, y^*) \;&= G(\Lambda w + y_0) + G^*(y^*) - (y^*, y_0 + \Lambda w)\\
&= D_G(\Lambda(w + u_0), y^*) = D_G(\Lambda v, y^*).
\end{aligned}$$

Similarly,

$$\begin{aligned}
\widehat{F}^*(-\Lambda^* y^*) =\;& \sup_{w\in V_0}\{\langle -\Lambda^* y^*, w\rangle - F(w + u_0)\}\\
=\;& \sup_{w\in V_0 + u_0}\{(-\Lambda^* y^*, w) + (y^*, y_0) - F(w)\}.
\end{aligned}$$

If F is an integral type functional, then one can usually estimate the value of this supremum by the quantity $F^*(-\Lambda^* y^*) + (y^*, y_0)$, where F^* is an integral functional whose integrand is given by a polar function. Then,

$$\begin{aligned}
D_{\widehat{F}}(w, -\Lambda^* y^*) \;&= \widehat{F}(w) + \widehat{F}^*(-\Lambda^* y^*) + \langle\Lambda^* y^*, w\rangle\\
&\le F(w + u_0) + F^*(-\Lambda^* y^*) + (y^*, \Lambda(w + u_0))\\
&= D_F(w + u_0, -\Lambda^* y^*) = D_F(v, -\Lambda^* y^*)
\end{aligned}$$

and the evaluation of the second term of the majorant is reduced to computing the functional D_F.

To show the effect of the above approach, we consider a particular nonlinear problem. Let Ω be a bounded connected domain in $\mathbb{R}^2$ with Lipschitz continuous boundary $\partial\Omega$ and A be a symmetric real-valued matrix subject to the condition

$$c_1|\xi|^2 \le A\xi \cdot \xi \le c_2|\xi|^2, \quad \forall \xi \in \mathbb{R}^2$$

with positive constants c_1 and c_2. Define the set

$$V_0 + u_0 = \{v \in H^1(\Omega) \,|\, v = w + u_0, \quad w \in V_0\},$$

where $V_0 := \overset{\circ}{H}{}^1(\Omega)$. The primal variational problem is as follows.

Problem $\mathcal{P}$. Find $u \in V_0 + u_0$ such that

$$J(u) = \inf \mathcal{P} := \inf_{v \in V_0 + u_0} J(v), \tag{7.3.3}$$

where

$$J(v) = \int_\Omega \left(\frac{1}{2} A\nabla v \cdot \nabla v + \frac{\lambda^k}{k}|v|^k - fv \right) dx,$$

$k > 1$, $\lambda > 0$, $f \in L^\infty(\Omega)$ and $u_0 \in H^1(\Omega)$.

To apply the general scheme, we set $\mathcal{V} = V_0$, $Y = L^2(\Omega, \mathbb{R}^2)$,

$$\widehat{G}(y) := \int_\Omega \frac{1}{2} A(y + y_0) \cdot (y + y_0)\, dx,$$

and

$$\widehat{F}(w) := \int_\Omega \phi(w + u_0)\, dx, \quad w \in V_0,$$

where the integrand ϕ is defined by the relation

$$\phi(\xi) = \frac{\lambda^k}{k}|\xi|^k - f\xi, \quad \forall \xi \in \mathbb{R}.$$

In this case,

$$\widehat{G}^*(y^*) = \int_\Omega \left(-y^* \cdot y_0 + \frac{1}{2} A^{-1}y^* \cdot y^* \right) dx,$$

$$(y^*, \Lambda w) = \int_\Omega y^* \cdot \nabla w\, dx.$$

Therefore,

$$D_{\widehat{G}}(\nabla w, y^*) = \int_{\Omega} \frac{1}{2} A(\nabla(w + u_0)) \cdot (\nabla(w + u_0)) \, dx$$

$$+ \int_{\Omega} \left(-y^* \cdot \nabla u_0 + \frac{1}{2} A^{-1} y^* \cdot y^*\right) dx - \int_{\Omega} \nabla w \cdot y^* \, dx.$$

To obtain $\widehat{F}^*(v^*)$, we must find

$$\sup_{w \in V_0} \{\langle v^*, w \rangle - \widehat{F}(w)\}.$$

If $v^* \in L^\infty(\Omega)$, then for any $k > 1$ this quantity can be evaluated explicitly.
Indeed,

$$\widehat{F}^*(v^*) = \sup_{w \in V_0} \int_{\Omega} (v^* w - \phi(w + u_0)) \, dx$$

$$\leq \sup_{v \in V_0 + u_0} \int_{\Omega} (v^* v - \phi(v)) \, dx - \int_{\Omega} v^* u_0 \, dx. \quad (7.3.4)$$

Now, we use the algebraic relation

$$\phi(\xi) + \phi^*(\xi^*) - \xi\xi^* \geq 0, \quad \forall \xi, \, \xi^* \in \mathbb{R}, \qquad (7.3.5)$$

in which

$$\phi^*(\xi^*) = \frac{1}{k^* \lambda^{k^*}} |v^* + f|^{k^*}, \quad k^* = \frac{k}{k-1},$$

is the respective conjugate function. Applying (7.3.5) to (7.3.4), we obtain

$$\widehat{F}^*(v^*) \leq \int_{\Omega} (-v^* u_0 + \phi^*(v^*)) \, dx.$$

Hence,

$$D_{\widehat{F}}(w, v^*) \leq \int_{\Omega} (\phi(w) + \phi^*(v^*) - (w + u_0)v^*) dx.$$

We see that the majorant takes the form (for $v \in V_0 + u_0$)

$$\mathcal{M}_\oplus(v, y^*) = \tfrac{1}{4} \int_{\Omega} (\nabla v - A^{-1} y^*) \cdot (A\nabla v - y^*) \, dx$$

$$+ \tfrac{1}{2} \int_{\Omega} \left(\frac{\lambda^k}{k} |v|^k - (\operatorname{div} y^* + f)v + \frac{\lambda^{-k^*}}{k^*} |\operatorname{div} y^* + f|^{k^*}\right) dx. \quad (7.3.6)$$

Note that the majorant may assume infinite values (e.g., if $\operatorname{div} y^* \notin L^{k^*}(\Omega)$). However, if $\operatorname{div} y^* \in L^\infty(\Omega)$, then $\mathcal{M}_\oplus(w, y^*) < +\infty$.

The functional $G(y)$ is uniformly convex in any ball (see Example 5.5.1) with forcing functional $\frac{1}{8} \int_\Omega Ay \cdot y \, dx$. Now (7.3.2) yields the estimate

$$\| e \|^2 \leq \int_\Omega (\nabla v - A^{-1} y^*) \cdot (A\nabla v - y^*) \, dx$$

$$+ 2 \int_\Omega \left(\frac{\lambda^k}{k} |v|^k - (\operatorname{div} y^* + f)v + \frac{\lambda^{-k^*}}{k^*} |\operatorname{div} y^* + f|^{k^*} \right) dx, \quad (7.3.7)$$

where

$$\| e \|^2 := \frac{1}{2} \int_\Omega A\nabla e \cdot \nabla e \, dx.$$

This estimate can be enforced if we recall that F is also uniformly convex (this fact follows from Clarkson's inequalities (see [201])). For example, if $k \geq 2$, then we have the inequality

$$\left| \tfrac{\phi + \psi}{2} \right|^k + \left| \tfrac{\phi - \psi}{2} \right|^k \leq \frac{1}{2} \left(|\phi|^k + |\psi|^k \right),$$

which holds for numbers as well as for real-valued functions. In this case, F is uniformly convex in any ball with

$$\gamma(v) = \lambda^k \int_\Omega |v|^k \, dx.$$

If $1 < k < 2$, then another inequality specifies the respective forcing functional.

Remark 7.3.1. A posteriori majorants for problems containing nonlinearities in higher terms (such as problems with power growth functionals) can be obtained in a similar way (see [171]).

Remark 7.3.2. The estimate (7.3.7) cannot be used if $\lambda = 0$. Therefore, the case

$$F(v) = \int_\Omega fv \, dx$$

requires a special consideration, which we present in the next section.

7.4 Error majorant for the case $F(v) = \langle l, v \rangle$

Considerable part of practically interesting variational problems is related to the case of a linear functional F. In this section, we show how one can derive functional type a posteriori estimates for variational problems with

$$F(v) = \langle l, v \rangle, \qquad l \in \mathcal{V}^*.$$

Now, Problem $\mathcal{P}$ is to minimize the functional

$$J(v) = G(\Lambda v) + \langle l, v \rangle \tag{7.4.1}$$

on the space $\mathcal{V}$.

In this case,

$$F^*(-\Lambda^* y^*) = \sup_{v \in \mathcal{V}} \langle -\Lambda^* y^* - l, v \rangle = \begin{cases} 0 & \text{if } y^* \in Q_l^*, \\ +\infty & \text{if } y^* \neq Q_l^*, \end{cases}$$

where

$$Q_l^* := \{ q^* \in Y^* \mid (q^*, \Lambda w) + \langle l, w \rangle = 0, \quad w \in \mathcal{V} \}.$$

Hence,

$$I^*(y^*) = \begin{cases} -G^*(y^*) & \text{if } y^* \in Q_l^*, \\ -\infty & \text{if } y^* \notin Q_l^* \end{cases}$$

and the respective Problem $\mathcal{P}^*$ is to find $p^* \in Q_l^*$ such that the functional $-G^*(p^*)$ attains its supremum on Q_l^*. Recall that G and $-G^*$ are convex, continuous and coercive on $\mathcal{V}$ and Y^*, respectively. Hence, we can prove the existence of a minimizer u and a maximizer p^* that satisfy the relation

$$I^*(p^*) = J(u). \tag{7.4.2}$$

Let v be an element of $\mathcal{V}$ viewed as an approximation of u. Consider the difference

$$J(v) - J(u) = J(v) - I^*(p^*) = G(\Lambda v) + G^*(p^*) + \langle l, v \rangle.$$

Since $p^* \in Q_l^*$, we have

$$\langle l, v \rangle = -(p^*, \Lambda v)$$

and

$$J(v) - J(u) = G(\Lambda v) + G^*(p^*) - (p^*, \Lambda v) = D_G(\Lambda v, p^*). \tag{7.4.3}$$

Next, if $q^* \in Q_l^*$, then

$$J(v) - I^*(q^*) = G(\Lambda v) + G^*(q^*) - (q^*, \Lambda v) = D_G(\Lambda v, q^*). \qquad (7.4.4)$$

From (7.4.3) and (7.4.4) (with $v = u$) it follows that

$$\boxed{D_G(\Lambda v, p^*) + D_G(\Lambda u, q^*) = J(v) - I^*(q^*).} \qquad (7.4.5)$$

In view of (7.4.4), this relation has another form:

$$\boxed{D_G(\Lambda v, p^*) + D_G(\Lambda u, q^*) = D_G(\Lambda v, q^*).} \qquad (7.4.6)$$

These relations are particular forms of (7.2.13) and (7.2.14). Indeed,

$$D_F(v, -\Lambda^* q^*) = \begin{cases} \langle l, v \rangle + \langle \Lambda^* q^*, v \rangle = 0 & \text{if } q^* \in Q_l^*, \\ +\infty & \text{if } q^* \notin Q_l^*. \end{cases}$$

Therefore,

$$\mathcal{M}_\oplus(v, q^*) = \tfrac{1}{2} D_G(v, \Lambda v)$$

and (7.4.5) follows from (7.2.13) and (7.4.6) from (7.2.14).

Also, the relations (6.1.10) and (6.1.11) of the linear theory are particular cases of (7.4.5) and (7.4.6). In this case,

$$G(y) = \tfrac{1}{2}(\mathcal{A}y, y), \qquad G^*(y^*) = \tfrac{1}{2}(\mathcal{A}^{-1}y^*, y^*)$$
$$p^* = \mathcal{A}\Lambda u, \qquad \Lambda u = \mathcal{A}^{-1}p^*.$$

By (7.4.5), we obtain the identity

$$D_G(\Lambda v, p^*) + D_G(\Lambda u, q^*) = D_G(\Lambda v, \mathcal{A}\Lambda u) + D_G(\mathcal{A}^{-1}p^*, q^*)$$
$$= \tfrac{1}{2} \parallel \Lambda(v - u) \parallel^2 + \tfrac{1}{2} \parallel p^* - q^* \parallel_*^2,$$

which implies (6.1.10) and (6.1.11).

Remark 7.4.1. By (7.4.5) and the relation

$$D_G(\Lambda u, p^*) = 0,$$

we deduce two important formulas:

$$D_G(\Lambda v, p^*) = J(v) - I^*(p^*), \qquad (7.4.7)$$
$$D_G(\Lambda u, q^*) = J(u) - I^*(q^*). \qquad (7.4.8)$$

They will be used in the sequel to derive two-sided estimates of the quantities $D_G(\Lambda v, p^*)$ and $D_G(\Lambda u, q^*)$.

7.5 Properties of compound functionals

This section is concerned with properties of compound functionals. We begin by recalling the fact proved in Chapter 6 (cf. Proposition 6.3.2): the compound functional

$$D_G(y, y^*) = G(y) + G^*(y^*) - (y^*, y)$$

is nonnegative and equals zero if and only if

$$y^* \in \partial G(y) \quad \text{and} \quad y \in \partial G^*(y^*).$$

For differentiable functionals, the latter relation reads

$$y^* = G'(y) \quad \text{and} \quad y = G^{*\prime}(y^*).$$

Therefore, the quantity $D_G(\Lambda v, p^*)$ can be viewed as a measure of the distance between the element $G'(\Lambda v) \in Y^*$ (or the set $\partial G(\Lambda v) \in Y^*$) and the exact solution $p^* \in Y^*$ of Problem $\mathcal{P}^*$. We recall that this distance is zero if v coincides with the minimizer u.

Analogously, for any $q^* \in Q_l^*$, the quantity $D_G(\Lambda u, q^*)$ can be viewed as a measure of the distance between the element $G^{*\prime}(q^*) \in Y$ (or the set $\partial G^*(q^*) \in Y$) and the element $\Lambda u \in Y$. This distance is zero if $q^* = p^*$.

In the general estimate (7.2.3), the majorant $\mathcal{M}_\oplus$ is formed by two compound functionals that measure mismatches in the duality relations (7.1.11) and (7.1.12). Similar functionals also arise in a posteriori error estimates for variational inequalities (see Chapter 8). For nonlinear variational problems, compound and forcing functionals probably represent quantities naturally adapted for the use in a posteriori error estimation. This fact motivates our interest to a deeper investigation of their properties.

First, we note that, in general, $D_G(y, y^*)$ is a nonconvex functional. This fact is easily observed in the simplest case $Y = \mathbb{R}$. Set $G(y) = \frac{1}{\alpha}|y|^\alpha$. Then, $G(y) = \frac{1}{\alpha^*}|y|^{\alpha^*}$ (see Example 5.2.2). For $\alpha = 3$ and $\alpha = 1.2$, the respective compound functionals $D(y, y^*)$ are pictured in Figs. 7.5.1 and 7.5.2. By the form of level lines, we observe that these functionals are not convex. However, they have an important property, which is to some extent similar to convexity.

Proposition 7.5.1. *For any $y_1, y_2 \in Y$ and $y_1^*, y_2^* \in Y^*$,*

$$D_G\left(\tfrac{y_1+y_2}{2}, \tfrac{y_1^*+y_2^*}{2}\right) \leq \tfrac{1}{4}\Big(D_G(y_1, y_1^*) + D_G(y_1, y_2^*) +$$

$$+ D_G(y_2, y_1^*) + D_G(y_2, y_2^*)\Big) \quad (7.5.1)$$

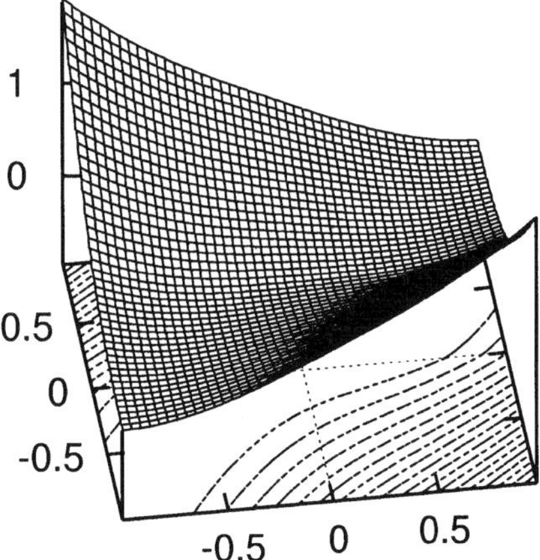

Figure 7.5.1: Compound functional and its level lines for $\alpha = 3$.

Proof. From the definition it follows that

$$D_G\left(y, \tfrac{y_1^* + y_2^*}{2}\right) = \ G(y) + G^*\left(\tfrac{y_1^* + y_2^*}{2}\right) - \left(\tfrac{y_1^* + y_2^*}{2}, y\right)$$
$$\leq \tfrac{1}{2}\left(D_G(y, y_1^*) + D_G(y, y_2^*)\right)$$

and

$$D_G\left(\tfrac{y_1 + y_2}{2}, y^*\right) = \ G\left(\tfrac{y_1 + y_2}{2}\right) + G^*(y^*) - \left(y^*, \tfrac{y_1 + y_2}{2}\right)$$
$$\leq \tfrac{1}{2}\left(D_G(y_1, y^*) + D_G(y_2, y^*)\right).$$

Therefore,

$$D_G\left(\tfrac{y_1 + y_2}{2}, \tfrac{y_1^* + y_2^*}{2}\right) \leq \tfrac{1}{2}\left(D_G(y_1, \tfrac{y_1^* + y_2^*}{2}) + D_G(y_2, \tfrac{y_1^* + y_2^*}{2})\right).$$

It is easy to see that (7.5.1) follows from this relation. $\qquad\square$

Henceforth, we assume that the forcing functionals that we use are even (i.e., $\Upsilon(y) = \Upsilon(-y)$).

Proposition 7.5.2. *1. If G satisfies the condition*

$$\Upsilon_\ominus\left(\tfrac{y_1 - y_2}{2}\right) \leq \tfrac{1}{2}(G(y_1) + G(y_2)) - G\left(\tfrac{y_1 + y_2}{2}\right), \tag{7.5.2}$$

where y_1 and y_2 are arbitrary elements of Y and $\Upsilon_\ominus$ is a functional in $\Gamma_0^+(Y)$ (cf. (5.5.1–(5.5.2)), then G^* satisfies the condition

$$\Upsilon_\ominus^*\left(\tfrac{y_1^* - y_2^*}{2}\right) \geq \tfrac{1}{2}(G^*(y_1^*) + G^*(y_2^*)) - G^*\left(\tfrac{y_1^* + y_2^*}{2}\right), \tag{7.5.3}$$

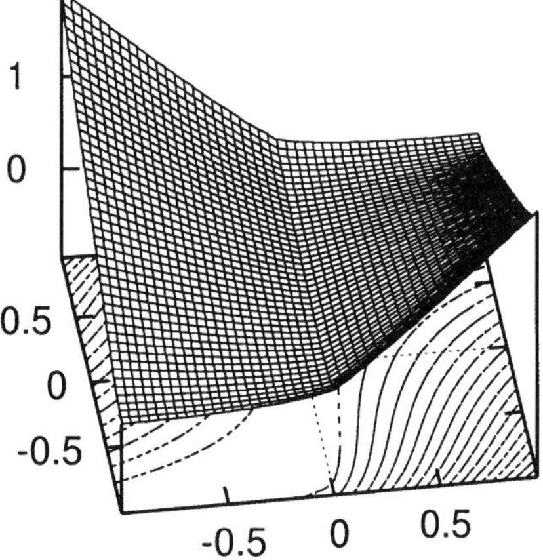

Figure 7.5.2: Compound functional and its level lines for $\alpha = 1.2$.

where y_1^* and y_2^* are arbitrary elements of Y^* and

$$\Upsilon_\ominus^*(y^*) = \sup_{y \in Y}\{(y^*, y) - \Upsilon_\ominus(y)\}$$

2. If G satisfies the condition

$$\Upsilon_\oplus\left(\tfrac{y_1-y_2}{2}\right) \geq \tfrac{1}{2}(G(y_1) + G(y_2)) - G\left(\tfrac{y_1+y_2}{2}\right), \qquad (7.5.4)$$

where y_1 and y_2 are arbitrary elements of Y and $\Upsilon_\oplus$ is a functional in $\Gamma_0^+(Y)$, then G^* satisfies the condition

$$\Upsilon_\oplus^*\left(\tfrac{y_1^*-y_2^*}{2}\right) \leq \tfrac{1}{2}(G^*(y_1^*) + G^*(y_2^*)) - G^*\left(\tfrac{y_1^*+y_2^*}{2}\right), \qquad (7.5.5)$$

where y_1^* and y_2^* are arbitrary elements of Y^* and

$$\Upsilon_\oplus^*(y^*) = \sup_{y \in Y}\{(y^*, y) - \Upsilon_\oplus(y)\}.$$

Proof. 1. Let y_1^* and y_2^* be two elements of Y^*. Consider the quantity

$$H_1 := \tfrac{1}{2}\left((y_1^*, y_1) - G(y_1)\right) + \tfrac{1}{2}\left((y_2^*, y_2) - G(y_2)\right) - G^*\left(\tfrac{y_1^*+y_2^*}{2}\right),$$

where y_1 and y_2 are arbitrary elements of Y. For any $y \in Y$, we have

$$\begin{aligned}
H_1 \leq \quad & \tfrac{1}{2}\left((y_1^*, y_1) - G(y_1)\right) + \tfrac{1}{2}\left((y_2^*, y_2) - G(y_2)\right) \\
& + G(y) - \tfrac{1}{2}(y_1^* + y_2^*, y).
\end{aligned}$$

Set $y_1 = y + z$ and $y_2 = y - z$, where z is an arbitrary element of Y. By virtue of (7.5.2) we have

$$G(y) \leq \tfrac{1}{2}G(y + z) + \tfrac{1}{2}G(y - z) - \Upsilon_\ominus(z),$$

and, consequently,

$$H_1 \leq \tfrac{1}{2}(y_1^* - y_2^*, z) - \Upsilon_\ominus(z) \leq \Upsilon_\ominus^*\big(\tfrac{y_1^* - y_2^*}{2}\big).$$

Thus,

$$\tfrac{1}{2}\big((y_1^*, y_1) - G(y_1)\big) + \tfrac{1}{2}\big((y_2^*, y_2) - G(y_2)\big) \leq \Upsilon_\ominus^*\big(\tfrac{y_1^* - y_2^*}{2}\big) + G^*\big(\tfrac{y_1^* + y_2^*}{2}\big).$$

Taking supremum with respect to y_1 and y_2, we arrive at (7.5.3).

2. Consider the quantity

$$H_2 := \tfrac{1}{2}G^*(y_1^*) + \tfrac{1}{2}G^*(y_2^*) + G(y) - \tfrac{1}{2}(y_1^* + y_2^*, y),$$

where $y_1^*, y_1^* \in Y^*$ and y is an arbitrary element of Y. Obviously,

$$G^*(y_1^*) \geq (y_1^*, y + z) - G(y + z), \quad \forall z \in Y,$$
$$G^*(y_2^*) \geq (y_2^*, y - z) - G(y - z), \quad \forall z \in Y.$$

By (7.5.4), we see that

$$G(y) = G\big(\tfrac{y + z + y - z}{2}\big) \geq \tfrac{1}{2}G(y + z) + \tfrac{1}{2}G(y - z) - \Upsilon_\oplus(z).$$

Therefore, we estimate H_2 as follows:

$$H_2 \geq \tfrac{1}{2}(y_1^* - y_2^*, z) - \Upsilon_\oplus(z).$$

This estimate holds for all z, so that

$$H_2 \geq \Upsilon_\oplus^*\big(\tfrac{y_1^* - y_2^*}{2}\big),$$

and, consequently,

$$\tfrac{1}{2}G^*(y_1^*) + \tfrac{1}{2}G^*(y_2^*) \geq \Upsilon_\oplus^*\big(\tfrac{y_1^* - y_2^*}{2}\big) + \tfrac{1}{2}(y_1^* + y_2^*, y) - G(y).$$

Take here supremum with respect to y. This gives the required estimate

$$\tfrac{1}{2}G^*(y_1^*) + \tfrac{1}{2}G^*(y_2^*) \geq \Upsilon_\oplus^*\big(\tfrac{y_1^* - y_2^*}{2}\big) + G^*\big(\tfrac{y_1^* + y_2^*}{2}\big).$$

$\square$

Quite analogously, one can obtain a similar proposition for the functional G^*.

Proposition 7.5.3.   *1. If G^* satisfies the condition*

$$_*\Upsilon_\ominus\left(\tfrac{y_1^*-y_2^*}{2}\right) \le \tfrac{1}{2}(G^*(y_1^*) + G^*(y_2^*)) - G^*\left(\tfrac{y_1^*+y_2^*}{2}\right), \qquad (7.5.6)$$

where y_1^* and y_2^* are arbitrary elements of Y^*, then G^{**} satisfies the condition

$$_*\Upsilon_\ominus^*\left(\tfrac{y_1-y_2}{2}\right) \ge \tfrac{1}{2}(G^{**}(y_1) + G^{**}(y_2)) - G^{**}\left(\tfrac{y_1+y_2}{2}\right), \qquad (7.5.7)$$

where y_1 and y_2 are arbitrary elements of Y and

$$_*\Upsilon_\ominus^*(y) = \sup_{y^*\in Y^*} \{(y^*,y) - \Upsilon_{*\ominus}(y^*)\}$$

2. If G^* satisfies the condition

$$_*\Upsilon_\oplus\left(\tfrac{y_1^*-y_2^*}{2}\right) \ge \tfrac{1}{2}(G^*(y_1^*) + G^*(y_2^*)) - G^*\left(\tfrac{y_1^*+y_2^*}{2}\right), \qquad (7.5.8)$$

where y_1^* and y_2^* are arbitrary elements of Y^*, then G^{**} satisfies the condition

$$_*\Upsilon_\oplus^*\left(\tfrac{y_1-y_2}{2}\right) \le \tfrac{1}{2}(G^{**}(y_1) + G^{**}(y_2)) - G^{**}\left(\tfrac{y_1+y_2}{2}\right), \qquad (7.5.9)$$

where y_1 and y_2 are arbitrary elements of Y and

$$_*\Upsilon_\oplus^*(y) = \sup_{y^*\in Y^*} \{(y^*,y) - {}_*\Upsilon_\oplus(y^*)\}.$$

Corollary 7.5.1. If $G \in \mathbb{CF}$ and G coincides with G^{**}, then (7.5.9) implies the estimate

$$_*\Upsilon_\ominus^*\left(\tfrac{y_1-y_2}{2}\right) \ge \tfrac{1}{2}(G(y_1) + G(y_2)) - G\left(\tfrac{y_1+y_2}{2}\right), \qquad (7.5.10)$$

where y_1 and y_2 are arbitrary elements of Y. Similarly (7.5.8) implies the estimate

$$_*\Upsilon_\oplus^*\left(\tfrac{y_1-y_2}{2}\right) \le \tfrac{1}{2}(G(y_1) + G(y_2)) - G\left(\tfrac{y_1+y_2}{2}\right). \qquad (7.5.11)$$

Remark 7.5.1. Define the functionals $\overline{\Upsilon}_\oplus(z)$ and $\overline{\Upsilon}_\ominus(z)$ by the relations

$$\overline{\Upsilon}_\ominus(z) := \inf_{y\in Y} \left\{\tfrac{1}{2}\left(G(y+z) + G(y-z)\right) - G(y)\right\},$$

$$\overline{\Upsilon}_\oplus(z) := \sup_{y\in Y} \left\{\tfrac{1}{2}\left(G(y+z) + G(y-z)\right) - G(y)\right\}.$$

For a convex functional G, these two functionals are nonnegative and vanish at 0_Y. Set $z = \frac{y_1 - y_2}{2}$ and $y = \frac{y_1 + y_2}{2}$. We see that $\overline{\Upsilon}_\oplus$ and $\overline{\Upsilon}_\ominus$ satisfy (7.5.2) and (7.5.4). Therefore, $\overline{\Upsilon}_\ominus$ is a forcing functional for G that provides the uniform convexity in any ball. Of course, for certain functionals $\overline{\Upsilon}_\ominus$ may be equal to zero. However, if the infimum is computed for $y \in \mathfrak{B} \subset Y$, then the respective forcing functional provides the uniform convexity of G in the ball $\mathfrak{B}$ with a nontrivial forcing functional. Note that the functionals $\overline{\Upsilon}_\oplus^*$ and $\overline{\Upsilon}_\ominus^*$ satisfy (7.5.3) and (7.5.5).

If G is the integral functional, i.e.,

$$G(y) = \int_\Omega g(y(x))\, dx,$$

where $g : \mathbb{R}^n \to \mathbb{R}$ is a given integrand, then $\overline{\Upsilon}_\oplus$ and $\overline{\Upsilon}_\ominus$ can be computed directly. Indeed, for any $\zeta \in \mathbb{R}^n$ we define the functions

$$\overline{\gamma}_\ominus(\zeta) := \inf_{\xi \in \mathbb{R}^n} \left\{ \tfrac{1}{2}\left(g(\xi + \zeta) + g(\xi - \zeta) \right) - g(\xi) \right\},$$

$$\overline{\gamma}_\oplus(\zeta) := \sup_{\xi \in \mathbb{R}^n} \left\{ \tfrac{1}{2}\left(g(\xi + \zeta) + g(\xi - \zeta) \right) - g(\xi) \right\}.$$

In this case,

$$\overline{\Upsilon}_\oplus(y) = \int_\Omega \overline{\gamma}_\oplus(y(x))\, dx,$$

$$\overline{\Upsilon}_\ominus(y) = \int_\Omega \overline{\gamma}_\ominus(y(x))\, dx,$$

so that the determination of $\overline{\Upsilon}_\oplus$ and $\overline{\Upsilon}_\ominus$ is reduced to an algebraic procedure. For example, if $g(\xi) = \tfrac{1}{2}\xi^2$, then

$$\tfrac{1}{2}\left(g(\xi + \zeta) + g(\xi - \zeta) \right) - g(\xi) = \tfrac{1}{2}\zeta^2$$

and $\overline{\Upsilon}_\ominus = \overline{\Upsilon}_\oplus = \tfrac{1}{2}\zeta^2$. In Fig. 7.5.3 we present scalar functions $\overline{\gamma}_\ominus(\xi)$ for the functional $g(\xi) = \frac{1}{\alpha}|\xi|^\alpha$ with $\alpha = 3$ and $\overline{\gamma}_\oplus(\xi)$ with $\alpha = 1.2$. Analogously, we define $_*\overline{\Upsilon}_\oplus(z)$ and $_*\overline{\Upsilon}_\ominus(z)$ by the relations

$$_*\overline{\Upsilon}_\ominus(z^*) := \inf_{y^* \in Y^*} \left\{ \tfrac{1}{2}\left(G^*(y^* + z^*) + G^*(y^* - z^*) \right) - G^*(y^*) \right\},$$

$$_*\overline{\Upsilon}_\oplus(z^*) := \sup_{y^* \in Y^*} \left\{ \tfrac{1}{2}\left(G^*(y^* + z^*) + G^*(y^* - z^*) \right) - G^*(y^*) \right\}.$$

It is clear that $_*\overline{\Upsilon}_\ominus$ and $_*\overline{\Upsilon}_\oplus$ satisfy (7.5.6) and (7.5.8) and, therefore, $_*\overline{\Upsilon}_\ominus^*$ and $_*\overline{\Upsilon}_\oplus^*$ satisfy (7.5.7) and (7.5.9).

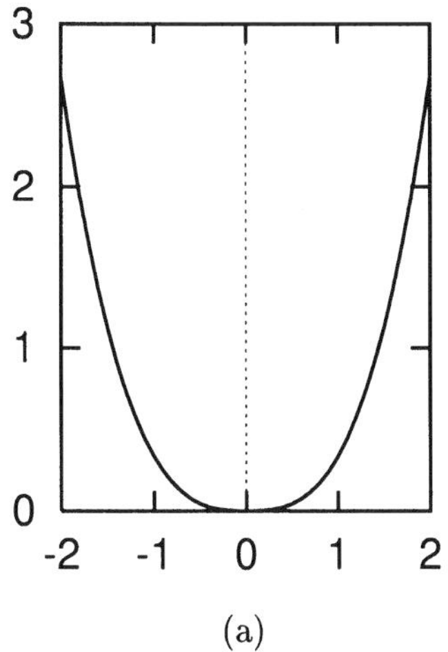

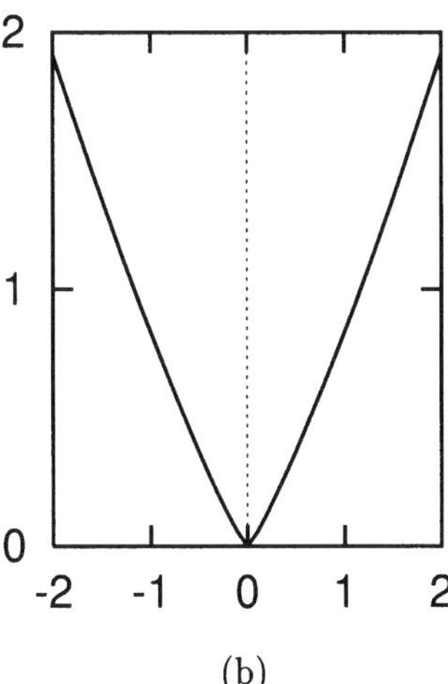

(a)   (b)

Figure 7.5.3: (a) $\overline{\gamma}_{\ominus}$ for $\alpha = 3$, (b) $\overline{\gamma}_{\oplus}$ for $\alpha = 1.2$.

Proposition 7.5.4.   *1. There the following relation holds:*

$$\overline{\Upsilon}_{\ominus}\left(\tfrac{y_1-y_2}{2}\right) + {}_*\overline{\Upsilon}_{\ominus}\left(\tfrac{y_1^*-y_2^*}{2}\right) \leq \left(\tfrac{y_1^*-y_2^*}{2}, \tfrac{y_1-y_2}{2}\right)$$
$$+ \tfrac{1}{2}\left(D_G(y_1, y_1^*) + D_G(y_2, y_2^*)\right), \quad (7.5.12)$$

where y_1 and y_2 are arbitrary elements of Y and y_1^* and y_2^* are arbitrary elements of Y^*.

2. The relation

$$\overline{\Upsilon}_{\oplus}\left(\tfrac{y_1-y_2}{2}\right) + {}_*\overline{\Upsilon}_{\oplus}\left(\tfrac{y_1^*-y_2^*}{2}\right) + D_G\left(\tfrac{y_1+y_2}{2}, \tfrac{y_1^*+y_2^*}{2}\right) \geq \left(\tfrac{y_1^*-y_2^*}{2}, \tfrac{y_1-y_2}{2}\right)$$
$$+ \tfrac{1}{2}\left(D_G(y_1, y_1^*) + D_G(y_2, y_2^*)\right), \quad (7.5.13)$$

holds, where y_1 and y_2 are arbitrary elements of Y and y_1^* and y_2^* are arbitrary elements of Y^*.

Proof. 1. We have

$$\overline{\Upsilon}_{\ominus}\left(\tfrac{y_1-y_2}{2}\right) + G\left(\tfrac{y_1+y_2}{2}\right) \leq \tfrac{1}{2}(G(y_1) + G(y_2)),$$
$${}_*\overline{\Upsilon}_{\ominus}\left(\tfrac{y_1^*-y_2^*}{2}\right) + G^*\left(\tfrac{y_1^*+y_2^*}{2}\right) \leq \tfrac{1}{2}(G^*(y_1^*) + G^*(y_2^*)),$$

and

$$G\left(\tfrac{y_1+y_2}{2}\right) + G^*\left(\tfrac{y_1^*+y_2^*}{2}\right) \geq \left(\tfrac{y_1^*+y_2^*}{2}, \tfrac{y_1+y_2}{2}\right).$$

Therefore,

$$\overline{\Upsilon}_{\ominus}\left(\tfrac{y_1-y_2}{2}\right) + {}_*\overline{\Upsilon}_{\ominus}\left(\tfrac{y_1^*-y_2^*}{2}\right) + \left(\tfrac{y_1^*+y_2^*}{2}, \tfrac{y_1+y_2}{2}\right)$$
$$\leq \tfrac{1}{2}\left(D_G(y_1,y_1^*) + D_G(y_2,y_2^*)\right) + \tfrac{1}{2}\left((y_1^*,y_1) + (y_2^*,y_2)\right),$$

which is equivalent to (7.5.15).

2. We have

$$\overline{\Upsilon}_{\oplus}\left(\tfrac{y_1-y_2}{2}\right) + G\left(\tfrac{y_1+y_2}{2}\right) \geq \tfrac{1}{2}(G(y_1) + G(y_2)),$$
$$_*\overline{\Upsilon}_{\oplus}\left(\tfrac{y_1^*-y_2^*}{2}\right) + G^*\left(\tfrac{y_1^*+y_2^*}{2}\right) \geq \tfrac{1}{2}(G^*(y_1^*) + G^*(y_2^*)),$$

and

$$\tfrac{1}{2}\left(G(y_1) + G(y_2) + G^*(y_1^*) + G^*(y_2^*)\right)$$
$$= \tfrac{1}{2}\left(D_G(y_1,y_1^*) + D_G(y_2,y_2^*) + (y_1^*,y_1) + (y_2^*,y_2)\right)$$

Hence,

$$\overline{\Upsilon}_{\oplus}\left(\tfrac{y_1-y_2}{2}\right) + {}_*\overline{\Upsilon}_{\oplus}\left(\tfrac{y_1^*-y_2^*}{2}\right) + D_G\left(\tfrac{y_1+y_2}{2}, \tfrac{y_1^*+y_2^*}{2}\right)$$
$$\geq \tfrac{1}{4}\left(y_1^* - y_2^*, y_1 - y_2\right) + \tfrac{1}{2}\left(D_G(y_1,y_1^*) + D_G(y_2,y_2^*)\right).$$

$$\square$$

Corollary 7.5.2. From (7.5.12) it follows that

$$_*\overline{\Upsilon}_{\ominus}\left(\tfrac{y_1^*-y_2^*}{2}\right) \leq \tfrac{1}{2}\left(D_G(y,y_1^*) + D_G(y,y_2^*)\right), \quad \forall y \in Y, \qquad (7.5.14)$$

$$\overline{\Upsilon}_{\ominus}\left(\tfrac{y_1-y_2}{2}\right) \leq \tfrac{1}{2}\left(D_G(y_1,y^*) + D_G(y_2,y^*)\right) \quad \forall y^* \in Y^*. \qquad (7.5.15)$$

Corollary 7.5.3. If $y_1^* \in \partial G(y_1)$ and $y_2^* \in \partial G(y_2)$, then

$$\overline{\Upsilon}_{\ominus}\left(\tfrac{y_1-y_2}{2}\right) + {}_*\overline{\Upsilon}_{\ominus}\left(\tfrac{y_1^*-y_2^*}{2}\right) \leq \left(\tfrac{y_1^*-y_2^*}{2}, \tfrac{y_1-y_2}{2}\right)$$
$$\leq {}_*\overline{\Upsilon}_{\oplus}\left(\tfrac{y_1^*-y_2^*}{2}\right) + \overline{\Upsilon}_{\oplus}\left(\tfrac{y_1-y_2}{2}\right) + D_G\left(\tfrac{y_1^*+y_2^*}{2}, \tfrac{y_1+y_2}{2}\right).$$

Proposition 7.5.5. *For any $y \in Y$ and $y^* \in Y^*$, the following estimates hold:*

$$_*\overline{\Upsilon}_{\ominus}\left(\tfrac{y^*-z^*}{2}\right) \leq \tfrac{1}{2}D_G(y,y^*), \qquad (7.5.16)$$
$$\overline{\Upsilon}_{\ominus}\left(\tfrac{y-z}{2}\right) \leq \tfrac{1}{2}D_G(y,y^*), \qquad (7.5.17)$$
$$(y^*-z^*, z-y) \geq D_G(y,y^*), \qquad (7.5.18)$$

where $z \in \partial G^*(y^*)$, and $z^* \in \partial G(y)$.

Proof. Estimates (7.5.16) and (7.5.17) follow from (7.5.14) and (7.5.15).

To prove (7.5.18), we note that $G^*(z^*) + G(y) - (z^*, y) = 0$. Therefore,

$$D_G(y, y^*) = G^*(y^*) - G^*(z^*) - (y^* - z^*, y) \leq (y^* - z^*, z) - (y^* - z^*, y)$$

which implies the estimate (7.5.18). $\qquad\Box$

If G and G^* are Gateaux differentiable, then the estimates (7.5.16)–(7.5.18) take a more transparent form:

$$_*\overline{\Upsilon}_\ominus \left(\tfrac{y^* - G'(y)}{2} \right) \ \leq \ \tfrac{1}{2} D_G(y, y^*), \qquad (7.5.19)$$

$$\overline{\Upsilon}_\ominus \left(\tfrac{y - G^{*\prime}(y^*)}{2} \right) \ \leq \ \tfrac{1}{2} D_G(y, y^*), \qquad (7.5.20)$$

$$(y^* - G'(y), G^{*\prime}(y^*) - y) \ \geq \ D_G(y, y^*). \qquad (7.5.21)$$

These estimates lead to two-sided estimates of the difference between the exact solutions of Problems $\mathcal{P}$ and $\mathcal{P}^*$ and their approximations.

7.6 Two-sided error estimates

Let us apply the estimates (7.5.19)–(7.5.21) to the problem (7.4.1). Set $y = \Lambda v$ and $z^* \in \partial G(\Lambda v)$. Then, we find that

$$_*\overline{\Upsilon}_\ominus \left(\tfrac{p^* - z^*}{2} \right) \leq \tfrac{1}{2} D_G(\Lambda v, p^*), \qquad (7.6.1)$$

$$\overline{\Upsilon}_\ominus \left(\tfrac{\Lambda(v - u)}{2} \right) \leq \tfrac{1}{2} D_G(\Lambda v, p^*), \qquad (7.6.2)$$

$$(p^* - z^*, \Lambda u - \Lambda v) \geq D_G(\Lambda v, p^*). \qquad (7.6.3)$$

If G is differentiable, then these relations can be represented as follows:

$$_*\overline{\Upsilon}_\ominus \left(\tfrac{G'(\Lambda u) - G'(\Lambda v)}{2} \right) \leq \tfrac{1}{2} D_G(\Lambda v, p^*), \qquad (7.6.4)$$

$$\overline{\Upsilon}_\ominus \left(\tfrac{\Lambda(v - u)}{2} \right) \leq \tfrac{1}{2} D_G(\Lambda v, p^*), \qquad (7.6.5)$$

$$(G'(\Lambda u) - G'(\Lambda v), \Lambda u - \Lambda v) \geq D_G(\Lambda v, p^*). \qquad (7.6.6)$$

By (7.4.7) and (7.6.5), we obtain estimates of the difference between Λv and Λu in terms of the functionals $\overline{\Upsilon}_\ominus$ and $_*\overline{\Upsilon}_\ominus$:

$$\overline{\Upsilon}_\ominus \left(\tfrac{\Lambda(v - u)}{2} \right) \leq \frac{1}{2}(J(v) - I^*(q^*)), \quad \forall q^* \in Q_l^*, \quad (7.6.7)$$

$$_*\overline{\Upsilon}_\ominus \left(\tfrac{G'(\Lambda u) - G'(\Lambda v)}{2} \right) \leq \frac{1}{2}(J(v) - I^*(q^*)). \qquad (7.6.8)$$

Now we set $y = \Lambda u$ and $y^* = q^* \in Q_l^*$, $z \in \partial G^*(q^*)$, and $z^* = p^* \in \partial G(\Lambda u)$. Then (7.5.16)–(7.5.18) yield

$$\overline{\Upsilon}_\ominus^* \left(\tfrac{q^*-p^*}{2} \right) \le \tfrac{1}{2} D_G(\Lambda u, q^*), \tag{7.6.9}$$

$$\overline{\Upsilon}_\ominus \left(\tfrac{\Lambda(u-z)}{2} \right) \le \tfrac{1}{2} D_G(\Lambda u, q^*), \tag{7.6.10}$$

$$(q^* - p^*, z - \Lambda u) \ge D_G(\Lambda u, q^*). \tag{7.6.11}$$

If G^* is differentiable, then we have

$$_*\overline{\Upsilon}_\ominus \big(\tfrac{q^*-p^*}{2} \big) \le \tfrac{1}{2} D_G(\Lambda u, q^*), \tag{7.6.12}$$

$$\overline{\Upsilon}_\ominus \left(\tfrac{G^{*\prime}(p^*)-G^{*\prime}(q^*)}{2} \right) \le \tfrac{1}{2} D_G(\Lambda u, q^*), \tag{7.6.13}$$

$$(G^{*\prime}(q^*) - G^{*\prime}(q^*), q^* - p^*) \ge \tfrac{1}{2} D_G(\Lambda u, q^*). \tag{7.6.14}$$

By (7.4.7) and (7.6.5), we obtain estimates of the difference between q^* and p^* in terms of the functionals $\overline{\Upsilon}_\ominus$ and $_*\overline{\Upsilon}_\ominus$:

$$_*\overline{\Upsilon}_\ominus \big(\tfrac{q^*-p^*}{2} \big) \le \frac{1}{2}(J(v) - I^*(p^*)), \quad \forall v \in V, \tag{7.6.15}$$

$$\overline{\Upsilon}_\ominus \left(\tfrac{G^{*\prime}(p^*)-G^{*\prime}(q^*)}{2} \right) \le \frac{1}{2}(J(v) - I^*(p^*)). \tag{7.6.16}$$

We see that for nonlinear problems, the differences between the values of the primal and dual functionals give upper bounds of the errors computed in terms of special functionals $\overline{\Upsilon}_\ominus$ and $_*\overline{\Upsilon}_\ominus$ that depend on the functional G. For the integral functionals they can explicitly be determined by the analysis of the respective integrands.

The estimates (7.4.7)–(7.4.8) and (7.6.1)–(7.6.12) lead to two-sided estimates of deviations from the exact solutions.

Let $\{V_{0k}\}$ and $\{Q_{lk}^*\}$ be two sequences of finite–dimensional subspaces, which are dense in V_0 and Q_l^*, respectively. Then

$$m_k \le D_G(\Lambda v, p^*) \le \mathcal{M}_k, \tag{7.6.17}$$

where

$$m_k := J(v) - \inf_{w \in V_{0k}+u_0} J(w),$$

$$\mathcal{M}_k := J(v) - \sup_{\eta^* \in Q_{lk}^*} I^*(\eta^*).$$

If J and I^* are continuous functionals, then the numbers m_k and $\mathcal{M}_k$ converge to the quantity $D_G(\Lambda v, p^*)$ that characterizes the distance between

Λv and Λu. By (7.6.4)–(7.6.6), we obtain

$$\overline{\Upsilon}_{\ominus}\left(\tfrac{G'(\Lambda u)-G'(\Lambda v)}{2}\right) \leq \tfrac{1}{2}\mathcal{M}_k \tag{7.6.18}$$

$$_*\Upsilon_{\ominus}\left(\tfrac{\Lambda(v-u)}{2}\right) \leq \tfrac{1}{2}\mathcal{M}_k, \tag{7.6.19}$$

$$(G'(\Lambda u) - G'(\Lambda v), \Lambda u - \Lambda v) \geq m_k. \tag{7.6.20}$$

Upper and lower bounds of the quantity $D_G(\Lambda u, q^*)$ are derived quite analogously:

$$m_k^* \leq D_G(\Lambda u, q^*) \leq \mathcal{M}_k^*, \tag{7.6.21}$$

where

$$m_k^* := \sup_{\eta^* \in Q_l^*} I^*(\eta^*) - I^*(q^*),$$

$$\mathcal{M}_k^* := \inf_{w \in V_{0k}+u_0} J(w) - I^*(q^*).$$

By (7.6.12)–(7.6.14), we obtain

$$\overline{\Upsilon}_{\ominus}\left(\tfrac{1}{2}\left(q^* - p^*\right)\right) \leq \tfrac{1}{2}\mathcal{M}_k^*, \tag{7.6.22}$$

$$_*\Upsilon_{\ominus}\left(\tfrac{1}{2}(\Lambda u - G^{*\prime}(q^*))\right) \leq \tfrac{1}{2}\mathcal{M}_k^*, \tag{7.6.23}$$

$$(q^* - p^*, G^{*\prime}(q^*) - \Lambda u) \geq m_k^*. \tag{7.6.24}$$

If it is known that u and its approximation v belong to a certain ball $\mathcal{B}_\delta \in Y$ (or p^* and its approximation q^* belong to $\mathcal{B}_\delta^*$), then a further revision of the actual accuracy can be obtained. For this purpose, the functionals

$$\overline{\Upsilon}_{\ominus}^{\delta}(z) = \inf_{y \in \mathcal{B}_\delta} \tfrac{1}{2}\left(G(y + z) + G(y - z)\right) - G(y)$$

and

$$_*\overline{\Upsilon}_{\ominus}^{\delta}(z) = \inf_{y^* \in \mathcal{B}_\delta^*} \tfrac{1}{2}\left(G^*(y^* + z^*) + G^*(y^* - z^*)\right) - G^*(y^*)$$

should be used. Obviously, the more a priori information on the properties of u (or p^*) we invoke, the larger $\overline{\Upsilon}_{\ominus}$ and $_*\overline{\Upsilon}_{\ominus}$ could be taken and, consequently, the sharper estimates could be obtained. At this point, a priori estimates of u (or p^*) known in the theory of partial differential equations (see, e.g., [132, 135]) can be effectively used. Such estimates imply the relations

$$u \in \mathcal{B}_\delta \subset V, \quad p^* \in \mathcal{B}_{\delta*}^* \subset Y^*,$$

where the numbers δ and δ^* are explicitly defined by the problem data. With account of this information, not only can one select the functionals $\overline{\Upsilon}_\ominus$ and $_*\overline{\Upsilon}_\ominus$, but also adjust the two-sided estimates as follows:

$$m_k := J(v) - \inf_{w \in (V_{0k}+u_0)\cap\mathcal{B}_\delta} J(w), \qquad (7.6.25)$$

$$\mathcal{M}_k := J(v) - \sup_{\eta^* \in Q^*_{lk}\cap\mathcal{B}^*_{\delta*}} I^*(\eta^*). \qquad (7.6.26)$$

To find $\mathcal{M}_k$ and m^*_k, we must use conforming approximations of the set Q^*_l. In general, such approximations are difficult to construct. In the next section, we suggest a way for overcoming this difficulty and obtain a computable estimate of $D_G(\Lambda v, p^*)$.

7.7   A class of nonlinear variational problems

Consider the problem (7.4.1) with

$$G(y) = \tfrac{1}{2}(\mathcal{A}y, y) + \Psi(y), \qquad (7.7.1)$$

where $\Psi : Y \to \mathbb{R}$ is a convex continuous functional. Note that if $\Psi \equiv 0$, then

$$\begin{aligned} D_G(\Lambda v, p^*) = \ & \tfrac{1}{2}(\mathcal{A}\Lambda v, \Lambda v) + \tfrac{1}{2}(\mathcal{A}\Lambda u, \Lambda u) - (\Lambda v, \mathcal{A}\Lambda u) \\ = \ & \tfrac{1}{2} \parallel \Lambda(v-u) \parallel^2, \end{aligned}$$

so that(7.4.7) leads to (6.1.10).

For the functional G, we have

$$\tfrac{1}{2}(G(y_1) + G(y_2)) - G\left(\tfrac{y_1+y_2}{2}\right) \geq \tfrac{1}{2} \parallel \tfrac{y_1-y_2}{2} \parallel^2, \qquad (7.7.2)$$

$$(G'(y_1) - G'(y_2), y_1 - y_2) \geq \parallel y_1 - y_2 \parallel^2 . \qquad (7.7.3)$$

In this case

$$\overline{\Upsilon}_\ominus(y) = \tfrac{1}{2} \parallel y \parallel^2, \quad {}_*\overline{\Upsilon}^*_\ominus(y^*) = \tfrac{1}{2} \parallel y^* \parallel^2_*,$$

and the estimate (7.6.7) has the form

$$\parallel \tfrac{\Lambda(v-u)}{2} \parallel^2 \leq J(v) - I^*(q^*). \qquad (7.7.4)$$

Here

$$\begin{aligned} J(v) - I^*(q^*) &= G(\Lambda v) + \langle l, v - u_0 \rangle + G^*(q^*) - (q^*, \Lambda u_0) \\ &= D_G(\Lambda v, y^*) + G^*(q^*) - G^*(y^*) + (y^* - q^*, \Lambda v), \quad (7.7.5) \end{aligned}$$

where y^* is an arbitrary element of Y^*. Since

$$G^*(q^*) - G^*(y^*) \le (q^* - y^*, G^{*\prime}(q^*)),$$

we obtain

$$\begin{aligned}
J(v) - I^*(q^*) \le \, & D_G(\Lambda v, y^*) \\
& + (q^* - y^*, G^{*\prime}(y^*) - \Lambda v) + (q^* - y^*, G^{\prime}(q^*) - G^{*\prime}(y^*)).
\end{aligned} \qquad (7.7.6)$$

By (7.7.4) and (7.7.6), we deduce the estimate

$$\|\!|\, \tfrac{\Lambda(v-u)}{2} \,|\!\|^2 \le m_D(\Lambda v, y^*) + m_R(y^*), \qquad (7.7.7)$$

in which

$$\begin{aligned}
m_D(\Lambda v, y^*) &= D_G(\Lambda v, y^*) + H(G^{*\prime}(y^*) - \Lambda v), \\
m_R(y^*) &= \inf_{q^* \in Q_l^*} \left((q^* - y^*, G^{*\prime}(q^*) - G^{*\prime}(y^*)) + H^*(q^* - y^*) \right),
\end{aligned}$$

$H \in \Gamma_0^+(Y)$, and $H^* \in \Gamma_0^+(Y^*)$ (H and H^* are two mutually conjugate functionals). It is easy to see that $m_D(\Lambda v, y^*) = 0$ if and only if Λv and y^* satisfy the *duality relation* $\Lambda v \in \partial G^*(y^*)$. Therefore, this term penalizes violations of this condition. The term $m_R(y^*)$ vanishes if and only if $y^* \in Q_l^*$. It penalizes violations of the second condition $\Lambda^* y^* + l = 0$. We remark that the functionals H and H^* may be chosen in many ways. Thus, (7.7.7) provides a series of various upper bounds of the error. Let us consider one particular variant. Take

$$H(y) = \overline{\Upsilon}_\ominus \left(\sqrt{\tfrac{\beta}{2}}\, y \right), \qquad H^*(y^*) = \overline{\Upsilon}_\ominus^* \left(\sqrt{\tfrac{2}{\beta}}\, y^* \right).$$

For any $\beta > 0$,

$$\begin{aligned}
(q^* - y^*, G^{*\prime}(y^*) - \Lambda v) \le \, & \Upsilon_\ominus \left(\sqrt{\tfrac{\beta}{2}}(G^{*\prime}(y^*) - \Lambda v) \right) \\
& + \Upsilon_\ominus^* \left(\sqrt{\tfrac{2}{\beta}}(q^* - y^*) \right).
\end{aligned}$$

Also, we have

$$\Upsilon_\ominus \left(\sqrt{\tfrac{\beta}{2}}(G^{*\prime}(y^*) - \Lambda v) \right) = 2\beta \Upsilon_\ominus \left(\tfrac{G^{*\prime}(y^*) - \Lambda v}{2} \right) \le \beta D_G(\Lambda v, y^*),$$

$$\Upsilon_\ominus^* \left(\sqrt{\tfrac{2}{\beta}}(q^* - y^*) \right) = \tfrac{1}{\beta} \|\!|\, q^* - y^* \,|\!\|_*^2 .$$

Let us estimate the quantity $(q^* - y^*, G^{*\prime}(q^*) - G^{*\prime}(y^*))$. From (7.7.3), we find that

$$\| G'(y_1) - G'(y_2) \|_* \geq \| y_1 - y_2 \| .$$

Set $y_1^* = G'(y_1)$ and $y_2^* = G'(y_2)$. If the functional Ψ has a nontrivial affine minorant, then $dom\, G^* = Y^*$ and G^* is differentiable at any point. In this case, we rewrite the inequality in the form

$$\| y_1^* - y_2^* \|_* \geq \| G^{*\prime}(y_1^*) - G^{*\prime}(y_2^*) \| .$$

Then

$$(q^* - y^*, G^{*\prime}(q^*) - G^{*\prime}(y^*)) \leq \| q^* - y^* \|_*^2,$$

and we conclude that

$$J(v) - I^*(q^*) \leq (1 + \beta) D_G(\Lambda v, y^*) + \left(1 + \tfrac{1}{\beta}\right) \| q^* - y^* \|_*^2 .$$

Now, (7.7.4) implies the estimate

$$\| \tfrac{\Lambda(v-u)}{2} \|^2 \leq (1 + \beta) D_G(\Lambda v, y^*) + \left(1 + \tfrac{1}{\beta}\right) \inf_{q^* \in Q_l^*} \| q^* - y^* \|_*^2 .$$

The second term on the right-hand side can be estimated by the norm of $\Lambda^* y^* + l$ in V^* (see Chaper 7), which leads to the estimate

$$\| \tfrac{1}{2}(\Lambda(v - u)) \|^2 \leq (1 + \beta) D_G(\Lambda v, y^*) + \left(1 + \tfrac{1}{\beta}\right) | \Lambda^* y^* + l |^2. \qquad (7.7.8)$$

It is easy to observe that this estimate and the estimate (6.2.3) are quite similar. Their right-hand sides consist of two terms associated with the errors in the duality relations and in the equation $\Lambda^* y^* + l = 0$. If $\Lambda^* y^* + l \in Y$, then the norm $|\cdot|$ can be estimated by the norm of Y, which leads to a fully computable upper bound.

It is worth noting that the general estimate (7.7.8) may fail to ensure the best upper bound of the error. To find a precise bound, we must analyze the functional of a particular problem and find proper functionals $\Upsilon_\ominus$ and H.

7.8   Comments

The method for deriving estimates of the deviations from exact solutions that is exposed in this chapter follows the lines of [168, 170, 174, 175]. Below, we briefly consider applications of this theory to some particular problems.

7.8.1 Power growth functionals

Let $G(y) = \frac{1}{\alpha} \int_\Omega |y|^\alpha \, dx$ and $F(v) = \int_\Omega fv dx$, where $\alpha > 1$. Then Problem $\mathcal{P}$ is to minimize the functional

$$I_\alpha(v) := \int_\Omega \left(\tfrac{1}{\alpha} |\nabla v|^\alpha + fv \right) dx$$

over the space $V = \{v \in H^\alpha(\Omega) \mid v = 0 \text{ on } \partial\Omega\}$. Problem $\mathcal{P}^*$ is to maximize the functional

$$I_{\alpha_*}^*(y^*) = -\tfrac{1}{\alpha_*} \int_\Omega |y^*|^{\alpha^*} \, dx$$

over the set

$$Q_l^* = \left\{ y^* \in Y^* := L^{\alpha_*}(\Omega, \mathbb{R}^n) \,\|\, \int_\Omega y^* \cdot \nabla w \, dx = \int_\Omega fw \, dx \; \forall w \in V \right\}.$$

The functional $G(y)$ is uniformly convex. For $\alpha \geq 2$ this fact follows from the first Clarkson's inequality (see [201])

$$\int_\Omega \left| \tfrac{y_1+y_2}{2} \right|^\alpha dx + \int_\Omega \left| \tfrac{y_1-y_2}{2} \right|^\alpha dx \leq \tfrac{1}{2} \int_\Omega \left(|y_1|^\alpha + |y_2|^\alpha \right) dx, \tag{7.8.1}$$

which is valid for all $y_1, y_2 \in Y$. Hence, we observe that in this case

$$\Upsilon_\ominus(z) = \frac{1}{\alpha} \|z\|_{\alpha,\Omega}^\alpha .$$

By (7.8.1), we obtain the basic estimate

$$\frac{1}{\alpha 2^\alpha} \int_\Omega |\nabla(v-u)|^\alpha \, dx \leq \frac{1}{2} \left(J_\alpha(v) - I_\alpha^*(q^*) \right), \; \forall q^* \in Q_l^*, \tag{7.8.2}$$

which is a particular form of (7.6.7). For this class of variational problems, computable majorants of approximation errors were derived in [171].

For $1 < \alpha \leq 2$, the functional J_α is also uniformly convex. This fact follows from the second Clarkson's inequality ([201])

$$\left(\int_\Omega \left(\frac{y_1+y_2}{2}\right)^\alpha dx\right)^{\frac{1}{\alpha-1}} + \left(\int_\Omega \left(\frac{y_1-y_2}{2}\right)^\alpha dx\right)^{\frac{1}{\alpha-1}}$$

$$\leq \left(\frac{1}{2}\int_\Omega (|y_1|^\alpha + |y_2|^\alpha)\,dx\right)^{\frac{1}{\alpha-1}}. \qquad (7.8.3)$$

However, in this case, the functional $\Upsilon_\ominus$ depends on the radius δ of a ball $\mathfrak{B}(0_Y,\delta)$ that contains y_1 and y_2 (see [146]), so that (7.2.1) holds with

$$\Upsilon_{\delta\ominus}(z) = \delta^{\frac{\alpha-2}{\alpha-1}}\kappa\,\|z\|_{\alpha,\Omega}^{\frac{\alpha}{\alpha-1}},$$

where $\kappa = \frac{1}{\kappa_0+1}$ and κ_0 is the integer part of $\frac{1}{\alpha-1}$.

7.8.2   Nonlinear elasticity

In Chapter 6, we derived a majorant of deviations from the exact solutions to boundary-value problems in linear elasticity (see (6.5.12)). Now, we consider a class of nonlinear variational problems asociated with the functional

$$J(v) = \int_\Omega g(\varepsilon(v))dx + \int_\Omega fv\,dx + \int_{\partial_2\Omega} Fv ds, \qquad (7.8.4)$$

where g (internal energy function) has the form

$$g(\varepsilon) = \delta \mathbb{L}\varepsilon : \varepsilon + \Phi(\varepsilon). \qquad (7.8.5)$$

In (7.8.5), δ is a positive real number and Φ is a convex function. A minimizer of J satisfies the relations (cf. (6.5.1)–(6.5.4))

$$\operatorname{div}\sigma^* = f, \qquad (7.8.6)$$

$$\sigma^* = \delta\mathbb{L}\varepsilon(u) + \Phi'(\varepsilon(v)), \qquad (7.8.7)$$

$$u = u_0 \qquad \text{on } \partial_1\Omega, \qquad (7.8.8)$$

$$\sigma^*\nu = F \qquad \text{on } \partial_2\Omega. \qquad (7.8.9)$$

Here, f and F are the volume and surface forces, respectively, ν is the outward unit normal vector to the boundary $\partial\Omega$, ε is the tensor of small strains. Equations of linear elasticity follow from (7.8.6)–(7.8.9) if we set $\delta = 1$ and $\Phi \equiv 0$. Other cases are connected with various models accepted nonlinear constitutive relations instead of the Hooke's law. For example, certain models arising in the deformation plasticity theory of isotropic solid bodies (see, e.g., [104, 100, 206]) are based on the constitutive relation

$$\sigma^* = K_0 \operatorname{tr}(\varepsilon)\mathbb{I} + \gamma(|\varepsilon^D(u(x))|)\,\varepsilon^D(u(x)), \qquad (7.8.10)$$

where

$$\gamma(t) = \begin{cases} 2\mu & \text{if } t \le t_0 = k_*/\sqrt{2}\mu, \\[2mm] (2\mu - \delta)t_0 t^{-1} + \delta & \text{if } t > t_0, \end{cases}$$

K_0 and μ are positive (elasticity) constants, $k_* > 0$ is a plasticity module, and $\delta > 0$ is a hardening module.

Assume that

$$f \in L^2(\Omega, \mathbb{R}^n), \quad F \in L^2(\Gamma_2\Omega, \mathbb{R}^n), \quad u_0 \in H^1(\Omega, \mathbb{R}^n).$$

The primal variational problem to find a displacement vector $u \in V_0 + u_0$ such that

$$J(u) = \inf_{v \in V_0 + u_0} J(v),$$

where the integrand g satisfies the relation

$$g(\varepsilon) = \frac{1}{2}a(\varepsilon, \varepsilon) + \mu_1 \phi(|\varepsilon^D|) \qquad (7.8.11)$$

with $\mu_1 = 1 - \delta/2\mu$,

$$a(\varepsilon_1, \varepsilon_2) = K_0 \operatorname{tr}(\varepsilon_1) \operatorname{tr}(\varepsilon_2) + \delta\varepsilon_1^D : \varepsilon_2^D, \quad \forall \varepsilon_1, \varepsilon_2 \in \mathbb{M}^{n \times n},$$

and

$$\phi(t) = \begin{cases} \mu\, t^2 & \text{if} |t| \le t_0 = k_*/\sqrt{2}\mu, \\[2mm] k_* \left(\sqrt{2}\,t - k_*/2\mu\right) & \text{if} |t| > t_0. \end{cases}$$

In view of (7.8.11), g has the form (7.8.5). It is easy to see that the functional (7.8.4), (7.8.5) is a particular case of the functional $G(\Lambda v) + F(v)$ and G lies in the class (7.7.1). Thus, deviations from the exact solutions of the nonlinear problems considered are subject to (7.7.8). The reader can find a more detailed analysis of a posteriori error estimates for nonlinear variational problems related to deformation theory of hardening elasto-plastic materials in [185, 186].

7.8.3 Nonconvex variational problems

Consider the functional

$$J(v) = \int_{\Omega} (g(\nabla v) - fv)\, dx, \qquad (7.8.12)$$

where the integrand $g : \mathbb{R}^n \to \mathbb{R}$ is a nonconvex function defined by the relations

$$g(\xi) = \min_{i=1,2}\{g_i(\xi)\}, \quad g_i(\xi) = \tfrac{1}{2}\mathrm{A}_i(\xi - a_i)\cdot(\xi - a_i) + \beta_i. \qquad (7.8.13)$$

In (7.8.13), $\mathrm{A}_i \in \mathbb{M}^{n\times n}$, $a_1, a_2 \in \mathbb{R}^n$, and β_1, β_2 are given real numbers.

It is assumed that the parameters a_1, a_2, β_1 and β_2 are defined in such a way that the function g coincides entirely neither with g_1 nor with g_2 and the matrices A_i, $i = 1, 2$, meet the conditions

$$\nu_i\,|\xi|^2 \le \mathrm{A}_i\xi\cdot\xi \le \mu_i\,|\xi|^2, \quad \forall \xi \in \mathbb{R}^n, \qquad (7.8.14)$$

where ν_i and μ_i are positive constants.

Variational problems with integrands of the type (7.8.13) arise in the theory of phase transitions in solids, where a specific nonconvex energy function is defined as a minimum of convex energy functions related to particular phases. In this case, the integrand g is a function of the small strain tensor (see, e.g., [21, 22, 120]).

Assume that $f \in L^2(\Omega)$ and $u_0 \in H^1(\Omega)$. Define the set of admissible functions

$$V_0 + u_0 = \{v \in V(\Omega) := H^1(\Omega) \mid v = u_0 + w, \ w \in V_0(\Omega)\},$$

where $V_0(\Omega) = \overset{\circ}{H}{}^1(\Omega)$. The problem of minimizing J over the set $V_0 + u_0$ is the primal Problem $\mathcal{P}$. Since the function g is nonconvex, the functional J is not sequentially lower semicontinuous on $V(\Omega)$ and, consequently, the set $V_0 + u_0$ may fail to contain a function u, which provides the exact lower bound of Problem $\mathcal{P}$. Therefore, in order to obtain a correct variational statement preserving this bound it is necessary and sufficient to make a lower semicontinuous regularization of the functional J. This procedure results in the so-called *relaxed variational problem*.

Problem $\mathcal{P}^{**}$. Find $u^{**} \in V_0 + u_0$ such that

$$J^{**}(u^{**}) = \inf_{v \in V_0 + u_0} J^{**}(v) := \inf \mathcal{P}^{**}, \qquad (7.8.15)$$

where

$$J^{**}(v) := \int_\Omega \left(g^{**}(\nabla v) - fv \right) dx.$$

$\mathcal{P}^{**}$ is a convex variational problem. However, J^{**} is not uniformly convex (see Example 5.2.7), so that one cannot apply our error estimation method directly to Problem $\mathcal{P}^{**}$. In [169], it was shown that in this case a duality error majorant can be derived for the dual variational problem $\mathcal{P}^*$: find a vector-valued function $p^* \in Y^*(\Omega) = L^2(\Omega, \mathbb{R}^n)$ such that

$$I^*(p^*) = \sup_{q^* \in Y^*(\Omega)} I^*(q^*) := \sup \mathcal{P}^*, \tag{7.8.16}$$

where

$$I^*(q^*) = \begin{cases} \int_\Omega \left(\nabla u_0 \cdot q^* - g^*(q^*) - f u_0 \right) dx & \text{if } q^* \in Q_f^* \\ -\infty & \text{if } q^* \notin Q_f^* \end{cases}$$

and Q_f^* is the set of functions that satisfy the equation

$$\operatorname{div} q^* + f = 0.$$

Here, g^* is the Fenchel conjugate of g^{**}, i.e., $g^* = (g^{**})^*$. For the problem discussed, it has the form

$$g^*(\xi) = \max\{g_1^*(\xi), g_2^*(\xi)\}, \tag{7.8.17}$$

where

$$g_i^*(\xi) = \tfrac{1}{2} \mathrm{A}_i^{-1} \xi \cdot \xi + \xi \cdot a_i - \beta_i.$$

The functional $-I^*$ is uniformly convex. This fact follows from the next assertion.

Proposition 7.8.1. *Let*

$$g^*(\xi) = \max_{i=1,2,\dots N} \{g_i^*(\xi)\},$$

where each $g_i^* : \mathbb{R}^n \to \mathbb{R}$ is uniformly convex with a certain forcing function ψ_i^*. Then,

$$g^*\left(\tfrac{\xi_1+\xi_2}{2} \right) + \psi^*(\tfrac{\xi_2-\xi_1}{2}) \le \tfrac{1}{2} \left(g^*(\xi_1) + g^*(\xi_2) \right), \tag{7.8.18}$$

where ξ_1 and ξ_2 are arbitrary vectors and

$$\psi^*(\xi) = \min_{i=1,2,\dots N} \{\psi_i^*(\xi)\}.$$

Proof. We have

$$g^*\left(\tfrac{\xi_1+\xi_2}{2}\right) = \max_{i=1,2,\ldots N}\left\{g_i^*\left(\tfrac{\xi_1+\xi_2}{2}\right)\right\}$$

$$\leq \max_{i=1,2,\ldots N}\left\{\tfrac{1}{2}g_i^*(\xi_1)+\tfrac{1}{2}g_i^*(\xi_2)-\psi_i^*(\tfrac{\xi_1-\xi_2}{2})\right\}$$

$$\leq \tfrac{1}{2}\left(\max_{i=1,2,\ldots N}\{g_i^*(\xi_1)\}+\max_{i=1,2,\ldots N}\{g_i^*(\xi_2)\}\right)$$

$$+\max_{i=1,2,\ldots N}\left\{-\psi_i^*(\tfrac{\xi_1-\xi_2}{2})\right\}.$$

Therefore,

$$g^*\left(\tfrac{\xi_1+\xi_2}{2}\right) \leq \tfrac{1}{2}\left(g^*(\xi_1)+g^*(\xi_2)\right)-\min_{i=1,2,\ldots N}\left\{\psi_i^*(\tfrac{\xi_1-\xi_2}{2})\right\},$$

and we arrive at (7.8.18). $\qquad\square$

The uniform convexity of the dual functional implies the estimate (see [169])

$$\Psi^*(y^*-p^*) \leq M^*(y^*), \quad \forall y^* \in Y^*(\Omega), \qquad (7.8.19)$$

in which the difference between p^* and a function y^* is measured by means of the functional

$$\Psi^*(y^*-p^*) = \int_\Omega \left(\min_{i=1,2}\{\tfrac{1}{8}A_i^{-1}(y^*-p^*)\cdot(y^*-p^*)\}\right)dx.$$

The majorant M^* is given by the relation

$$M^*(y^*) := \left[\sqrt{m_D^2(y^*)+\alpha_1\,m_f^2(y^*)+\alpha_2\,m_f(y^*)}+\alpha_3\,m_f(y^*)\right]^2.$$

Here, the constants α_i, $i=1,2,3$ depend on the problem data and can easily be computed. The terms m_D and m_f are defined by the relations

$$m_D^2(y^*) := \inf_{v\in V_0+u_0}\int_\Omega D(\nabla v,y^*)\,dx,$$

$$m_f^2(y^*) := \inf_{q^*\in Y_f^*(\Omega)}\int_\Omega |y^*-q^*|^2\,dx,$$

where

$$D\left(\xi,\xi^*\right) := g^{**}(\xi)+g^*(\xi^*)-\xi\cdot\xi^*.$$

Note that

$$m_D(y^*) = 0 \quad \text{if and only if} \quad y^* = g^{**\prime}(\nabla v), \ v \in V_0 + u_0,$$
$$m_f(y^*) = 0 \quad \text{if and only if} \quad y^* \in Q_f^*(\Omega).$$

Therefore, the majorant vanishes if and only if y^* coincides with the exact solution p^*. Moreover, it is proved that it possesses proper continuity properties, namely,

$$M^*(y_m^*) \ \to \ 0 \quad \text{as} \quad y_m^* \ \to \ p^* \ \text{in} \ Y^*(\Omega)\,.$$

In [169], this fact is used to justify a certain numerical procedure that, in principle, is able to solve variational problems of this type with any a priori given tolerance.

Chapter 8

A posteriori estimates for variational inequalities

8.1 Variational inequalities

Mathematical models related to variational inequalities often arise in various applied problems (see, e.g., [64]). In this chapter, we present a method of deriving estimates of deviations from exact solutions of stationary variational inequalities. It can be regarded as a modification of the duality technique used in Chapters 6 and 7 for variational problems with uniformly convex functionals. In Chapter 8, we consider three classical problems related to variational inequalities, namely, a problem with obstacles, an elasto-plastic torsion problem, and a problem with friction type boundary conditions. Also, we analyze some problems arising in the theory of viscous incompressible fluids.

Let $a : V \times V \to \mathbb{R}$ be a bilinear V-elliptic form and $j : V \to \mathbb{R}$ be a given convex continuous functional. Consider the following problem: find $u \in K$ such that the inequality

$$a(u, w - u) + j(w) - j(u) \geq \langle f, w - u \rangle \qquad (8.1.1)$$

holds for any $w \in K$, where K is a convex closed subset of V and $f \in V^*$.

The solution u of (8.1.1) is a minimizer of the following variational problem $\mathcal{P}$: Find $u \in K$ such that

$$J(u) = \inf_{w \in K} J(w), \qquad (8.1.2)$$

$$J(w) = \frac{1}{2} a(w, w) + j(w) - \langle f, w \rangle.$$

245

By (8.1.1), we easily obtain the inequality

$$J(v) - J(u) \;\; = \;\; \frac{1}{2}a(v - u, v - u) + a(u, v - u) - \langle f, v - u \rangle + j(v) - j(u)$$
$$\geq \;\; \frac{1}{2}a(v - u, v - u),$$

which implies the basic "deviation" estimate

$$\boxed{\frac{1}{2} \parallel v - u \parallel^2 \;\leq\; J(v) - J(u), \quad \forall v \in K\,,} \qquad (8.1.3)$$

where $\parallel v \parallel := (a(v, v))^{1/2}$. Recall that for linear problems we have the estimate (cf. (6.1.10))

$$\frac{1}{2} \parallel v - u \parallel^2 \;\leq\; J(v) - I^*(p^*),$$

which by the duality relation $J(u) = I^*(p^*)$ coincides with (8.1.3). In Chapter 6, we obtained an estimate of $J(v) - J(u)$ invoking the dual problem and using its properties in order to find a directly computable and physically meaningful upper bound of $J(v) - I^*(q^*)$. For variational inequalities, deviation estimates are obtained in a similar way, but with some complications caused by the fact that the problem dual to $\mathcal{P}$ has a more cumbersome form. Below, we show how we can circumvent this difficulty by using the so-called *perturbed functionals* and obtain explicitly computable majorants of $\parallel v - u \parallel$ for several types of variational inequalities.

8.2   Problems with obstacles

8.2.1   Variational formulations

Consider a bilinear form $a : V_0 \times V_0 \to \mathbb{R}$ defined by the relation

$$a(v, w) := \int_\Omega A \nabla v \cdot \nabla w \, dx\,, \qquad (8.2.1)$$

where Ω is a bounded domain in $\mathbb{R}^2$ with Lipschitz continuous boundary $\partial\Omega$, $V_0 := \overset{\circ}{H}{}^1(\Omega)$, and $A = \{a_{ij}\}$ is a symmetric matrix satisfying the conditions

$$\nu_1 \, |\xi|^2 \;\leq\; A\xi \cdot \xi \leq \nu_2 \, |\xi|^2\,, \quad \forall \xi \in \mathbb{R}^n, \quad \nu_2 \geq \nu_1 > 0. \qquad (8.2.2)$$

Let

$$K = K_{\varphi\psi} := \{v \in V_0 \;\parallel\; \phi(x) \leq v(x) \leq \psi(x) \text{ a.e. in } \Omega\}\,,$$

where $\phi, \psi \in H^2(\Omega)$ are two given functions such that

$$\phi(x) \leq \psi(x), \quad \forall x \in \Omega,$$

Set $j \equiv 0$ and $\langle f, v \rangle = \int_\Omega fv\, dx$. Then $\mathcal{P}$ is the classical obstacle problem (see, e.g., [75, 119]). A solution u minimizes the functional

$$J(v) = \int_\Omega A\nabla v \cdot \nabla v\, dx - \int_\Omega fv\, dx$$

on the set $K_{\varphi\psi}$. In general, it divides Ω into three sets:

$$\begin{aligned}
\Omega^u_\oplus &:= \left\{ x \in \Omega \;\|\; u(x) = \psi(x) \right\}, \\
\Omega^u_\ominus &:= \left\{ x \in \Omega \;\|\; u(x) = \phi(x) \right\}, \\
\Omega^u_0 &:= \left\{ x \in \Omega \;\|\; \phi(x) < u(x) < \psi(x) \right\}.
\end{aligned}$$

The sets $\Omega^u_\oplus$ and $\Omega^u_\ominus$ are called *upper* and *lower coincidence sets* and Ω^u_0 is an open set, where a solution satisfies the differential equation. Thus, we see that this problem involves free boundaries, which are unknown a priori.

Differentiability properties of solutions to linear and quasiliniear problems with obstacles were investigated by many authors (see, e.g., [7, 50, 75, 119, 208]). In particular, it was proved that, under natural assumptions on external data $u \in H^2(\Omega)$ and even for very smooth data, solutions have a limited regularity (which is $W^{2,\infty}$). Henceforth, we assume that these assumptions are fulfilled and the solutions are H^2-regular.

8.2.2 Perturbed problem

To estimate the difference $J(v) - J(u)$ we introduce the *perturbed functional*

$$J_\lambda(v) := J(v) - \int_\Omega \lambda \cdot (\mathrm{v} - \Phi)\, dx,$$

where $\Phi = (\phi, -\psi)$ and $\mathrm{v} = (v, -v)$,

$$\lambda \in \aleph_\oplus := \left\{ (\lambda_1, \lambda_2) \;\|\; \lambda_i \in L^2(\Omega),\ \lambda_i(x) \geq 0\ \text{a.e. in}\, \Omega,\ i = 1, 2 \right\}.$$

It is easy to see that

$$\begin{aligned}
\sup_{\lambda \in \aleph_\oplus} J_\lambda(v) &= J(v) - \inf_{\lambda \in \aleph_\oplus} \int_\Omega \lambda \cdot (\mathrm{v} - \Phi)\, dx \\
&= \begin{cases} J(v) & \text{if } v \in K_{\varphi\psi}, \\ +\infty & \text{if } v \notin K_{\varphi\psi}. \end{cases}
\end{aligned} \qquad (8.2.3)$$

With J_λ we associate a *perturbed variational problem.*

Problem $\mathcal{P}_\lambda$.   Find $u_\lambda \in V_0$ such that

$$J_\lambda(u_\lambda) = \inf_{v \in V_0} J_\lambda(v) := \inf \mathcal{P}_\lambda. \qquad (8.2.4)$$

Since

$$\inf_{v \in V_0} J_\lambda(v) \leq \inf_{v \in K_{\varphi\psi}} J_\lambda(v) = \inf_{v \in K_{\varphi\psi}} J(v) = \inf \mathcal{P},$$

we see that

$$\inf \mathcal{P}_\lambda \leq \inf \mathcal{P}, \quad \forall \lambda \in \aleph_\oplus. \qquad (8.2.5)$$

It follows from (8.1.3) and (8.2.5) that

$$\frac{1}{2} \parallel v - u \parallel^2 \leq J(v) - \inf \mathcal{P}_\lambda, \quad \lambda \in \aleph_\oplus. \qquad (8.2.6)$$

To estimate the right-hand side of (8.2.5), we introduce a dual counterpart of Problem $\mathcal{P}_\lambda$.

8.2.3   Dual perturbed problem

By the Lagrangian

$$L(v, \tau^*, \lambda) := \int_\Omega \left(\tau^* \cdot \nabla v - \frac{1}{2} A^{-1} \tau^* \cdot \tau^* - fv - \lambda \cdot (\mathbf{v} - \Phi) \right) dx,$$

we define the perturbed functional as follows:

$$J_\lambda(v) = \sup_{\tau^* \in Y^*} L(v, \tau^*, \lambda),$$

where $Y^* := L^2(\Omega, \mathbb{R}^2)$. Now, we can define a new variational problem, which is dual to Problem $\mathcal{P}_\lambda$.

Problem $\mathcal{P}_\lambda^*$.   Find τ_λ^* such that

$$J_\lambda^*(\tau_\lambda^*) = \sup_{q^* \in Q_{f\lambda}^*} J_\lambda^*(q^*),$$

where

$$J_\lambda^*(q^*) = \int_\Omega \left(-\frac{1}{2} A^{-1} q^* \cdot q^* + \lambda \cdot \Phi \right) dx,$$

$$Q_{f\lambda}^* := \left\{ q^* \in Y^* \mid \int_\Omega q^* \cdot \nabla v \, dx = \int_\Omega (fv + \lambda \cdot \mathbf{v}) \, dx \; \forall v \in V_0 \right\}.$$

Note that the set $Q^*_{f\lambda}$ is a closed affine manifold in Y^* and the functional $-J^*_\lambda$ is convex and continuous on Y^*. Therefore, Problem $\mathcal{P}^*_\lambda$ has a solution and

$$\inf \mathcal{P}_\lambda \;=\; \sup \mathcal{P}^*_\lambda. \tag{8.2.7}$$

8.2.4 Estimates of the deviation

From (8.1.3) and (8.2.7) we deduce the estimate

$$\frac{1}{2}\parallel v - u \parallel^2 \le J(v) - \sup\mathcal{P}^*_\lambda \le J(v) - J^*_\lambda(q^*)\,, \tag{8.2.8}$$

which holds for any $v \in K_{\varphi\psi}$, $q^* \in Q^*_{f\lambda}$ and $\lambda \in \aleph_\oplus$. Rewriting the right-hand side of (8.2.8), we find that

$$\frac{1}{2}\parallel v - u \parallel^2 \le \int_\Omega \left(\frac{1}{2}A\nabla v \cdot \nabla v - fv\right)\,dx + \int_\Omega \frac{1}{2}A^{-1}\tau^* \cdot \tau^*\,dx$$

$$+ \frac{1}{2}\int_\Omega \left(A^{-1}q^* \cdot q^* - A^{-1}\tau^* \cdot \tau^*\right)\,dx - \int_\Omega \lambda \cdot \Phi\,dx.$$

In view of the relation

$$A^{-1}a \cdot a - A^{-1}b \cdot b = A^{-1}(a - b) \cdot (a - b) + 2A^{-1}b \cdot (a - b)$$

and the integral identity

$$\int_\Omega fv\,dx = \int_\Omega q^* \cdot \nabla v\,dx - \int_\Omega \lambda \cdot \mathbf{v}\,dx, \quad \forall q^* \in Q^*_{f\lambda}\,,$$

we obtain

$$\frac{1}{2}\parallel v - u \parallel^2 \le \frac{1}{2}\int_\Omega (A\nabla v - \tau^*) \cdot (\nabla v - A^{-1}\tau^*)\,dx$$

$$+ \int_\Omega \lambda \cdot (\mathbf{v} - \Phi)dx + \frac{1}{2}\int_\Omega A^{-1}(q^* - \tau^*) \cdot (q^* - \tau^*)$$

$$+ \int_\Omega (\nabla v - A^{-1}\tau^*) \cdot (\tau^* - q^*)\,dx. \tag{8.2.9}$$

The last integral can be estimated as follows:

$$\int\limits_{\Omega} (\nabla v - A^{-1}\tau^*) \cdot (\tau^* - q^*)dx$$

$$\leq \tfrac{\beta}{2} \int\limits_{\Omega} A(\nabla v - A^{-1}\tau^*) \cdot (\nabla v - A^{-1}\tau^*)\, dx$$

$$+ \tfrac{1}{2\beta} \int\limits_{\Omega} A^{-1}(q^* - \tau^*) \cdot (q^* - \tau^*)\, dx,$$

where β is any positive number.

Introduce the quantity

$$d^2(\tau^*, Q^*_{f\lambda}) := \inf_{q^* \in Q^*_{f\lambda}} \int\limits_{\Omega} A^{-1}(q^* - \tau^*) \cdot (q^* - \tau^*)\, dx, \qquad (8.2.10)$$

which is the distance between τ^* and the set $Q^*_{f\lambda}$. Now, (8.2.9) has the form

$$\| v - u \|^2$$

$$\leq (1+\beta)d^2(\tau^*, Q^*_{f\lambda}) + (1+\beta^{-1}) \| \nabla v - A^{-1}\tau^* \|^2$$

$$+ 2\int\limits_{\Omega} \lambda \cdot (\mathbf{v} - \Phi)\, dx. \quad (8.2.11)$$

If $\tau^* \in Q^* := \{\tau^* \in Y^* \ \| \ \operatorname{div}\tau^* \in L^2(\Omega)\}$, then $d(\tau^*, Q^*_{f\lambda})$ can be estimated by the procedure used in Chapter 6, which gives

$$d(\tau^*, Q^*_{\lambda}) \leq C_{\Omega,A} \|\operatorname{div}\tau^* + f + \lambda_1 - \lambda_2\|, \qquad (8.2.12)$$

where $C_{\Omega,A}$ is a constant in the inequality

$$\|w\|_{2,\Omega} \leq C_{\Omega,A} \| w \|, \quad \forall w \in V_0.$$

Thus, for $y^* \in Q^*$ we obtain

$$\| v - u \|^2 \leq C_{\Omega,A}^2 (1+\beta) \|\operatorname{div}\tau^* + f + \lambda_1 - \lambda_2\|^2$$

$$+ (1+\beta^{-1}) \| \nabla v - A^{-1}\tau^* \|)^2 + 2\int\limits_{\Omega} \lambda \cdot (\mathbf{v} - \Phi)\, dx. \quad (8.2.13)$$

To obtain a rigorous upper bound of $\| v - u \|$, one should find a suitable λ that makes the right-hand side of (8.2.13) close to its minimal value.

The first possible choice of λ is as follows. For a continuous function v we define three sets (some of which may be empty)

$$\begin{aligned}
\Omega_0^v &:= \{x \in \Omega \ \| \ \phi(x) < v(x) < \psi(x)\}, \\
\Omega_\ominus^v &:= \{x \in \Omega \ \| \ \quad v(x) = \phi(x)\}, \\
\Omega_\oplus^v &:= \{x \in \Omega \ \| \ \quad v(x) = \psi(x)\}
\end{aligned}$$

and set

$$\begin{aligned}
\lambda_1 = \lambda_2 = 0 & \qquad \text{a.e. in } \ \Omega_0^v, \\
\lambda_1 = -\langle \operatorname{div} \tau^* + f \rangle_\ominus, \ \lambda_2 = 0 & \qquad \text{a.e. in } \ \Omega_\ominus^v, \\
\lambda_1 = 0, \ \lambda_2 = \langle \operatorname{div} \tau^* + f \rangle_\oplus, & \qquad \text{a.e. in } \ \Omega_\oplus^v.
\end{aligned}$$

Then, we arrive at the estimate

$$\| v - u \|^2 \le \mathcal{M}_1(v, \tau^*, \beta) := \left(1 + \frac{1}{\beta}\right) \| \nabla v - A^{-1}\tau^* \|^2$$
$$+ C_{\Omega,A}^2 (1 + \beta) \left[\int_{\Omega_0^v} |r(\tau^*)|^2 \, dx + \int_{\Omega_\oplus^v} \langle r(\tau^*) \rangle_\ominus^2 \, dx + \int_{\Omega_\ominus^v} \langle r(\tau^*) \rangle_\oplus^2 \, dx \right],$$
$$\tag{8.2.14}$$

where $r(\tau^*) = \operatorname{div} \tau^* + f$ and $\langle \cdot \rangle_\ominus$ and $\langle \cdot \rangle_\oplus$ denote the negative and positive parts of a quantity, respectively.

The majorant $\mathcal{M}_1$ consists of four terms. The first term penalizes the error in the duality relation $\nabla v = A^{-1}\tau^*$. Other terms penalize "improper" behavior of $r(\tau^*)$ on the sets Ω_0^v, $\Omega_\oplus^v$, and $\Omega_\ominus^v$, respectively. It is easy to see that the majorant $\mathcal{M}_1$ is a nonnegative function of its arguments, which vanishes if and only if

$$\begin{aligned}
\nabla v(x) &= A^{-1}\tau^*(x) & \text{for a.e. } x \in \Omega, & \tag{8.2.15} \\
\operatorname{div} \tau^*(x) + f(x) &\le 0, & \text{for a.e. } x \in \Omega_\ominus^v, & \tag{8.2.16} \\
\operatorname{div} \tau^*(x) + f(x) &= 0, & \text{for a.e. } x \in \Omega_0^v, & \tag{8.2.17} \\
\operatorname{div} \tau^*(x) + f(x) &\ge 0 & \text{for a.e. } x \in \Omega_\oplus^v. & \tag{8.2.18}
\end{aligned}$$

If (8.2.15)–(8.2.18) hold, then for any $w \in K_{\varphi\psi}$, we have

$$
\int_\Omega A\nabla v \cdot \nabla(w - v)\, dx \;-\; \int_\Omega f(w - v)\, dx
$$

$$
= \int_\Omega (\operatorname{div}\tau^* + f)(v - w)\, dx = \int_{\Omega_\ominus^v} (\operatorname{div}\tau^* + f)(\phi - w)\, dx
$$

$$
+ \int_{\Omega_0^v} (\operatorname{div}\tau^* + f)(v - w)\, dx + \int_{\Omega_\oplus^v} (\operatorname{div}\tau^* + f)(\psi - w)\, dx \geq 0.
$$

This inequality means that v coincides with a solution of the variational inequality

$$
a(u, w - u) \geq \int_\Omega f(w - u)\, dx, \quad \forall w \in K_{\varphi\psi}. \tag{8.2.19}
$$

Hence, we arrive at the following assertion.

Proposition 8.2.1. *For any $\beta > 0$, $\mathcal{M}_1(v, \tau^*, \beta)$ is a nonnegative functional that majorizes $\|\, v - u\, \|^2$ and vanishes if and only if $v = u$ and $\tau^* = A\nabla u$, where u is a solution of (8.2.19).*

To obtain a more rigorous upper bound of $\|\, v - u\, \|$, one should find λ by minimizing the right-hand side of (8.2.13). Such a minimization procedure results in the estimate

$$
\|\, v - u\, \|^2 \leq \mathcal{M}_2(v, \tau^*, \beta) := \left(1 + \frac{1}{\beta}\right) \|\, \nabla v - A^{-1}\tau^*\, \|^2
$$

$$
+ \int_\Omega R(v, \operatorname{div}\tau^* + f, \beta)\, dx, \tag{8.2.20}
$$

where

$$
R(v, r, \beta) = \begin{cases} -\dfrac{(\phi - v)^2}{c_\beta} + 2r(\phi - v) & \text{if } c_\beta r + v \leq \phi, \\[2mm] c_\beta r^2 & \text{if } \phi < c_\beta r + v < \psi, \\[2mm] -\dfrac{(\psi - v)^2}{c_\beta} + 2r(\psi - v) & \text{if } c_\beta r + v \geq \psi \end{cases}
$$

and $c_\beta = C^2_{\Omega, A}(1 + \beta)$ The functional $\mathcal{M}_2$ is defined for any $v \in K_{\varphi\psi}$, $\tau^* \in Q^*$, and $\beta > 0$. It is clear that

$$
\mathcal{M}_2(v, \tau^*, \beta) \leq \mathcal{M}_1(v, \tau^*, \beta).
$$

This fact implies the following assertion.

Proposition 8.2.2. *For any $\beta > 0$, $\mathcal{M}_2(v, \tau^*, \beta)$ is a nonnegative functional that majorizes $\| v - u \|^2$ and vanishes if and only if $v = u$ and $\tau^* = A\nabla u$, where u is a solution of (8.2.19).*

In (8.2.20), the first term is the same as in (8.2.14), while the second one replaces three other terms in (8.2.14) and, in a sense, inherits all their properties. This fact becomes obvious if we look at the shape of the integrand R regarded as a function of its two arguments for the case. In Fig. 8.2.1, it is presented for $\phi = 0$ and $\psi = 1$. We see that the term $R(v, r, \beta)$

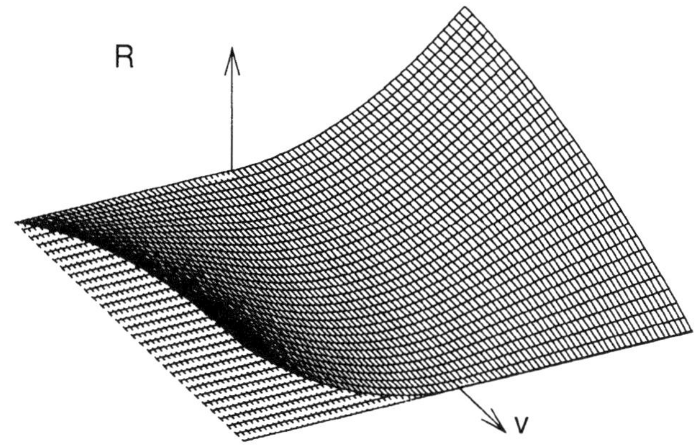

Figure 8.2.1: Generalized residual term of the Duality Majorant.

is equal to zero in the following three cases:

- $r = 0$,

- $v = \phi$ and $r < 0$,

- $v = \psi$ and $r > 0$.

Remark 8.2.1. One can prove that for any positive β the majorant $\mathcal{M}_2(v, \tau^*, \beta)$ possesses necessary continuity properties with respect to the first and second arguments. Namely,

$$\mathcal{M}_2(v_k, \tau_k^*, \beta) \rightarrow 0$$

if $v_k \rightarrow u$ in V_0 and $\tau_k^* \rightarrow A\nabla u$ in Q^*.

Remark 8.2.2. If $\psi = +\infty$ (i.e., if only one obstacle exists), then the function R has a more compact form:

$$R(v, \tau^*, \beta) = c_\beta \left[|r|^2 - \langle \frac{v - \phi}{c_\beta} + r \rangle_\ominus^2 \right].$$

For a membrane, this case was analyzed in [40], where one can also find some numerical examples.

Remark 8.2.3. Estimates (8.2.14) and (8.2.20) contain "free" parameters $\beta > 0$ and $\tau^* \in Q^*$. Therefore, as in the case of linear problems, the determination of rigorous upper bounds of errors requires some minimization over β and τ^*.

8.3　The elasto-plastic torsion problem

Let Ω be a bounded domain in $\mathbb{R}^2$ with Lipschitz continuous boundary $\partial\Omega$. We consider the classical torsion problem for a long elasto-plastic bar whose cross-section is the domain Ω. If such a bar is made of an isotropic material, then the torsion problem is reduced to the following variational inequality (see, e.g., [64, 89, 91]): find $u \in K$ such that

$$\int_\Omega \nabla u \cdot \nabla(v - u)\,dx \geq \mu \int_\Omega (v - u)\,dx, \quad \forall v \in K, \qquad (8.3.1)$$

where μ is a positive parameter,

$$K := \{v \in V_0 \ \| \ |\nabla v| \leq 1 \quad \text{a.e. in } \Omega\},$$

and V_0 is defined as in the previous section.

Mathematical properties of this problem have been investigated by many authors. It has a unique solution that contains a free boundary that separates the sets

$$\Omega_e := \{x \in \Omega \ \| \ |\nabla u| < 1\}$$

and

$$\Omega_p := \{x \in \Omega \ \| \ |\nabla u| = 1\},$$

which are called *elastic* and *plastic* sets, respectively. It is known that, under natural assumptions on external data, a solution u has second generalized derivatives summable with any power (see, e.g., [75]).

If Ω is a 1-connected domain, then u coincides with a solution of the following obstacle problem (see, e.g., [75, 91]): find $u \in K_d$ such that

$$J(u) = \inf_{v \in K_d} J(v), \quad J(v) = \frac{1}{2} \int_\Omega (|\nabla v|^2 - \mu v)dx, \quad (8.3.2)$$

where

$$K_d := \{v \in V_0 \ \| \ |v| \le d(x, \partial\Omega) \text{ for a.e. } x \in \Omega\}$$

and $d(x, \partial\Omega)$ denotes the distance between x and $\partial\Omega$.

It is easy to see that (8.3.2) is an obstacle problem considered in the previous section. Therefore, we can use estimates (8.2.14) or (8.2.20) with $\phi = -d(x, \partial\Omega)$ and $\psi = d(x, \partial\Omega)$. In particular, if v has a fixed sign in Ω (e.g., v is nonnegative in Ω), then (8.2.14) implies the estimate

$$\| v - u \|^2 \le \left(1 + \frac{1}{\beta}\right) \| \nabla v - \tau^* \|^2$$

$$+ C_\Omega^2 (1 + \beta) \left[\int_{\Omega_e(v)} (\operatorname{div} \tau^* + \mu)^2 \, dx + \int_{\Omega_p(v)} \langle \operatorname{div} \tau^* + \mu \rangle_\ominus^2 \, dx \right]. \quad (8.3.3)$$

In (8.3.3), $\Omega_e(v)$ and $\Omega_p(v)$ are *elastic* and *plastic* sets defined by the function $v \in K$ and C_Ω is a constant in the Friedrichs–Poincaré inequality.

In the case of a multiply connected domain, the elasto-plastic torsion problem can also be represented as an obstacle problem (see [75]). Therefore, the estimates (8.2.14) and (8.2.20) are valid for such problems.

8.4 A model problem with a friction type boundary condition

Let $\partial\Omega$ consist of two measurable nonoverlapping parts Γ_0 and Γ_1,

$$j(v) = \gamma \int_{\Gamma_1} |v| \, dx, \quad \gamma > 0,$$

and

$$K = V_0 := \{v \in H^1(\Omega) \ \| \ v = 0 \text{ on } \Gamma_0\}.$$

In this case, Problem $\mathcal{P}$ can be viewed as a model problem with a friction type boundary condition on Γ_1. It has a variational form: find $u \in V_0$ such that

$$J(u) \ = \ \inf_{v \in V_0} J(v) \ = \ \inf \mathcal{P} \, ,$$

where

$$J(v) = \int\limits_{\Omega} (\frac{1}{2} |\nabla v|^2 - fv) \, dx + j(v), \quad f \in L^2(\Omega).$$

This functional is strictly convex and coercive on V_0, therefore Problem $\mathcal{P}$ has a unique minimizer. Hereafter, we assume that external data are such that this solution has square summable second derivatives. Then, directly from the variational inequality (8.1.1), it follows that u almost everywhere satisfies the relations

$$\Delta u + f \ = \ 0 \qquad \text{in } \Omega, \tag{8.4.1}$$
$$u \ = \ 0 \qquad \text{on } \Gamma_0, \tag{8.4.2}$$
$$|u_{,\nu}| \ < \ \gamma \ \Rightarrow \ u = 0 \quad \text{on } \Gamma_1, \tag{8.4.3}$$
$$|u_{,\nu}| \ = \ \gamma \ \Rightarrow \ u \leq 0 \quad \text{on } \Gamma_1, \tag{8.4.4}$$
$$|u_{,\nu}| \ = \ -\gamma \ \Rightarrow \ u \geq 0 \quad \text{on } \Gamma_1. \tag{8.4.5}$$

Henceforth, ν denotes the unit outward normal to Γ_1. It is not difficult to verify that conditions (8.4.3)–(8.4.5) are equivalent to the relations

$$|u_{,\nu}| \ \leq \ \gamma \qquad \text{on } \Gamma_1 \, , \tag{8.4.6}$$
$$u \, u_{,\nu} + \gamma \, |u| \ = \ 0 \qquad \text{on } \Gamma_1 \, , \tag{8.4.7}$$

which present "friction" conditions on Γ_1. As in previous sections, we derive an a posteriori estimate by using an auxiliary problem. For this purpose we introduce the functional

$$J_\lambda(v) = \int\limits_{\Omega} \left(\frac{1}{2} |\nabla v|^2 - fv \right) dx + \gamma \int\limits_{\Gamma_1} \lambda v \, d\Gamma \, , \tag{8.4.8}$$

where

$$\lambda \in \aleph := \{\lambda \in L^\infty(\Omega) \ \| \ |\lambda| \leq 1 \text{ a.e. on } \Gamma_1\} \, .$$

Consider the (perturbed) Problem $\mathcal{P}_\lambda$: find $u_\lambda \in V_0$ such that

$$J_\lambda(u_\lambda) \ = \ \inf_{v \in V_0} J_\lambda(v) \ = \ \inf \mathcal{P}_\lambda \, .$$

It is a minimization problem for a quadratic functional, which has a unique solution for any $\lambda \in \aleph$.

It is easy to see that

$$J(v) = \sup_{\lambda \in \aleph} J_\lambda(v)$$

and, consequently,

$$\inf \mathcal{P}_\lambda = \inf_{v \in V_0} J_\lambda(v) \leq \inf_{v \in V_0} J(v) = \inf \mathcal{P}. \qquad (8.4.9)$$

Problem $\inf \mathcal{P}_\lambda$ has a dual counterpart-type Problem $\mathcal{P}_\lambda^*$. The latter consists of finding $\tau_\lambda^* \in Q_{f\gamma\lambda}^*$ such that

$$I_\lambda^*(\tau_\lambda^*) = \sup_{q^* \in Q_{f\gamma\lambda}^*} I_\lambda^*(q^*),$$

where

$$I_\lambda^*(q^*) = -\frac{1}{2} \int_\Omega |q^*|^2 \, dx$$

and $Q_{f\gamma\lambda}^*$ is a subset of $Y^* := L^2(\Omega, \mathbb{R}^n)$ that consists of functions satisfying the relation

$$\int_\Omega q^* \cdot \nabla v dx = \int_\Omega fv dx - \int_{\Gamma_1} \gamma\lambda v \, d\Gamma, \quad \forall v \in V_0.$$

Problem $\mathcal{P}_\lambda$ also has a unique solution and

$$\inf \mathcal{P}_\lambda = \sup \mathcal{P}_\lambda^*.$$

In view of (8.1.3), we obtain

$$\frac{1}{2} \|\nabla(v - u)\|^2 \leq J(v) - \sup \mathcal{P}_\lambda^*$$

$$\leq J(v) - I_\lambda^*(q^*), \quad \forall q^* \in Q_{f\gamma\lambda}^*. \qquad (8.4.10)$$

Let $\tau^* \in Y^*$ and $q^* \in Q_{f\gamma\lambda}^*$, then

$$J(v) - I_\lambda^*(q^*) = \tfrac{1}{2} \int_\Omega |\nabla v - \tau^*|^2 \, dx + \int_\Omega \tfrac{1}{2} |\tau^* - q^*|^2 \, dx$$

$$+ \int_\Omega (\tau^* - q^*) \cdot (\nabla v - \tau^*) dx + j(v) - \int_{\Gamma_1} \gamma\lambda v \, d\Gamma.$$

Now, we estimate the third term and find that

$$\|\nabla(v - u)\|^2 \leq \left(1 + \frac{1}{\beta}\right) \|\nabla v - \tau^*\|^2$$

$$+ (1 + \beta) \inf_{q^* \in Q^*_{f\gamma\lambda}} \|q^* - \tau^*\|^2 + \gamma \int_{\Gamma_1} (|v| - \lambda v)\, dx, \quad (8.4.11)$$

where β is a positive number.

If $\tau^* \in H(\mathrm{div}, \Omega)$ and $\tau_\nu^* := \tau^* \cdot \nu \in L^2(\Gamma_1)$, then

$$\inf_{q^* \in Q^*_{f\gamma\lambda}} \|q^* - \tau^*\|^2$$

$$\leq C_{\Omega\Gamma}^2 \left(\int_\Omega |\mathrm{div}\, \tau^* + f|^2\, dx + \int_{\Gamma_1} |\gamma\lambda + \tau_\nu^*|^2\, d\Gamma \right), \quad (8.4.12)$$

where $C_{\Omega\Gamma}$ is a constant in the inequality

$$\int_\Omega |w|^2\, dx + \int_{\Gamma_1} |w|^2\, d\Gamma \leq C_{\Omega\Gamma}^2 \int_\Omega |\nabla w|^2\, dx, \quad \forall w \in V_0. \qquad (8.4.13)$$

Combining (8.4.11) and (8.4.12), we obtain

$$\|\nabla(v - u)\|^2 \leq \left(1 + \frac{1}{\beta}\right) \|\nabla v - \tau^*\|^2$$

$$+ (1 + \beta)C_{\Omega\Gamma}^2 \left(\int_\Omega |\mathrm{div}\, \tau^* + f|^2\, dx + \int_{\Gamma_1} |\gamma\lambda + \tau_\nu^*|^2\, d\Gamma \right)$$

$$+ \gamma \int_{\Gamma_1} (|v| - \lambda v)\, dx. \quad (8.4.14)$$

Let us choose λ as follows:

$$\lambda(x) = \begin{cases} \tau_\nu^*(x)/\gamma & \text{if } |\tau_\nu^*(x)| < \gamma, \\ 1 & \text{if } \tau_\nu^*(x) \leq -\gamma, \\ -1 & \text{if } \tau_\nu^*(x) \geq \gamma. \end{cases}$$

In this case the estimate has the form

$$\|\nabla(v - u)\|^2 \leq \mathcal{M}(v, \tau^*, \beta) := \left(1 + \frac{1}{\beta}\right) \|\nabla v - \tau^*\|^2$$

$$+ (1 + \beta)\, C_{\Omega\Gamma}^2 \int_\Omega |\mathrm{div}\, \tau^* + f|^2\, dx + \int_{\Gamma_1} \Theta_1(v, \tau_\nu^*, \beta)\, dx, \quad (8.4.15)$$

where

$$
\Theta_1(v,\tau_\nu^*,\beta) = \begin{cases} (1+\beta)C_{\Omega\Gamma}^2 \, |\tau_\nu^* + \gamma|^2 + \gamma(|v| - v) & \text{if } \tau_\nu^* < -\gamma, \\ \tau_\nu^* v + \gamma\,|v| & \text{if } |\tau_\nu^*| \le \gamma, \\ (1+\beta)C_{\Omega\Gamma}^2 \, |\tau_\nu^* - \gamma|^2 + \gamma(|v| + v) & \text{if } \tau_\nu^* > +\gamma. \end{cases}
$$

In this relation, zero values may arise only in the second branch. Thus, for any $\beta > 0$ the functional $\mathcal{M}(v,\tau^*,\beta)$ is nonnegative and vanishes if and only if

$$
\nabla v = \tau^* \qquad\qquad \text{a.e. in } \Omega, \qquad (8.4.16)
$$
$$
\operatorname{div}\tau^* + f = 0 \qquad\qquad \text{a.e. in } \Omega, \qquad (8.4.17)
$$
$$
|\tau_\nu^*| \le \gamma \quad \text{and} \quad \tau_\nu^* v + \gamma\,|v| = 0 \quad \text{a.e. on } \Gamma_1. \qquad (8.4.18)
$$

Since $\nabla v \cdot \nu = v_{,\nu}$, the above relations hold if and only if the pair of functions (v,τ^*) meets the relations (8.4.16), (8.4.17), and the conditions (8.4.6) and (8.4.7).

We summarize the results in the following assertion.

Proposition 8.4.1. *For any $\beta > 0$, $\mathcal{M}(v,\tau^*,\beta)$ is a nonnegative functional that vanishes if and only if $v = u$ and $\tau^* = \nabla u$.*

To obtain a sharper estimate, the right-hand side of (8.4.14) should be minimized over $\lambda \in [-1,1]$. Then, we obtain a more rigorous estimate:

$$
\|\nabla(v - u)\|^2 \le \left(1 + \frac{1}{\beta}\right) \|\nabla v - \tau^*\|^2
$$
$$
+ (1+\beta)\,C_{\Omega\Gamma}^2 \int_\Omega |\operatorname{div}\tau^* + f|^2 \, dx + \int_{\Gamma_1} \Theta_2(v,\tau_\nu^*,\beta)\,dx\,,
$$

where

$$
\Theta_2(v,\tau_\nu^*,\beta) = \gamma\,|v| + \inf_{\xi \in [-1,1]} \left\{ (1-\beta)C_{\Omega\Gamma}^2 (\gamma\xi + \tau_\nu^*)^2 + \gamma v \xi \right\}.
$$

8.5 Variational problems in the theory of viscous fluids

8.5.1 Preliminaries

In this section, we consider slow stationary flows of generalized Newtonian fluids with nonlinear viscosity. The classical statement of this problem is as follows: find a vector-valued function u (velocity), a scalar-valued function p (pressure), and a tensor-valued function σ^* (stress deviator) such that

$$- \operatorname{div} \sigma^* = f - \nabla p \qquad \text{in } \Omega, \qquad (8.5.1)$$

$$\operatorname{div} u = 0 \qquad \text{in } \Omega, \qquad (8.5.2)$$

$$\sigma^* \in \partial g(\varepsilon(u)) \qquad \text{in } \Omega, \qquad (8.5.3)$$

$$u = u_0 \qquad \text{on } \partial\Omega, \qquad (8.5.4)$$

where Ω is a connected bounded domain in $\mathbb{R}^n$ with Lipschitz continuous boundary $\partial\Omega$, f and u_0 are given functions and g is the *dissipative potential* that defines physical properties of a fluid considered.

Henceforth, it is assumed that

$$f \in L^2(\Omega, \mathbb{R}^n), \quad u_0 \in H^1(\Omega, \mathbb{R}^n), \quad \operatorname{div} u_0 = 0 \qquad (8.5.5)$$

and

$$g(\varepsilon) = \frac{\mu}{2} |\varepsilon|^2 + \psi(\varepsilon), \qquad (8.5.6)$$

where μ is a positive constant (viscosity) and $\psi : \mathbb{M}^{n \times n} \to \mathbb{R}_+$ is a convex function that meets the conditions

$$\psi(0) = 0, \quad \psi(\varepsilon) \le c_1 |\varepsilon|^2 + c_2, \quad c_1 > 0. \qquad (8.5.7)$$

The cases

$$\psi \equiv 0 \qquad (8.5.8)$$

and

$$\psi(\varepsilon) = k_* |\varepsilon| \qquad (8.5.9)$$

correspond to Newtonian and Bingham models, respectively. In (8.5.9), $k_* > 0$ is the *plasticity yield* of the fluid. Other sensible examples are given by power constitutive laws

$$\psi(\varepsilon) = k_* |\varepsilon|^\alpha \qquad 1 < \alpha \le 2, \qquad (8.5.10)$$

$$\psi(\varepsilon) = k_* \left(1 + |\varepsilon|^2\right)^{\alpha/2} \qquad (8.5.11)$$

and the Powell–Eyring model, in which

$$\psi(\varepsilon) = k_* |\varepsilon| \ln \left(1 + |\varepsilon|\right). \qquad (8.5.12)$$

In all the models mentioned above, the stress tensor can be split into two terms associated with two parts of the dissipative potential, so that the basic system admits the following form:

$$- \operatorname{div}(\sigma_1^* + \sigma_2^*) = f - \nabla p \qquad \text{in } \Omega, \qquad (8.5.13)$$

$$\operatorname{div} u = 0 \qquad \text{in } \Omega, \qquad (8.5.14)$$

$$\sigma_1^* = \mu \varepsilon(u) \qquad \text{in } \Omega, \qquad (8.5.15)$$

$$\sigma_2^* \in \partial \psi(\varepsilon(u)) \qquad \text{in } \Omega, \qquad (8.5.16)$$

$$u = u_0 \qquad \text{on } \partial\Omega, \qquad (8.5.17)$$

Investigations of mathematical properties of solutions to such type problems started in [131, 133]. Various types of dissipative potentials associated with generalized Newtonian fluids, their physical justifications, and properties of the respective mathematical problems were considered by many authors. At this point we refer, e.g., to [8, 64, 140, 146]. The regularity of solutions to such problems was investigated in [80, 81, 192]. These and other results are systematically exposed in the book [79].

As in previous sections, our goal is to obtain computable majorants of the quantity $\|\varepsilon(v - u)\|$ for any function v from the natural functional space containing the exact solution u. For this purpose we again apply the approach based upon a certain perturbed variational problem. The analysis presented below follows the general lines of [178, 179].

Let us begin by defining the notation. In this section we assume that $V = H^1(\Omega, \mathbb{R}^n)$, $V_0 = \overset{\circ}{H}{}^1(\Omega, \mathbb{R}^n)$, and $Y = L^2(\Omega, \mathbb{M}^{n \times n})$. By $\overset{\circ}{L}_2$ we denote the space of square-summable scalar functions with zero mean. The set of smooth solenoidal fields is denoted by $\overset{\cdot}{J}{}^{\infty}$, i.e.,

$$\overset{\cdot}{J}{}^{\infty}(\Omega) := \{ v \in C_0^{\infty}(\Omega, \mathbb{R}^n) \;\;\|\; \operatorname{div} v = 0,\; \operatorname{supp} v \subset\subset \Omega \}.$$

The space $\mathcal{H}^1$ is the closure of the set $\overset{\cdot}{J}{}^{\infty}$ in the norm $\|\nabla v\|$. The set $\mathcal{H}^1 + u_0$ contains functions $w + u_0$, where $w \in \mathcal{H}^1$. All tensor-valued functions that represent stresses are marked by stars. The functions $\varepsilon(x)$ and $\sigma^*(x)$ are viewed as elements of two dual spaces Y and Y^*, respectively. By Y_{div}^* we denote a subspace of Y^* that contains functions with square summable divergence; Y_{div}^* is a Hilbert space with respect to the scalar product

$$(\tau^*, \eta^*) := \int_{\Omega} (\tau^* : \eta^* + \operatorname{div} \tau^* \cdot \operatorname{div} \eta^*) \, dx.$$

Also, we will use the following two affine subsets of Y^*. The first subset

$Y_f^*(\Omega, \mathcal{H}^1)$ contains functions satisfying the condition

$$\int_\Omega \varepsilon(w) : \tau^* \, dx = \int_\Omega f \cdot w \, dx, \quad \forall \, w \in \mathcal{H}^1.$$

Analogously, the set $Y_f^*(\Omega, V_0)$ is defined by the relation

$$\int_\Omega \varepsilon(w) : \tau^* \, dx = \int_\Omega f \cdot w \, dx, \quad \forall \, w \in V_0.$$

Since $\mathcal{H}^1 \subset V_0$, we see that $Y_f^*(\Omega, V_0) \subset Y_f^*(\Omega, \mathcal{H}^1)$. In what follows, we also use the norm

$$|g| := \sup_{w \in V_0} \frac{\displaystyle\int_\Omega gw \, dx}{\|\varepsilon(w)\|},$$

which is a norm in the space V_0^*.

8.5.2 Variational problems

We define a generalized solution of (8.5.13)– (8.5.17) as a function u that satisfies the variational inequality

$$\int_\Omega \mu\varepsilon(u) : \varepsilon(v - u) \, dx$$

$$+ \, \Psi(\varepsilon(v)) - \Psi(\varepsilon(u)) \geq \int_\Omega f \cdot (v - u) \, dx, \quad \forall v \in \mathcal{H}^1 + u_0, \quad (8.5.18)$$

where

$$\Psi(\varepsilon) := \int_\Omega \psi(\varepsilon) \, dx.$$

This problem has a variational form.

Problem $\mathcal{P}$.   Find $u \in \mathcal{H}^1 + u_0$ such that

$$J(u) = \inf \mathcal{P} := \inf_{v \in \mathcal{H}^1 + u_0} J(v),$$

where

$$J(v) = \int_\Omega g(\varepsilon(v)) \, dx - \int_\Omega f \cdot v \, dx.$$

By usual arguments it is not difficult to prove that Problem $\mathcal{P}$ is uniquely solvable. Now, the estimate (8.1.3) has the form

$$\int_\Omega \frac{\mu}{2} \, |\varepsilon(v-u)|^2 \, dx \leq J(v) - J(u), \quad \forall v \in \mathcal{H}^1 + u_0. \tag{8.5.19}$$

To estimate the value of $J(u)$, we introduce a collection of perturbed variational problems whose functionals are defined on a wider set of functions. For $q \in \overset{\circ}{L}_2$ and $\tau_2^* \in Y^*$, we define the functional

$$\bar{J}(v) := \int_\Omega \left(\frac{\mu}{2} \, |\varepsilon(v)|^2 + \tau_2^* : \varepsilon(v) - \psi^*(\tau_2^*) - f \cdot v - q \, \mathrm{div}(v - u_0) \right) \, dx$$

and consider the following auxiliary problem.

Problem $\overline{\mathcal{P}}$. Find $\bar{u} \in V_0 + u_0$ such that

$$\bar{J}(\bar{u}) = \inf \overline{\mathcal{P}} := \inf_{v \in V_0 + u_0} \bar{J}(v).$$

Obviously, the mininimizer $\bar{u}$ depends on q and τ_2^*, so that it would be more correct to write $\bar{u}_{q,\tau_2^*}$, but we want to avoid cumbersome notation and omit these subscripts. It is not difficult to verify that Problem $\overline{\mathcal{P}}$ has a unique solution and

$$\inf \overline{\mathcal{P}} \leq \inf \mathcal{P}. \tag{8.5.20}$$

Indeed, for any $v \in \mathcal{H}^1 + u_0$ we have

$$\bar{J}(v) = \int_\Omega \left(\frac{\mu}{2} \, |\varepsilon(v)|^2 + \tau_2^* : \varepsilon(v) - v\psi^*(\tau_2^*) \right) \, dx - \int_\Omega f \cdot v dx \leq J(v).$$

Therefore,

$$\inf_{v \in V_0 + u_0} \bar{J}(v) \leq \inf_{v \in \mathcal{H}^1 + u_0} \bar{J}(v) \leq \inf_{v \in \mathcal{H}^1 + u_0} J(v) = \inf \mathcal{P}.$$

Introduce a problem dual to $\overline{\mathcal{P}}$. For this purpose, we use the Lagrangian

$$\bar{L}(v; \tau_1^*, \tau_2^*) := \int_\Omega \left(\varepsilon(v) : (\tau_1^* + \tau_2^*) - \frac{1}{2\mu} \, |\tau_1^*|^2 - \psi^*(\tau_2^*) \right) \, dx$$

$$- \int_\Omega f \cdot v \, dx - \int_\Omega q \cdot \mathrm{div} \cdot (v - u_0) dx.$$

It is easy to see that

$$\bar{J}(v) = \sup_{\tau_1^* \in Y^*} \bar{L}(v; \tau_1^*, \tau_2^*)$$

and

$$\inf_{v \in V_0 + u_0} \bar{L}(v; \tau_1^*, \tau_2^*) = \begin{cases} I^*(\tau_1^*, \tau_2^*), & \text{if } (\tau_1^* + \tau_2^*) \in Y^*_{f - \nabla q}(\Omega), \\ -\infty & \text{if } (\tau_1^* + \tau_2^*) \notin Y^*_{f - \nabla q}(\Omega), \end{cases}$$

where

$$I^*(\tau_1^*, \tau_2^*) = \int_\Omega \left(\varepsilon(u_0) : (\tau_1^* + \tau_2^*) - \frac{1}{2\mu}|\tau_1^*|^2 - \psi^*(\tau_2^*) - f \cdot u_0 \right) dx.$$

Here, $\eta^* \in Y^*$ belongs to $Y^*_{f - \nabla q} \subset Y^*$ if it satisfies the integral identity

$$\int_\Omega \varepsilon(w) : \eta^* dx = \int_\Omega (f \cdot w + q \operatorname{div} w) dx, \quad \forall\, w \in V_0.$$

Thus, we arrive at a problem dual to $\overline{\mathcal{P}}$ with respect to the Lagrangian $\bar{L}$.

Problem $\overline{\mathcal{P}}^*$. Find $\hat{\sigma}_1^* \in Y^*$ such that $(\hat{\sigma}_1^* + \tau_2^*) \in Y^*_{f - \nabla q}(\Omega)$ and

$$I^*(\hat{\sigma}_1^*, \tau_2^*) = \sup \overline{\mathcal{P}}^* := \sup_{\tau_1^*} I^*(\tau_1^*, \tau_2^*),$$

where $\tau_1^* \in Y^*$ must satisfy the condition $(\tau_1^* + \tau_2^*) \in Y^*_{f - \nabla q}(\Omega)$.

Proposition 8.5.1. *Problem $\overline{\mathcal{P}}^*$ has a unique solution $\hat{\sigma}_1^*$ and*

$$\inf \overline{\mathcal{P}} = \bar{J}(\bar{u}) = \sup \overline{\mathcal{P}}^* = I^*(\hat{\sigma}_1^*, \tau_2^*), \tag{8.5.21}$$

$$\mu\varepsilon(\bar{u}) = \hat{\sigma}_1^*. \tag{8.5.22}$$

Proof. The function $\bar{u}$ meets the integral identity

$$\int_\Omega (\mu\varepsilon(\bar{u}) : \varepsilon(w) + \tau_2^* : \varepsilon(w))\, dx = \int_\Omega (f \cdot w + q \operatorname{div} w) dx, \quad \forall w \in V_0.$$

Hence,

$$\int_\Omega f \cdot \bar{u}\, dx = \int_\Omega (\mu\,|\varepsilon(\bar{u})|^2 - \mu\varepsilon(\bar{u}) : \varepsilon(u_0) + \tau_2^* : \varepsilon(\bar{u} - u_0)$$

$$+ f \cdot u_0 + q \operatorname{div}(\bar{u} - u_0)) dx.$$

Therefore,

$$
\begin{aligned}
\inf \overline{\mathcal{P}} \;=\;& \bar{J}(\bar{u}) \\
=\;& \int_{\Omega} \left(\frac{\mu}{2}\,|\varepsilon(\bar{u})|^{2} + \tau_{2}^{*} : \varepsilon(\bar{u}) - \psi^{*}(\tau_{2}^{*}) \right) dx + \int_{\Omega} (q\,\mathrm{div}(\bar{u}-u_{0}) - f\cdot\bar{u})dx \\
=\;& \int_{\Omega} \left(\tau_{2}^{*} : \varepsilon(u_{0}) - \frac{\mu}{2}\,|\varepsilon(\bar{u})|^{2} - \psi^{*}(\tau_{2}^{*}) + \mu\varepsilon(u_{0}) : \varepsilon(\bar{u})) \right) dx - \int_{\Omega} f\cdot u_{0}dx.
\end{aligned}
$$

Since $\mu\bar{u} + \tau_{2}^{*} \in Y_{f-\nabla q}^{*}$, we find that

$$
I^{*}(\mu\bar{u},\tau_{2}^{*}) \leq \sup \overline{\mathcal{P}}^{*}
$$

and, consequently,

$$
I^{*}(\mu\varepsilon(\bar{u}),\tau_{2}^{*}) = \int_{\Omega} \left(\varepsilon(u_{0}) : (\mu\varepsilon(\bar{u})+\tau_{2}^{*}) - \frac{\mu}{2}\,|\varepsilon(\bar{u})|^{2} - \psi^{*}(\tau_{2}^{*}) - f\cdot u_{0} \right) dx
$$

$$
\leq \sup \overline{\mathcal{P}}^{*} \leq \inf \overline{\mathcal{P}}
$$

$$
= \int_{\Omega} \left(\tau_{2}^{*} : \varepsilon(u_{0}) - \frac{\mu}{2}\,|\varepsilon(\bar{u})|^{2} - \psi^{*}(\tau_{2}^{*}) + \mu\varepsilon(u_{0}) : \varepsilon(\bar{u})) \right) dx - \int_{\Omega} f\cdot u_{0}\, dx.
$$

Thus, we observe that $\mu\varepsilon(\bar{u}) = \hat{\sigma}_{1}^{*}$ and $\inf \overline{\mathcal{P}} = \sup \overline{\mathcal{P}}^{*}$. $\qquad\square$

8.5.3 Estimates for solenoidal fields

Assume that $v \in \mathcal{H}^{1} + u_{0}$ (i.e., $\mathrm{div}\,v = 0$). From (8.5.19), (8.5.20), and (8.5.21) it follows that

$$
\int_{\Omega} \frac{\mu}{2}\,|\varepsilon(v-u)|^{2}\, dx \leq J(v) - \inf \bar{\mathcal{P}} =
$$

$$
= J(v) - \sup \bar{\mathcal{P}}^{*} \leq J(v) - I^{*}(\tau_{1}^{*},\tau_{2}^{*}). \quad (8.5.23)
$$

This estimate holds for any $v \in \mathcal{H}^{1}+u_{0}$ and $(\tau_{1}^{*}+\tau_{2}^{*}) \in Y_{f-\nabla q}^{*}(\Omega)$. Transform the right-hand side of (8.5.23).

$$
J(\bar{v}) - I^{*}(\tau_{1}^{*},\tau_{2}^{*}) \leq \int_{\Omega} \left(\frac{\mu}{2}\,|\varepsilon(v)|^{2} + \frac{1}{2\mu}\,|\tau_{1}^{*}|^{2} - \varepsilon(u_{0}) : \tau_{1}^{*} \right) dx
$$

$$
+ \int_{\Omega} (\psi(\varepsilon(v)) + \psi^{*}(\tau_{2}^{*}) - \varepsilon(u_{0}) : \tau_{2}^{*})\, dx + \int_{\Omega} f\cdot(u_{0}-v)dx.
$$

Since $(\tau_1^* + \tau_2^*) \in Y_{f-\nabla q}^*(\Omega)$ and $v \in \mathcal{H}^1 + u_0$, we have

$$\int_\Omega f \cdot (u_0 - v)dx = \int_\Omega \left(f \cdot (u_0 - v) + q \operatorname{div}(u_0 - v) \right) dx$$

$$= \int_\Omega \varepsilon(u_0 - v) : (\tau_1^* + \tau_2^*)dx.$$

By this relation, we obtain the first principal estimate

$$\frac{\mu}{2} \|\varepsilon(v - u)\|^2 \leq D_1(\varepsilon(v), \tau_1^*) + D_2(\varepsilon(v), \tau_2^*). \tag{8.5.24}$$

The right-hand side of (8.5.24) is the majorant $\mathcal{M}_1(v, \tau_1^*, \tau_2^*)$. Here,

$$D_1(\varepsilon(v), \tau_1^*) := \int_\Omega \left(\tfrac{\mu}{2} |\varepsilon(v)|^2 + \tfrac{1}{2\mu} |\tau_1^*|^2 - \varepsilon(v) : \tau_1^* \right) dx = \tfrac{1}{2\mu} \|\mu\varepsilon(v) - \tau_1^*\|^2$$

and

$$D_2(\varepsilon(v), \tau_2^*) := \int_\Omega \left(\psi(\varepsilon(v)) + \psi^*(\tau_2^*) - \varepsilon(v) : \tau_2^* \right) dx.$$

The compound functionals D_1 and D_2 are nonnegative. Moreover,

$$D_1(\varepsilon(v), \tau_1^*) = 0 \quad \Longleftrightarrow \quad \tau_1^* = \mu\varepsilon(v),$$
$$D_2(\varepsilon(v), \tau_2^*) = 0 \quad \Longleftrightarrow \quad \tau_1^* \in \partial\psi(\varepsilon(v)).$$

Now, the meaning of (8.5.24) is quite transparent. Since $v \in \mathcal{H}^1 + u_0$ and $(\tau_1^* + \tau_2^*) \in Y_{f-\nabla q}^*(\Omega))$, we observe that the relations (8.5.13), (8.5.14), and (8.5.17) hold. In this case, the estimate (8.5.24) shows that the error is controlled by the functionals $D_1(\varepsilon(v), \tau_1^*)$ and $D_2(\varepsilon(v), \tau_2^*)$ that penalize a possible violation of the two remaining relations (8.5.15) and (8.5.16). It is obvious that $\mathcal{M}_1(v, \tau_1^*, \tau_2^*)$ is equal to zero if and only if v coincides with u, τ_1^* with σ_1^*, and τ_2^* with σ_2^*.

Certainly, the condition $(\tau_1^* + \tau_2^*) \in Y_{f-\nabla q}^*(\Omega)$ is rather inconvenient. To avoid it, we may proceed as follows. Take a pair of functions $(\eta_1^*, \tau_2^*) \in Y^* \times Y^*$ for which the condition $(\eta_1^* + \tau_2^*) \in Y_{f-\nabla q}^*(\Omega)$ may be violated. Since

$$D_1(\varepsilon(v), \tau_1^*) = D_1(\varepsilon(v), \eta_1^*)$$

$$+ \int_\Omega \left(\tfrac{1}{2\mu} |\tau_1^* - \eta_1^*|^2 - 2\sqrt{D_1(\varepsilon(v), \eta_1^*)} : \sqrt{\tfrac{1}{2\mu}}(\tau_1^* - \eta_1^*) \right) dx$$

$$\leq (1 + \beta)D_1(\varepsilon(v), \eta_1^*) + \tfrac{1}{2\mu}(1 + \tfrac{1}{\beta}) \|\tau_1^* - \eta_1^*\|^2,$$

where β is an arbitrary positive number, we have

$$\frac{\mu}{2}\|\varepsilon(v-u)\|^2 \le (1+\beta)D_1(\varepsilon(v),\eta_1^*) + D_2(\varepsilon(v),\tau_2^*) + \frac{1+\beta}{2\beta\mu}\|\tau_1^* - \eta_1^*\|^2.$$

Note that $\|\tau_1^* - \eta_1^*\| = \|\tau_1^* + \tau_2^* - \eta_1^* - \tau_2^*\|$, where $\tau_1^* + \tau_2^* \in Y^*_{f-\nabla q}(\Omega)$. Therefore,

$$\frac{\mu}{2}\|\varepsilon(v-u)\|^2 \le (1+\beta)D_1(\varepsilon(v),\eta_1^*)$$
$$+ D_2(\varepsilon(v),\tau_2^*) + \frac{1+\beta}{2\beta\mu}\inf_{\eta^* \in Y^*_{f-\nabla q}(\Omega)}\|\eta_1^* + \tau_2^* - \eta^*\|^2. \quad (8.5.25)$$

By arguments similar to those used in Chapter 6, we estimate the value of the infimum in (8.5.25) by the norm

$$\|\operatorname{div}(\eta_1^* + \tau_2^*) + f - \nabla q\|.$$

This observation leads to the estimate

$$\frac{\mu}{2}\|\varepsilon(v-u)\|^2 \le (1+\beta)D_1(\varepsilon(v),\eta_1^*)$$
$$+ D_2(\varepsilon(v),\tau_2^*) + \tfrac{1+\beta}{2\mu\beta}\|\operatorname{div}(\eta_1^* + \tau_2^*) + f - \nabla q\|^2, \quad (8.5.26)$$

in which $v \in \mathcal{H}^1 + u_0$, $(\eta_1^*,\tau_2^*) \in Y^* \times Y^*$, and $\beta > 0$. We denote the right-hand side of (8.5.25) by $\mathcal{M}_2(\beta,v,\eta_1^*,\tau_2^*,q)$. If $(\eta_1^* + \tau_2^*) \in Y^*_{\operatorname{div}}(\Omega)$ and $q \in \overset{\circ}{L}_2(\Omega) \cap \mathcal{H}^1(\Omega)$, then the norm $\|\ \|$ is estimated by the L^2 norm with a multiplier given by a constant in the inequality

$$\|w\| \le C_\Omega \|\varepsilon(w)\|, \quad \forall w \in V_0.$$

Assume that $\mathcal{M}_2(\beta,v,\eta_1^*,\tau_2^*,q) = 0$. In this case,

$$-\operatorname{div}(\eta_1^* + \tau_2^*) = f - \nabla q,$$
$$\eta_1^* = \mu\varepsilon(v),$$
$$\tau_2^* \in \partial\psi(\varepsilon(v)).$$

Since $v \in \mathcal{H}^1 + u_0$, we conclude that v coincides with u, η_1^* with σ_1^*, and τ_2^* with σ_2^*. The above results can be summarized as follows.

Proposition 8.5.2. *For all*

$$\beta \in \mathbb{R}_+, \ v \in \mathcal{H}^1 + u_0, \ \eta_1^* \in Y^*, \ \tau_2^* \in Y^*, \ q \in \overset{\circ}{L}_2,$$

the functional $\mathcal{M}_2(\beta,v,\eta_1^*,\tau_2^*,q)$ majorizes the norm $\|\varepsilon(v-u)\|^2$.

For any $\beta \in \mathbb{R}_+$, the exact lower bound of this functional on

$$(\mathcal{H}^1 + u_0) \times Y^* \times Y^* \times \overset{\circ}{L}_2$$

is zero. It is attained if and only if $v = u$, $\eta_1^* = \sigma_1^*$, $\tau_2^* = \sigma_2^*$, and $q = p$.

8.5.4 Special cases

Stokes problem

The Stokes problem corresponds to the case $\psi(\varepsilon) \equiv 0$. Set $\tau_2^* = 0$. Then

$$D_2(\varepsilon(v), \tau_2^*) \equiv 0$$

and (8.5.26) yields the estimates

$$\mu^2 \|\varepsilon(u - v)\|^2 \le (1 + \beta)\|\mu\varepsilon(v) - \tau^*\|^2$$
$$+ \frac{1 + \beta}{\beta} |\operatorname{div}\tau^* + f - \nabla q|^2, \quad \forall \beta > 0, \quad (8.5.27)$$

and

$$\mu\|\varepsilon(u - v)\| \le \|\mu\varepsilon(v) - \tau^*\| + |\operatorname{div}\tau^* + f - \nabla q|. \quad (8.5.28)$$

It should be remarked that if τ^* belongs to the set Y_{div}^* and q has square summable derivatives, then the last term of (8.5.27) and (8.5.28) is estimated by the L^2-norm, which is easily computable.

Bingham problem

In this case, $\psi(\varepsilon) = k_*|\varepsilon|$ and

$$\psi^*(\tau^*(x)) = \begin{cases} 0 & \text{if } |\tau^*(x)| \le k_*, \\ +\infty & \text{if } |\tau^*(x)| \ge k_*. \end{cases}$$

Therefore,

$$D_2(\varepsilon(v), \tau_2^*) = \begin{cases} \int_\Omega (k_*|\varepsilon(v)| - \varepsilon(v) \cdot \tau_2^*)dx & \text{if } |\tau_2^*(x)| \le k_*, \\ +\infty & \text{if } |\tau_2^*(x)| \ge k_*, \end{cases}$$

and we obtain

$$\tfrac{\mu}{2}\|\varepsilon(v - u)\|^2$$
$$\le \int_\Omega \left(\tfrac{(1+\beta)}{2\mu} |\mu\varepsilon(v) - \eta_1^*|^2 + k_*|\varepsilon(v)| - \varepsilon(v) : \tau_2^* \right) dx$$
$$+ \tfrac{1+\beta}{2\beta\mu} |\operatorname{div}(\eta_1^* + \tau_2^*) + f - \nabla q|^2, \quad (8.5.29)$$

where $v \in \mathcal{H}^1 + u_0$ and $|\tau_2^*(x)| \le k_*$. If we take η_1^*, τ_2^*, and q such that

$$\operatorname{div}(\eta_1^* + \tau_2^*) \in L^2, \qquad q \in \overset{\circ}{L}_2 \cap H^1,$$

then we arrive at an estimate, whose right-hand side is given by a directly computable integral.

It is well-known that a solution of the Bingham problem have different properties in two subdomains Ω_0 (where $\varepsilon(u) \equiv 0$) and Ω_1 (where $|\varepsilon(u)| > 0$). The estimate (8.5.29) reflects this fact. It is easy to see that in Ω_0 the right-hand side of (8.5.29) vanishes if $\eta_1^* = 0$ and $\mathrm{div}\,\tau_2^* + f - \nabla q = 0$ for a tensor-valued function τ_2^* satisfying the condition $|\tau_2^*(x)| \leq k_*$. The part associated with Ω_1 vanishes if

$$\tau_2^* = k_* \frac{\varepsilon(v)}{|\varepsilon(v)|}, \quad \eta_1^* = \mu\varepsilon(v), \quad \mathrm{div}(\eta_1^* + \tau_2^*) + f - \nabla q = 0.$$

Finally, we consider the problem of anti-plane slow flow of the Bingham fluid in a pipe. It can be reduced to the variational problem (see, e.g., [64, 146])

$$\inf_{\widehat{v} \in \widehat{V}_0} \int_\Omega \left(\frac{\mu}{2} |\nabla\widehat{v}|^2 + k_* |\nabla\widehat{v}| - f\widehat{v} \right) dx,$$

where $\Omega \in \mathbb{R}^2$ is a cross-section of the pipe, $\widehat{V}_0$ is the subspace of $H^1(\Omega)$ that contains functions with zero traces on $\partial\Omega$, and $f = const$ is a linear decay of pressure. In this case, the condition $\mathrm{div}\,v = 0$ is satisfied because of the plane structure of the motion, v has only one nonzero component $\widehat{v}$, so that $\varepsilon(v)$ is replaced by $\nabla\widehat{v}$ and $\psi(\xi) = k_*|\xi|$ for any $\xi \in \mathbb{R}^2$. By (8.5.29), we obtain

$$\frac{\mu}{2} \|\nabla(\widehat{v} - \widehat{u})\|^2 \leq \int_\Omega \left(\tfrac{(1+\beta)}{2\mu} (\mu\nabla\widehat{v} - \eta_1^*)^2 + k_* |\nabla\widehat{v}| - \nabla\widehat{v} \cdot \tau_2^* \right) dx$$

$$+ \tfrac{1+\beta}{2\mu\beta} c_\Omega^2 \|\mathrm{div}(\eta_1^* + \tau_2^*) + f\|^2. \quad (8.5.30)$$

In (8.5.30), η_1^* and τ_2^* are vector-valued functions that satisfy the conditions

$$\eta_1^*, \tau_2^* \in L^2(\Omega, \mathbb{R}^2), \qquad \mathrm{div}(\eta_1^* + \tau_2^*) \in L^2(\Omega),$$
$$|\tau_2^*(x)| \leq k_* \text{ for a.e. } x \in \Omega.$$

8.5.5 Estimates taking into account violations of the incompressibility condition

Now, our goal is to show a way of deriving error majorants for approximations that may not satisfy the incompressibility condition. Assume that an approximate solution is represented by the function $\widehat{v} \in V_0 + u_0$. To find a majorant of $\|u - \hat{v}\|$, we take an arbitrary function $v \in \mathcal{H}^1 + u_0$ and use the inequality

$$\|\varepsilon(\hat{v} - u)\|^2 \leq \tfrac{(1+\gamma)}{\gamma} \|\varepsilon(\hat{v} - v)\|^2 + (1 + \gamma) \|\varepsilon(v - u)\|^2, \tag{8.5.31}$$

where γ is a positive constant. Since $v \in \mathcal{H}^1 + u_0$, the second term is estimated by (8.5.26). For example, for the Stokes problem we have

$$\|\varepsilon(\hat{v} - u)\|^2 \leq \tfrac{(1+\gamma)}{\gamma} \|\varepsilon(\hat{v} - v)\|^2$$
$$+ (1+\gamma)(\tfrac{1+\beta}{\mu^2} \|\mu\varepsilon(v) - \tau^*\|^2 + \tfrac{1+\beta}{\beta\mu^2}|\operatorname{div}\tau^* + f - \nabla q|^2). \tag{8.5.32}$$

From this estimate, we obtain

$$\mu\|\varepsilon(\hat{v} - u)\| \leq 2\mu\|\varepsilon(\hat{v} - v)\|$$
$$+ \|\mu\varepsilon(\hat{v}) - \tau^*\| + |\operatorname{div}\tau^* + f - \nabla q|. \tag{8.5.33}$$

Since v is an arbitrary function from $\mathcal{H}^1 + u_0$, we can replace the first term on the right-hand side in (8.5.33) by the quantity

$$\inf_{v \in \mathcal{H}^1 + u_0} \|\varepsilon(\hat{v} - v)\| = \inf_{w \in \mathcal{H}^1} \|\varepsilon(\hat{v} - u_0 - w)\|.$$

To estimate this quantity we recall an assertion that was established in [134]. It reads: for any $f \in \overset{\circ}{L}_2$, there exists a vector-valued function $\bar{u} \in \overset{\circ}{H}^1(\Omega, \mathbb{R}^n)$ such that

$$\operatorname{div}\bar{u} = f \qquad \text{and} \qquad \|\nabla\bar{u}\| \leq \mathsf{C}_\Omega\|f\|, \tag{8.5.34}$$

where C_Ω is a positive constant dependent on Ω. Set

$$f = \operatorname{div}(\hat{v} - u_0) = \operatorname{div}\hat{v}.$$

In view of (8.5.34), we can find a function $\bar{u}$ such that

$$\operatorname{div}\bar{u} = \operatorname{div}\hat{v} \qquad \text{and} \qquad \|\nabla\bar{u}\| \leq \mathsf{C}_\Omega\|\operatorname{div}\hat{v}\|. \tag{8.5.35}$$

Note that the function $w_0 = \hat{v} - u_0 - \bar{u}$ belongs to $\mathcal{H}^1(\Omega)$, so that (8.5.35) means that

$$\|\nabla(\hat{v} - u_0 - w_0)\| \leq \mathsf{C}_\Omega\|\operatorname{div}\hat{v}\|. \tag{8.5.36}$$

From here, we obtain the required upper bound. It is worth remarking that (see [179])

$$\mathsf{C}_\Omega = 1/\mathbb{C}_\Omega,$$

where $\mathbb{C}_\Omega$ is a constant in the well-known *Ladyzhenskaya–Babuska–Brezzi* condition ([131, 13, 37])

$$\inf_{\substack{q\in\overset{\circ}{L}_2 \\ q\neq 0}} \quad \sup_{\substack{w\in\overset{\circ}{H}^1 \\ w\neq 0}} \quad \frac{\displaystyle\int_\Omega q\,\operatorname{div} w\,dx}{\|q\|\,\|\nabla w\|} \geq \mathbb{C}_\Omega\,. \tag{8.5.37}$$

Thus, for the Stokes problem we have

$$\mu\,\|\varepsilon(\hat{v}-u)\| \leq 2\mu\mathbb{C}_\Omega\,\|\operatorname{div}\hat{v}\|$$
$$+\,\|\mu\varepsilon(\hat{v})-\tau^*\|+|\operatorname{div}\tau^*+f-\nabla q|. \tag{8.5.38}$$

This estimate shows that the error majorant is a sum of three quantities that penalize violations of the *incompressibility condition, constitutive law,* and *flow equation*, respectively. In [178], a similar estimate is obtained for approximations of the pressure function.

For problems with dissipative potentials of the type (8.5.6), estimates of the deviations from exact solutions were derived in [179]. In general, it follows the same lines as for the Stokes problem, but contains a number of technical details. Here we omit them and give only the final estimate

$$\frac{\mu}{2}\,\|\varepsilon(\hat{v}-u)\|^2 \leq c_1 D_1(\varepsilon(\hat{v}),\eta_1^*)$$
$$+\,c_2\big(D_2(\varepsilon(\hat{v}),\tau_2^*)+\tfrac{\delta}{2\mu}\|\hat{\eta}_2^*-\tau_2^*\|^2\big)$$
$$+\,c_3|\operatorname{div}(\eta_1^*+\tau_2^*)+f-\nabla q|^2+c_4 R_{\operatorname{div}}(\hat{v}). \tag{8.5.39}$$

Here,

$$c_1=(1+\gamma)(1+\alpha_1+\frac{1}{\alpha_3}),\quad c_2=c_4=(1+\gamma),$$
$$c_3=(1+\alpha_2+\alpha_3)\frac{1+\gamma}{2\mu},$$

and γ, δ, α_1, α_2, and α_3 are arbitrary positive constants. The term R_{div} penalizes violations of the incompressibility condition. It has the form

$$R_{\operatorname{div}}(\hat{v})=\inf_{v\in\mathcal{H}^1}\left\{\tfrac{\kappa}{2}\,\|\varepsilon(\hat{v}-v)\|^2+\mathcal{R}(v,\hat{v})\right\},$$

where

$$\mathcal{R}(v,\hat{v})=\int_\Omega(\psi'(v)-\psi'(\hat{v})):\varepsilon(v-\hat{v})dx$$

and $\kappa = \mu\left(1 + \frac{1}{\gamma} + \frac{1}{\alpha_1} + \frac{1}{\alpha_2} + \frac{1}{\delta}\right)$. For certain types of dissipative potentials, the term $\mathcal{R}(v, \hat{v})$ can be estimated, which leads to an explicitly computable form of the error majorant (see [179]).

8.6　Examples

Example 8.6.1. Consider a model 1-dimensional obstacle problem with homogeneous boundary conditions. The problem is solved on the interval $(0, 1)$. An approximate solution was computed for a uniform mesh with 60 intervals. In this example, $f = -2.0$, $\phi(x) = -0.16$, and the coincidence set is [.400, .600]. The minimal value of the functional is $-.149$. Exact and approximate solutions are presented in Table 8.6.1.

In this case, the error is 0.000118. The value of $M_\oplus$ computed for $y^* = G_h(\nabla u_h)$ (when the dual variable y^* is computed by a simple gradient averaging procedure) gives the first upper bound 0.000647. Thus, without

Table 8.6.1:

x	Exact solution	Approximate solution	Obstacle
.03	$-.025556$	$-.025183$	$-.160000$
.08	$-.059722$	$-.058874$	$-.160000$
.13	$-.088889$	$-.087732$	$-.160000$
.18	$-.113056$	$-.111755$	$-.160000$
.23	$-.132222$	$-.130945$	$-.160000$
.28	$-.146389$	$-.145300$	$-.160000$
.33	$-.155556$	$-.154822$	$-.160000$
.38	$-.159722$	$-.159520$	$-.160000$
.43	$-.160000$	$-.160000$	$-.160000$
.48	$-.160000$	$-.160000$	$-.160000$
.53	$-.160000$	$-.160000$	$-.160000$
.58	$-.160000$	$-.160000$	$-.160000$
.63	$-.158889$	$-.158508$	$-.160000$
.68	$-.153056$	$-.152209$	$-.160000$
.73	$-.142222$	$-.141077$	$-.160000$
.78	$-.126389$	$-.125110$	$-.160000$
.83	$-.105556$	$-.104309$	$-.160000$
.88	$-.079722$	$-.078674$	$-.160000$
.93	$-.048889$	$-.048206$	$-.160000$
.98	$-.013056$	$-.012881$	$-.160000$

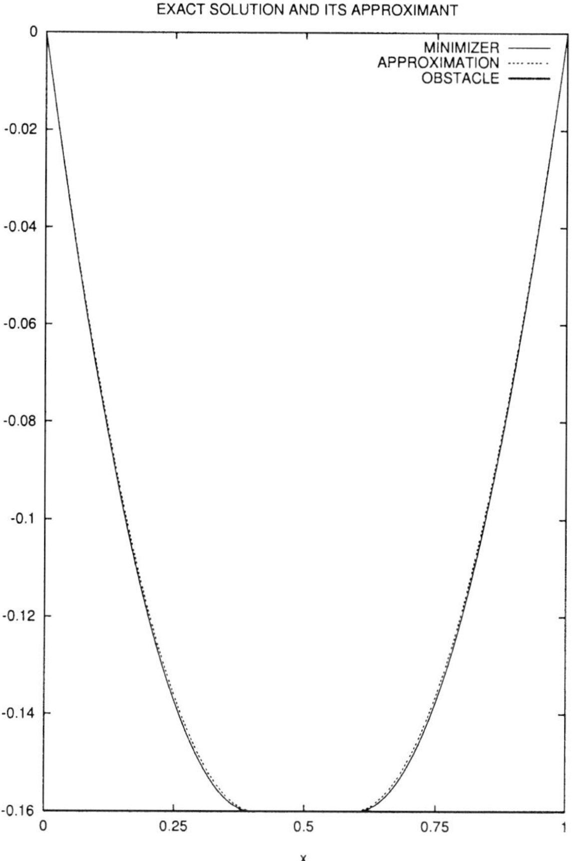

Figure 8.6.1: Example 8.6.1. Approximate and exact solutions of an obstacle problem.

serious additional expenditures, we obtain an estimate with $I_{eff} = 5.473$. Here, two parts of the majorant have the following values: 0.000164 (duality term) and .000483 (generalized residual term).

Then, $M_{\oplus}$ was minimized with respect to the dual variable. In Table 8.6.2, we present values of the majorant obtained in this process. Computational expenditures are measured by the "time unit", which is the time required for computing the approximate solution. Distributions of subinterval errors and errors computed by the Duality Error Majorant are depicted in Fig. 8.6.2. It shows that $M_{\oplus}$ provides not only reliable and efficient

Table 8.6.2:

Iteration	The majorant	I_{eff}	Expenditures
1	.000214	1.804802	.448
2	.000163	1.379279	.660
3	.000152	1.281185	.787
4	.000146	1.232812	.881
5	.000143	1.209496	.977
6	.000140	1.185637	1.073
7	.000137	1.158350	1.169
8	.000136	1.148871	1.243
9	.000134	1.134931	1.336

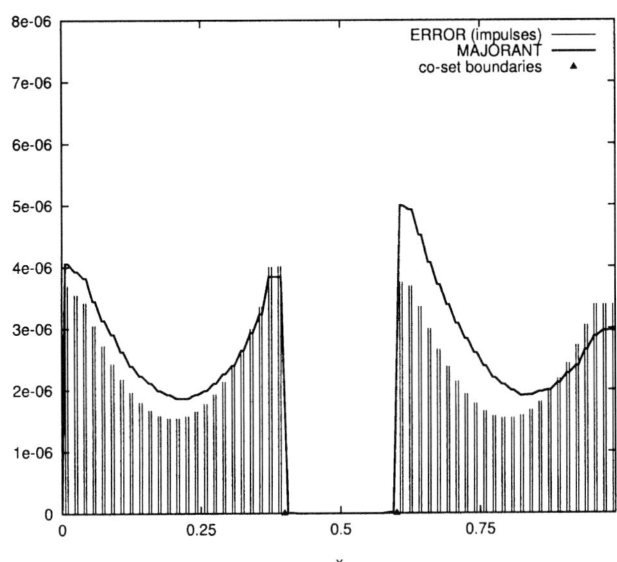

Figure 8.6.2: Example 8.6.1. Error and Duality Majorant.

upper bound for the global error but also provides a good representation of local errors.

Example 8.6.2. Consider another 1-dimensional problem with homogeneous boundary conditions on the unit interval. In this example, $f = -2.0$ and the obstacle is given by the function

$$\phi(x) = -0.3x^2 - 0.06.$$

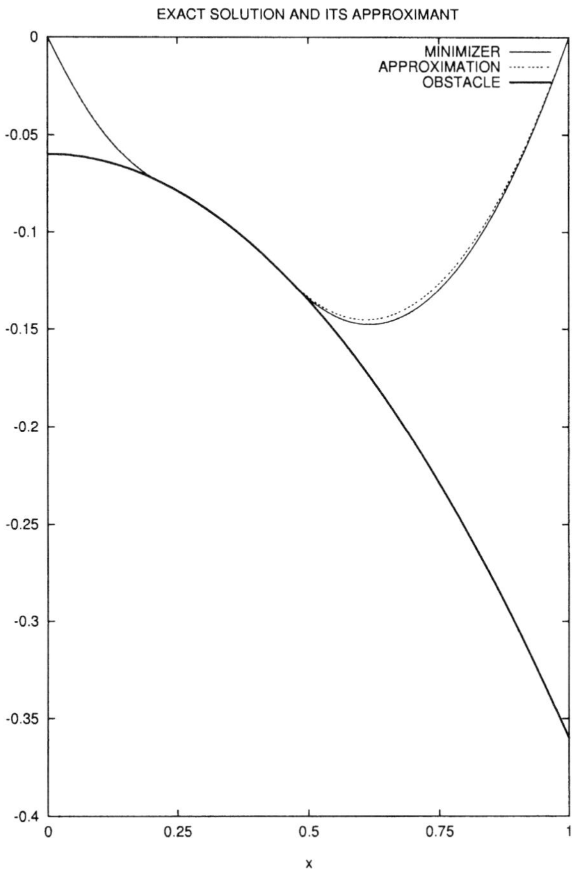

Figure 8.6.3: Example 8.6.2. Approximate solution of an obstacle problem.

The respective coincidence set is $[\,.215, .474]$ and the lower bound of the primal variational problem is equal to $-.125$. Again, an approximate solution was computed for the uniform mesh with 60 subintervals.

Exact and approximate solutions are presented in Fig. 8.6.3. In this example, the error is 0.000158. The value of $M_{\oplus}$ computed for $y^* = G_h(\nabla u_h)$ gives the first (rough) upper bound of the error 0.000861. Thus, without serious additional expenditures, we obtain an estimate with $I_{eff} = 5.457$. Two parts of this majorant are as follows: .000156 (duality term) and .000704 (generalized residual term). Then, the majorant was minimized by optimizing the dual variable. The respective results are presented in

Table 8.6.3:

Iteration	The majorant	I_{eff}	Expenditures
1	.000296	1.877766	.625
2	.000245	1.551655	.842
3	.000240	1.521251	.938
4	.000235	1.492785	1.033
5	.000227	1.439252	1.165
6	.000218	1.379380	1.287
7	.000212	1.343913	1.383
8	.000210	1.334278	1.457
9	.000209	1.322333	1.542

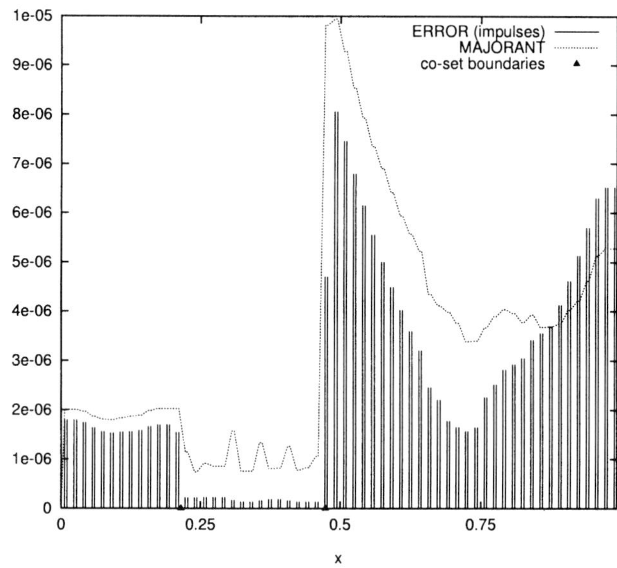

Figure 8.6.4: Example 8.6.2. Error and Duality Majorant.

Table 8.6.3, where computational expenditures are measured by the time unit, which is the time required for finding the approximate solution. In Fig. 8.6.4, the distribution of actual errors on the intervals is compared with the distribution computed by the Duality Majorant. It is clear that in this example the majorant also provides reliable and efficient upper bound for the global error and also correctly shows the local distribution of errors on intervals.

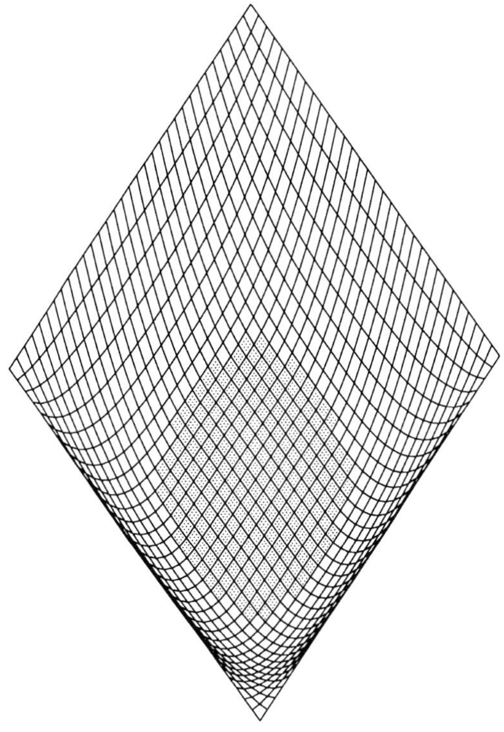

Figure 8.6.5: Example 8.6.3. Approximate solution of an obstacle problem.

Example 8.6.3. Now we consider a 2-dimensional obstacle problem with a plane obstacle and take Ω as a unit square. In Fig. 8.6.5, we present an approximate solution computed by the finite element method on a uniform mesh. Here, the elements that belong to the coincidence set are colored black. Figure 8.6.6 shows the distribution of local errors (which was found by comparing the approximate solution with a reference one computed on a much finer mesh) and the distribution given by the integrand of the majorant. Figure 8.6.7 depicts the distribution of errors computed on the same mesh but with help of a more accurate minimization procedure. We observe that if the expenditures incidental to the evaluation of the majorant enlarge, then the majorant provides a correct distribution of local errors. In Fig. 8.6.8, we show the behavior of the global effectivity index with respect to the costs of the error estimation.

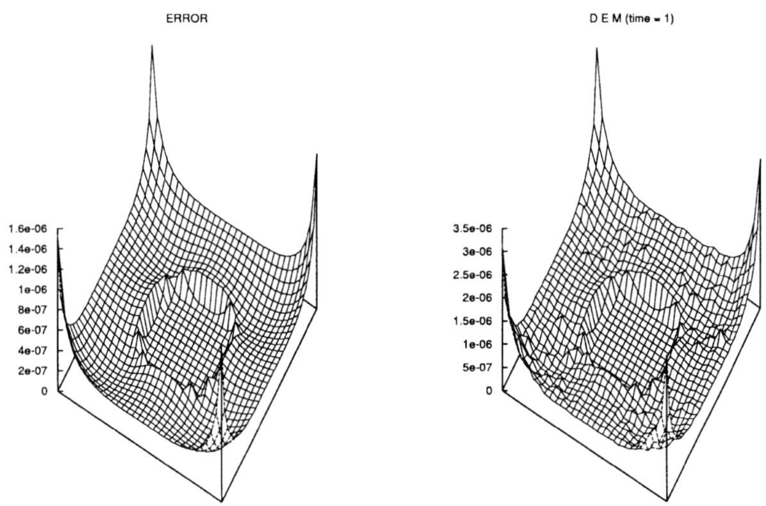

Figure 8.6.6: Example 8.6.3. Local errors and the distribution computed by DEM for $t = 1$.

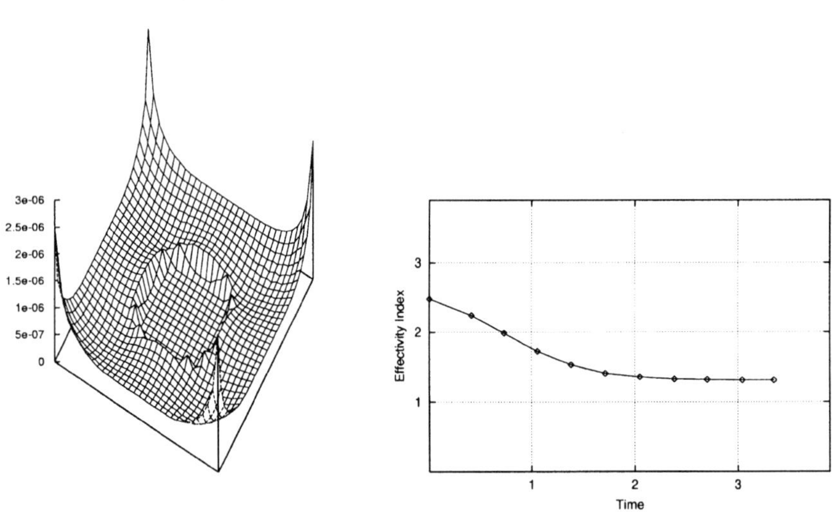

Figure 8.6.7: Example 8.6.3. Local errors computed by DEM for $t = 3$.

Figure 8.6.8: Example 8.6.3. Effectivity index.

8.7 Notes for the Chapter

Various mathematical models related to variational inequalities are presented in the book by G. Duvaut and J.-L. Lions [64]. The existence, uniqueness, and differentiability properties of their solutions were analyzed by many authors (see, e.g., C. Baiocchi and A. Capelo [20], H. Brezis [34], H. Brezis and D. Kinderlehrer [35], H. Brezis and M. Sibony [36], M. Chipot [50], A. Friedman [75], D. Kinderlehrer and G. Stampacchia [119], N. Uraltseva [208] and the references therein).

Much effort has been directed toward creating numerical methods for variational inequalities. Nowadays three main groups of them are presented by *regularization* methods, *penalty* methods and methods based upon *minimax* approaches. At this point we refer to, e.g., R. Glowinski [89] and R. Glowinski, J.-L. Lions and R. Trémolierès [91], H. Kavarada [116]. A priori error estimate of the projection type was established by R. S. Falk [72]. A posteriori error estimates of the errors caused by regularizations of nondifferentiable terms of variational inequalities were obtained in papers by W. Han [93, 94], W. Han and D. C. Reddy [95, 96], and W. Han, S. Jensen and I. Shimansky [97].

Functional type a posteriori error estimates for the main types of stationary variational inequalities were derived in S. Repin [177].

Bibliography

[1] M. Ainsworth and J. T. Oden, A unified approach to a posteriori error estimation using element residual methods, *Numer. Math.*, 65 (1993) 23–50.

[2] M. Ainsworth and J. T. Oden, A posteriori error estimation for the Stokes and Oseen equations, *SIAM J. Numer. Anal.*, 34 (1997) 228–245.

[3] M. Ainsworth and J. T. Oden, *A posteriori error estimation in finite element analysis*, Wiley and Sons, New York, 2000.

[4] M. Ainsworth, J. T. Oden and C. Y. Lee, Local a posteriori error estimators for variational inequalities, *Numer. Methods Partial Differential Equations*, 9 (1993), 23–33.

[5] M. Ainsworth, J. Z. Zhu, A. W. Craig and O. C. Zienkiewicz, Analysis of the Zienkiewicz–Zhu a posteriori error estimator in the finite element method, *Int. J. Numer. Methods Engrg.*, 28 (1989), 2161–2174.

[6] M. Amara, M. Ben Younes and C. Bernardi, Error indicators for the Navier–Stokes equations in stream function and vorticity formulation, *Numer. Math.*, 80 (1998), 181–206.

[7] A. Arkhipova, Limit smootheness of the solution of a problem with two–sided obstacle *Vestnik Leningr. Univ. Mat. Mekh. Astronom.*, 2 (1984), 5–9 (in Russian).

[8] G. Astarita and G. Marrucci, *Principles of non Newtonian fluid mechanics*. McGraw Hill, London, 1974.

[9] J.P.Aubin, *Approximation of elliptic boundary value problems*. Wiley-Interscience, New York-London-Sydney, 1972.

[10] G. Auchmuty, A posteriori error estimates for linear equations, *Numer. Math.*, 61 (1992), 1–6.

[11] O. Axelsson, *Iterative solution methods.* Cambridge University Press, Cambridge, 1994.

[12] I. Babuška, Courant element: before and after, in *Fifty years of Courant element, Eds., M. Křižek, P. Neittaanmäki and R. Stenberg , Marcel Dekker, 1994*, 37–51.

[13] I. Babuška, The finite element method with Lagrangian multipliers, *Numer. Math.*, 20 (1972/73), 179–192.

[14] I. Babuška, F. Ihlenburg, T. Strouboulis and S. K. Gangaraj, A posteriori error estimation for finite element solutions on Helmholtz' equation–Part II: estimation of the pollution error, *Int. J. Numer. Meth. Engrg.*, 40 (1997), 3883–3900.

[15] I. Babuška and W. C. Rheinboldt, A-posteriori error estimates for the finite element method, *Internat. J. Numer. Meth. Engrg.*, 12 (1978) 1597–1615.

[16] I. Babuška and W. C. Rheinboldt, Error estimates for adaptive finite element computations, SIAM J.Numer. Anal., 15 (1978), 736–754.

[17] I. Babuška and R. Rodriguez, The problem of the selection of an a posteriori error indicator based on smoothing techniques, *Internat. J. Numer. Meth. Engrg.*, 36 (1993), 539–567.

[18] I. Babuška and T. Strouboulis, *The finite element method and its reliability*, Clarendon Press, Oxford, 2001.

[19] I. Babuška, T. Strouboulis, C. S. Upadhyay and S. K. Gangaraj, A posteriori estimation and adaptive control of the pollution-error in the h-version of the FEM, *Int. J. Numer. Meth. Engrg.*, 38 (1995), 4207–4235.

[20] C. Baiocchi and A. Capelo, *Variational and quasivariational inequalities. Applications to free boundary problems*, Wiley and Sons, New York, 1984.

[21] J. M. Ball and B. D. James, Fine phase mixtures as minimizers of energy, *Arch. Rational Mech. Anal.*, 100 (1987), 13–52.

[22] P. Baumann and D. Phillips, A nonconvex variational problem related to change of phase, *Appl. Math. Optim.*, 21 (1990), 113–138.

[23] R. E. Bank and R. K. Smith, Mesh smoothing using a posteriori error estimates, *SIAM J. Numer. Anal.*, 34 (1997), 3, 979–997.

[24] R. E. Bank and A. Weiser, Some a posteriori error estimators for elliptic partial differential equations, *Math. Comput.*, 44 (1985), 283–301.

[25] R. E. Bank and B. D. Welfert, A posteriori error estimates for the Stokes problem, *SIAM J. Numer. Anal.*, 28 (1991), 591–623.

[26] W. Bangerth and R. Rannacher, *Adaptive finite element methods for differential equations.*, Birkhäuser, Berlin, 2003.

[27] R. Becker and R. Rannacher, A feed–back approach to error control in finite element methods: Basic approach and examples, *East–West J. Numer. Math.*, 4 (1996), 237–264.

[28] R. Becker, H. Kapp and R. Rannacher, Adaptive finite element methods for optimal control of partial differential equations: Basic concept. *SIAM J. Control Optim.*, 39 (2000),1, 113–132.

[29] B. Boroomand and O. C. Zienkiewicz, Recovery by equilibrium in patches (REP), *Int. J. Numer. Meth. Engrg.*, 40 (1997), 137–164.

[30] B. Boroomand and O. C. Zienkiewicz, An improved REP recovery and the effectivity robustness test, *Int. J. Numer. Meth. Engrg.*, 40 (1997), 3247–3277.

[31] S. W. Brady and A. R. Elcrat, Some results on a posteriori error estimation for approximate solution of second order elliptic problems, *SIAM J. Numer. Anal.*, 16 (1979), 6, 877–889.

[32] D. Braess, *Finite elements.*, Cambridge University Press, Cambridge, 1997.

[33] J.H. Bramble, R.D. Lazarov and J. E. Pasciak, Least-squares for second-order elliptic problems, *Comput. Methods Appl. Mech. Engrg.* , 152 (1998), 195–210.

[34] H. Brezis, Problémes unilatéraux, *J. Math. Pures Appl.*, 9 (1971), 1–168.

[35] H. Brezis and D. Kinderlehrer, The smoothness of solutions to nonlinear variational inequalities, *Indiana Univ. Math J.*, 23 (1974), 831–844.

[36] H. Brezis and M. Sibony, Equivalence de deux inéquations variationnelles et applications, *Arch. Rat. Mech. Anal.*, 41 (1971), 254–265.

[37] F. Brezzi, On the existence, uniqueness and approximation of saddlepoint problems arising from Lagrange multipliers, *R.A.I.R.O., Annal. Numer.*, 8 (1974), R-2, 129–151.

[38] F. Brezzi and M. Fortin, *Mixed and hybrid finite element methods*, Springer Series in Computational Mathematics 15, New York, 1991.

[39] I. G. Bubnov, *Selected Works.* Sudpromgiz, Leningrad, 1956 (in Russian).

[40] H. Buss and S. Repin, A posteriori error estimates for boundary-value problems with obstacles, in *Proceedings of 3d European Conference on Numerical Mathematics and Advanced Applications, Jyuvaskyla, 1999, Eds., P. Neittaanmki, T. Tiihonen and P. Tarvainen, World Scientific, 2000*, 162–170.

[41] G. F. Carey and D. L. Humphrey, Mesh refinement and iterative solution methods for finite element computations, *Int. J. Numer. Meth. Engrg.*, 17 (1981), 1717–1734.

[42] C. Carstensen, Quasi-interpolation and a posteriori error analysis of finite element methods, *Mathematical Modelling in Numerical Analysis*, 33 (1999), 6, 1187–1202.

[43] C. Carstensen and S. Bartels, Each averaging technique yields reliable a posteriori error control in FEM on unstructured grids. I: Low order conforming, nonconforming, and mixed FEM, *Math. Comp.*, 71 (2002), 239, 945–969.

[44] C. Carstensen and S. A. Funken, Fully reliable localized error control in the FEM, *SIAM J. Sci. Comput.*, 21 (2000), 1465–1484.

[45] C. Carstensen and S. A. Funken, Constants in Clément-interpolation error and residual based a posteriori error estimates in Finite Element Methods, *East–West J. Numer. Math.*, 8 (2000), 153–175.

[46] C. Carstensen and R. Verfürth, Edge residuals dominate a posteriori error estimates for low order finite element methods, *SIAM J. Numer. Anal.*, 36 (1999), 1571–1587.

[47] A. Charbonneau, K. Dossou and R. Pierre, A residual-based a posteriori error estimator for Ciarlet–Raviart formulation of the first biharmonic problem, *Numer. Meth. for PDE's*, 13 (1997), 93–111.

[48] Z. Chen and R. H. Nochetto, Rezidual type a posteriori error estimates for elliptic obstacle problems, *Numer. Math.*, 84 (2000), 527–548.

[49] X. Cheng, W. Han and H. Huang, Analysis of some mixed elements for the Stokes problem, *Journal of Computational and Applied Mathematics*, 85 (1997), 19–35.

[50] M. Chipot, Regularity for the two obstacles problem, in *Free boundary problems, Vol. II (Pavia, 1979), Ist. Naz. Alta Mat. Francesco Severi, Rome, 1980*, 135–140.

[51] S. S. Chow, Finite element error estimates for non-linear elliptic equations of monotone type, *Numer. Math.*, 54 (1989), 373–393.

[52] S.-S. Chow, G. F. Carey and R. D. Lazarov, Natural and post-processed superconvergence in semilinear problems, *Numer. Meth. PDE*, 7 (1991), 245–259.

[53] P. G. Ciarlet, *The finite element method for elliptic problems*, North Holland, New York, Oxford, 1978.

[54] Ph. Clément, Approximations by finite element functions using local regularization, *RAIRO Anal. Numér.*, 9 (1975), R-2, 77–84.

[55] L. Collatz, *Funktionanalysis und numerische mathematik*, Springer-Verlag, Berlin, 1964.

[56] P. Coorevits, J.-P. Dumeau and J.-P. Pelle, Error estimator and adaptivity for three-dimensional finite element analysis, in *Advances in Adaptive Computational Methods in Mechanics, Eds., P. Ladevéze and J. T. Oden, Elsevier, New York, 1998*, 443–457.

[57] R. Courant, Variational methods for some problems of equilibrium and vibration, *Bulletin of AMS*, 49 (1943), 1–23.

[58] Computational Methods in the mechanics of Fracture, Ed. S.N. Atluri, North–Holland, New York, 1986.

[59] P. Destuynder and B. Métivet, Explicit error bounds of a conforming finite element method, *Math. Comput.*, 68 (1999), 1379–1396.

[60] W. Dörfler and M. Rumpf, An adaptive strategy for elliptic problems including a posteriori controlled boundary approximation, *Math. Comput.*, 67 (1998), 224, 1361–1362.

[61] J. Douglas, Jr. T. Dupont and L. B. Wahlbin, The stability in L^q of the L^2-projection into finite element function spaces, *Numer. Math.*, 23 (1975), 193–197.

[62] R. Duran, M. A. Muschietti and R. Rodriguez, On the asymptotic exactness of error estimators for linear triangle elements, *Numer. Math.*, 59 (1991), 107–127.

[63] R. Duran and R. Rodriguez, On the asymptotic exactness of Bank–Weiser's estimator, *Numer. Math.*, 62 (1992), 297–303.

[64] G. Duvant and J.-L. Lions, *Les inequations en mecanique et en physique*, Dunod, Paris, 1972.

[65] D. C. Drucker and W. Prager, Soil mechanics and plastic analysis or limit design, *Quart. Appl. Math.*, 10 (1952), 157–165.

[66] I. Ekeland and R. Temam, *Convex analysis and variational problems*, North-Holland, Amsterdam, 1976.

[67] H.W. Engl and O. Scherzer, Convergence rates results for iterative methods for solving nonlinear ill-posed problems, In *Surveys on solution methods for inverse problems*, 7–34, Springer-Verlag, Vienna, 2000.

[68] K. Eriksson, D. Estep, P. Hansbo and C. Johnson, *Computational differential equations*, Cambridge University Press, Cambridge, 1996.

[69] K. Eriksson and C. Johnson, An adaptive finite element method for linear elliptic problems, *Math. Comput.*, 50 (1988), 361–383.

[70] K. Eriksson and C. Johnson, Adaptive finite element methods for parabolic problems, I. A linear model problem, *SIAM J. Numer. Anal.*, 28 (1991), 1, 43–77.

[71] R. E. Ewing, R. D. Lazarov and J. Wang, Superconvergence of the velocity along the Gauss lines in mixed finite element methods, *SIAM J. Numer. Anal.*, 18 (1991), 1015–1029.

[72] R. S. Falk, Error estimates for the approximation of a class of variational inequalities, *Math. Comput.*, 28 (1974), 963–971.

[73] M. Feistauer, *Mathematical methods in fluid dynamics*, Longman, Harlow, 1993.

[74] W. Fenchel,On the conjugate convex functions, *Canad. J. Math.* 1 (1949), 73–77.

[75] A. Friedman, *Variational principles and free-boundary problems*, Wiley and Sons, New York, 1982.

[76] D. A. Field, Laplacian smoothing and Delanay triangulations, *Commum. Appl. Numer. Methods*, 4 (1988), 709–712.

[77] M. Frolov, P. Neittaanmäki and S. Repin, On the reliability, effectivity and robustness of a posteriori error estimation methods, Reports of the Department of Mathematical Information Technology of the University of Jyväskylä, No. B14/2002.

[78] M. Frolov, P. Neittaanmäki and S. Repin, On computational properties of a posteriori error estimates based upon the method of duality error majorants (to appear in Proc. 5th European Conference on Numerical Mathathematics and aplications, Praha,2003).

[79] M. Fuchs and G. A. Seregin, *Variational methods for problems from plasticity theory and for generalized Newtonian fluids*, Lect. Notes in Mathematics 1749, Springer-Verlag, Berlin (2000).

[80] M. Fuchs and G. A. Seregin, Regularity results for the quasi static Bingham variational inequality in dimension two and three, *Math. Z.*, 227 (1998), 525–541.

[81] M. Fuchs and G. A. Seregin, A regularity theory for variational integrals with $LlnL$ growth, *Calc. of Var.*, 6 (1998) 2, 171–187.

[82] H. Gaevskiĭ, H. K. Gröger and K. Zacharias, *Nichtlineare Operatorgliechungen and Operatordifferentialgleiichungen*, Akademie-Verlag, Berlin, 1974.

[83] B. G. Galerkin, Beams and plates. Series in some questions of elastic equilibrium of beams and plates, *Vestnik Ingenerov*, 19 (1915), 897–908 (in Russian).

[84] L. Gallimard, P. Ladeveze and J. P. Pelle, Error estimation and adaptivity in elastoplasticity, *Int. J. Numer. Meth. Engrg.*, 39 (1996), 189–217.

[85] E. G. Geisler, A. A. Tal and D.P. Garg, On a-posteriori error bounds for the solution of ordinary nonlinear differential equations, in *Computers and mathematics with applications*, 407–416, Pergamon, Oxford, 1976.

[86] C. I. Goldstain, Variational crimes and L^∞ error estimates in the finite element method, *Math. Comput.*, 35 (1980), 152, 1131–1157.

[87] D. Gilbarg and N.S. Trudinger, *Elliptic partial differential equations of second order*, Springer-Verlag, Berlin, 1977.

[88] V. Girault and P. A. Raviart, *"Finite element approximation of the Navier–Stokes equations"*, Springer-Verlag, Berlin, 1986.

[89] R. Glowinski, *Numerical methods for nonlinear variational problems*, Springer-Verlag, New-York, 1982.

[90] R. Glowinski and O. Pironneau, Numerical Methods for the First Biharmonic Equation and for the Two-Dimensional Stokes Problem, *SIAM Review*, (21)1999, 2, 167–212.

[91] R. Glovinski, J.-L.Lions, R. Trémolierès, *Analyse numérique des inéquations variationnelles*, Dunod, Paris, 1976.

[92] R. Glowinski, T.-W. Pan and J. Periaux, A fictitious domain method for Dirichlet problem and applications, *Comput. Methods Appl. Mech. Engrg.*, 111 (1994), 283–303.

[93] W. Han, Finite element analysis of a holonomic elasto-plastic problem, *Numer. Math.*, 60 (1992), 493–508.

[94] W. Han, A posteriori error analysis for linearization of nonlinear elliptic problems and their discretization, *Math. Meth. Appl. Sci.*, 17 (1994), 487–508.

[95] W. Han and D. C. Reddy, On the finite element method for mixed variational inequalities arising in elastoplasticity, *SIAM J. Numer. Anal.*, 32 (1995), 1776–1807.

[96] W. Han and D. Reddy, Qualitative and numerical analysis of quasi-static problems in elastoplasticity, *SIAM J. Numer. Anal.*, 34 (1997), 143–177.

[97] W. Han, S. Jensen and I. Shimansky, The Kačanov method for some nonlinear problems, *Applied Numerical Mathematics*, 24 (1997), 57–79.

[98] P. Hansbo, Generalized Laplacian smoothing of unstructured grids, *Commun. Numer. Methods Eng.*, 11 (1995), 455–464.

[99] Y. Hayashi, On a posteriori error estimation in the numerical solution of systems of ordinary differential equations, *Hiroshima Math. J*, 9 (1979), 201–243.

[100] R.Hill, *The mathematical theory of plasticity*, Oxford, 1983.

[101] I. Hlaváček and M. Křížek, On a superconvergence finite element scheme for elliptic systems. I. Dirichlet boundary conditions, *Aplikace Matematiky*, 32 (1987), 131–154.

[102] R. H. W. Hoppe and R. Kornhuber, Adaptive multilevel methods for obstacle problems, *SIAM J. Numer. Anal.*, 31 (1994), 301–323.

[103] P. Houston, R. Rannacher, E. Süli, A posteriori error analysis for stabilised finite element approximations of transport problems, *Comput. Methods Appl. Mech. Engrg.*, 190 (2000), 1483–1508.

[104] A. A. Iljushin, *Plasticite. (Deformations elastico-plastiques)*, Eyrolles, Paris, 1956.

[105] A.D. Ioffe and V.M. Tikhomirov, *Theory of extremal problems. Studies in Mathematics and its Applications, 6*, North-Holland, Amsterdam-New York, 1979.

[106] J.L. Jensen, Sur les fonctions convexes et les inegalités entre les valeurs moyennes, *Acta Math.*, 30 (1906), 175–193.

[107] Jin Qi-nian, Error estimates of some Newton-type methods for solving nonlinear inverse problems in Hilbert scales, *Inverse Problems*, 16 (2000), 187–197.

[108] C. Johnson, *Numerical Solution of Partial Differential Equations by the Finite Element Methods*, Cambridge University Press, Cambridge, 1987.

[109] C. Johnson, Adaptive finite element methods for diffusion and convection problems, *Comput. Methods Appl. Mech. Engrg.*, 82 (1990), 301–322.

[110] C. Johnson and P. Hansbo, Adaptive finite elements in computational mechanics, *Comput. Methods Appl. Mech. Engrg.* 101 (1992), 143–181.

[111] C.Johnson and B.Mercier, Some equilibrium finite element methods two-dimensional elasticity problems, *Numer. Math.*, 30 (1978), 101–116.

[112] C. Johnson and R. Rannacher, On error control in computational fluid dynamics (CFD), Preprint 1994–07, Chalmers University of Technology, Göteborg.

[113] C. Johnson, R. Rannacher and M. Boman, Numerics in hydrodynamic stability. Towards error control in CFD, *SIAM J. Numer. Anal.*, 32 (1995), 1058–1079.

[114] C. Johnson and A. Szepessy, Adaptive finite element methods for conservation laws based on a posteriori error estimates, *Commun. Pure and Appl. Math.*, vol. XLVIII (1995), 199–234.

[115] B.-O. Heimsund, X.-C. Tai and J. Wang, Superconvergence for the gradient of finite element approximations by L^2 projections, *SIAM J. Numer. Anal.*, 40 (2002), 4, 1263–1280.

[116] H. Kavarada, *Numerical problems for free surface problems by means of penalty*, Lecture Notes in Mathematics 704, Springer-Verlag, 1979.

[117] H. Kardestuncer and D. H. Norrie Eds., *Finite Element Handbook*, McGraw–Hill, New York, 1987.

[118] D. W. Kelly, The self equilibration of residuals and complementary error estimates in the finite element method, *Internat. J. Numer. Meth. Engrg.*, 20 (1984) 1491–1506.

[119] D. Kinderlehrer and G. Stampacchia, *An introduction to variational inequalities and their applications*, Academic Press, New York, 1980.

[120] R. V. Kohn, The relaxation of a double-well energy, *Continuum Mech. Thermodynamics*, 3 (1991), 3, 193–236.

[121] A. N. Kolmogorov and S. V. Fomin, *Introductory real analysis.* Dover Publications, Inc., New York, 1975.

[122] R. Kornhuber, A posteriori error estimates for elliptic variational inequalities, *Comput. Math. Appl.*, 31 (1996), 49–60.

[123] S. Korotov, P. Neittaanmaki and S. Repin, A posteriori error estimation of goal-oriented quantities by the superconvergence patch recovery, *J. Numer. Math.*, 11 (2003), 33–59.

[124] S. Korotov, P. Neittaanmaki and S. Repin. A posteriori error estimation in terms of linear functionals for elliptic problems (to appear in Proc. 5th European Conference on Numerical Mathathematics and aplications, Praha, 2003).

[125] M. Křížek, P. Neittaanmäki and R. Stenberg Eds., *Finite Element Methods. Superconvergence, Post-Processing and A Posteriori Error Estimates*, Lecture Notes in Pure and Applied Mathematics 196, Marcel Dekker, New York, 1998.

[126] M. Křížek and P. Neittaanmäki, *Finite Element Approximations of Variational Problems and Applications*, Wiley and Sons, New York, 1990.

[127] M. Křížek and P. Neittaanmäki, Superconvergence phenomenon in the finite element method arising from averaging of gradients, *Numer. Math.*, 45 (1984), 105–116.

[128] P. Ladevéze and D. Leguillon, Error estimate procedure in the finite element method and applications, *SIAM J. Numer. Anal.*, 20 (1983), 485–509.

[129] P. Ladevéze, J.-P. Pelle and Ph. Rougeot, Error estimation and mesh optimization for classical finite elements, *Engineering Computations*, 8 (1991), 69–80.

[130] P. Ladevéze and Ph. Rougeot, New advances on a posteriori error on constitutive relation in f.e. analysis, *Comput. Methods Appl. Mech. Engrg.*, 150 (1997), 239–249.

[131] O. A. Ladyzhenskaya, *Mathematical problems in the dynamics of a viscous incompressible fluid*, Nauka, Moscow, 1970 (in Russian).

[132] O. A. Ladyzhenskaya, *The boundary value problems of mathematical physics*, Springer-Verlag, New York, 1985.

[133] O. A. Ladyzhenskaya, On modified Navier–Stokes equations for large velocity gradients, *Zapiski Nauchnych Seminarov LOMI*, 7 (1968), 126–154.

[134] O. A. Ladyzhenskaja and V. A. Solonnikov, Some problems of vector analysis, and generalized formulations of boundary value problems for the Navier–Stokes equation, *Zap. Nauchn. Sem. Leningrad. Otdel. Mat. Inst. Steklov. (LOMI)*, 59 (1976), 81–116 (in Russian).

[135] O. A. Ladyzhenskaya and N. N. Uraltseva, *Linear and Quasilinear Elliptic equations*, Academic Press, New York, 1968.

[136] R. D. Lazarov, Superconvergence of the gradient for triangular and tetrahedral finite elements of a solution of linear problems in elasticity

theory, *Computational Processes and Systems*, 6 (1988), 180–191 (in Russian).

[137] A.S. Leonov, Some a posteriori stopping rules for iterative methods for solving linear ill-posed problems, *Comput. Math. Math. Phys.*, 34 (1994), 121–126.

[138] J.-L. Lions and E. Magenes, *Problmés aux limites non homogénes et applications*, Dunod, Paris, 1968.

[139] R. Löhner, K. Morgan and O. C. Zienkiewicz, Adaptive grid refinement for the compressible Euler equations, in *Accuracy Estimates and Adaptive Refinements in Finite Element Computations, Eds., I. Babuška, O. C. Zienkiewicz, J. Gago and E. R. de Olivieira, Wiley and Sons, 1986*, 281–297.

[140] J. Malek, J. Nečas, J. Rokuta and M. Ružička, *Weak and measure valued solutions to evolution partial differential equations, Applied Mathematic and Mathematical Computation 13.*, Chapman and Hall, 1996.

[141] J. Medina, M. Picasso and J. Rappaz, Error estimates and adaptive finite elements for nonlinear diffusion-convection problems, Ecole Polytechnique Federale de Lausanne, Preprint CH-1015, 1995.

[142] P. Meyer, A unifying theorem on Newton's method, *Numer. Funct. Anal. Optim.* 13 (1992), 5–6, 463–473.

[143] S. G. Mikhlin, *Variational methods in mathematical physics*, Pergamon, Oxford, 1964.

[144] S. G. Mikhlin, *Error Analysis in Numerical Processes*, Wiley and Sons, Chichester–New York, 1991.

[145] S. G. Mikhlin, *Constants in some inequalities of analysis*, Wiley and Sons, Chichester–New York, 1986.

[146] P. Mosolov and V. Myasnikov, *Mechanics of rigid plastic bodies*, Nauka, Moscow, 1981 (in Russian).

[147] A. Muzalevsky and S. Repin, On two-sided error estimates for approximate solutions of problems in the linear theory of elasticity, *Russian J. Numer. Anal. Math. Modelling*, 18 (2003), 65–85.

[148] P. Neittaanmäki and M. Křížek, On $0(h^4)$ superconvergence of piecewise bilinear FE-approximations, in Teubner-Texte Math. 107, Teubner, 1988, 250–255.

[149] P. Neittaanmäki and S. Repin, A posteriori error estimates for boundary-value problems related to the biharmonic operator, *East-West J. Numer. Math.*, 9 (2001), 157–178.

[150] J. T. Oden, L. Demkowicz, T. Strouboulis and P. Devloo, Adaptive methods for problems in solid and fluid mechanics, in *Accuracy Estimates and Adaptive Refinements in Finite Element Computations*, Eds., I. Babuška, O. C. Zienkiewicz, J. Gago and E. R. de Oliveira, Wiley and Sons, 1996.

[151] J. T. Oden, L. Demkowicz, W. Rachowicz and T. A. Westermann, Towards a universal $h - p$ adaptive finite element strategy. Part 2. A posteriori error estimation, *Comput. Methods Appl. Mech. Engrg.*, 77 (1989) 113–180.

[152] J. T. Oden, S. Prudhomme, Goal-oriented error estimation and adaptivity for the finite element method, *Comput. Math. Appl.*, 41 (2001), 735–756.

[153] J. T. Oden and J. N. Reddy, *Mathematical theory of finite elements*, Wiley and Sons, 1976.

[154] J. T. Oden, W. Wu and M. Ainsworth, An a posteriori error estimate for finite element approximations of the Navier–Stokes equations, *Comput. Methods. Appl. Mech. Engrg.*, 111 (1994), 185–202.

[155] L. A. Oganesjan and L. A. Ruchovec, An investigation of the rate of convergence of variation-difference schemes for second order elliptic equations in a two-dimensional region with smooth boundary, *Z. Vychisl. Mat. i Mat. Fiz.*, 9 (1969), 1102–1120 (in Russian).

[156] A. Ostrowski, Les estimations des erreurs a posteriori dans les procédés itératifs, *C.R. Acad.Sci. Paris Sér. A–B*, 275 (1972), A275–A278.

[157] D.R.J.Owen and E.Hinton, *Finite Elements in Plasticity Theory: Theory and Practice.* Prinerdge Press, Swansea, 1980.

[158] C. Padra, A posteriori error estimators for nonconforming approximation of some quasi-newtonian flows, *SIAM J. Numer. Anal.*, 34 (1997), 1600–1615.

[159] J. Peraire and A. T. Patera, Bounds for linear-functional outputs of coercive partial differential equations: Local indicators and adaptive

refinement, in *Advances in Adaptive Computational Methods in Mechanics, Eds., P. Ladevéze and J. T. Oden, Elsevier, New York, 1998*, 199–228.

[160] F. Potra, Sharp error bounds for a class of Newton-like methods, *Libertas Math.* 5 (1985), 71–84

[161] J. Pousin and J. Rappaz, Consistance, stabilité, erreurs a priori et a posteriori pour des problemes non linéaries, *C. R. Acad. Sci. Paris*, 312 (1991), 699–703.

[162] J. Pousin and J. Rappaz, Consistency, stability, a priori and a posteriori errors for Petrov–Galerkin methods applied to nonlinear problems, *Numer. Math.*, 69 (1994), 213–231.

[163] W. Prager and J. L. Synge, Approximation in elasticity based on the concept of function space, *Quart. Appl. Math.*, 5 (1947), 241–269.

[164] R. Rannacher, Zur L^∞-Konvergenz linear finite elemente beim Dirichlet Problem, *Math. Z.*, 149 (1976), 69–77.

[165] R. Rannacher and F. T. Suttmeier, A feed–back approach to error control in finite element methods: application to linear elasticity, IWR, Preprint 96–42 (SFB 359), Heidelberg 1996.

[166] R. Rannacher and F.T. Suttmeier, A posteriori error control and mesh adaptation for F.E. models in elasticity and elasto-plasticity, in *Advances in Adaptive Computational Methods in Mechanics, Eds., P. Ladevéze and J. T. Oden, Elsevier, New York, 1998*, 275–292.

[167] W. C. Rheinboldt, On a theory of mesh-refinement processes, *SIAM J. Numer. Anal.*, 17 (1980), 766–778.

[168] S. Repin, A posteriori estimates for approximate solutions of variational problems with strongly convex functionals, *Problems of Mathematical Analysis*, 17 (1997), 199–226. (in Russian, translated in *Journal of Mathematical Sciences*, 97 (1999), 4311–4328).

[169] S. Repin, A posteriori estimates of the accuracy of variational methods for problems with nonconvex functionals, *Algebra i Analiz*, 11 (1999), 151–182 (in Russian, translated in *St.-Petersburg Mathematical Journal*, 11 (2000), 4, 651–672).

[170] S. Repin, A posteriori error estimation for nonlinear variational problems by duality theory, *Zapiski Nauchnych Seminarov POMI*, 243 (1997) 201–214.

[171] S. Repin, A posteriori error estimation for variational problems with power growth functionals based on duality theory, *Zapiski Nauchnych Seminarov POMI*, 249 (1997), 244–255.

[172] S. Repin, A posteriori error estimates for approximate solutions of variational problems, in *Proceedings of 2nd European Conference on Numerical Mathematics and Advanced Applications, Heidelberg, 1997, Eds., H. G. Bock, G. Kanschat, R. Rannacher, F. Brezzi, R. Glowinski, Yu. Kuznetsov and J. Priaux, World Sci. Publishing, River Edge, New York, 1998*, 524–531.

[173] S. Repin, A unified approach to a posteriori error estimation based on duality error majorants, *Mathematics and Computers in Simulation*, 50 (1999), 313–329.

[174] S. Repin, A posteriori error estimation for variational problems with uniformly convex functionals, *Math. Comput.*, 69 (2000), 481–500.

[175] S. Repin, Two-sided estimates for deviation from an exact solution to uniformly elliptic equation. *Trudi St.-Petersburg Math. Society*, 9 (2001), 148–179 (in Russian, translated in *American Mathematical Translations Series 2*, 209 (2003), 143–171).

[176] S.Repin, Estimates of deviation from exact solutions of initial-boundary value problems for the heat equation, *Rend. Mat. Acc. Lincei*, 13 (2002), 121–133.

[177] S. Repin, Estimates of deviations from exact solutions of elliptic variational inequalities, *Zapiski Nauchn. Semin. V.A. Steklov Mathematical Institute in St.-Petersburg (POMI)*, 271 (2000), 188–203.

[178] S. Repin, Aposteriori estimates for the Stokes problem, *J. Math. Sci. (New York)* 109 (2002), 5, 1950–1964.

[179] S. Repin, Estimates of deviations for generalized Newtonian fluids, *Zapiski Nauchn. Semin. V.A. Steklov Mathematical Institute in St.-Petersburg (POMI)*, 288 (2002), 178–203.

[180] S. Repin, Estimates for errors in two-dimensional models of elasticity theory, *J. Math. Sci. (New York)*, 106 (2001), 3, 3027–3041.

[181] S. Repin and M. Frolov, A posteriori error estimates for approximate solutions of elliptic boundary value problems, *Zh. Vychisl. Mat. Mat. Fiz.*, 42 (2002), 12, 1774–1787 (in Russian).

[182] S. Repin, S. Sauter and A. Smolianski, A posteriori error estimation for the Dirichlet problem with account of the error in the approximation of boundary conditions, *Computing*, 70 (2003), 205–233.

[183] S. Repin, S. Sauter and A. Smolianski. Duality Based A Posteriori Error estimator for the Dirichlet Problem, *Proc.Appl. Math. Mech.*, 2 (2003), 513–514.

[184] S. Repin, S. Sauter and A. Smolianski. A posteriori error estimation of dimension reduction errors (to appear in Proc. 5th European Conference on Numerical Mathathematics and aplications, Praha,2003).

[185] S. I. Repin and L. S. Xanthis, A posteriori error estimation for elastoplastic problems based on duality theory, *Comput. Methods Appl. Mech. Engrg.*, 138 (1996), 317–339.

[186] S. I. Repin and L. S. Xanthis, A posteriori error estimation for nonlinear variational problems, *Comptes Rendus de l'Académie des Sciences, Mathématique*, 324 (1997), 1169–1174.

[187] W. Ritz, Über eine neue Methode zur Lözing gewisser Variationsprobleme der Mathematischen Physics, *J. Reine Angew. Math.*, 135 (1909), 1–61.

[188] R. T. Rockafellar, *Convex analysis*, Princeton Univ. Press, 1970.

[189] R. Rodrigues, Some remarks on Zienkiewicz–Zhu estimator, *Numer. Methods for PDE*, 10 (1994), 625–635.

[190] W. Rudin, *Functional analysis*, McGraw-Hill, 1973.

[191] J. L. Segerlind, *Applied finite element analysis*, Wiley and Sons, 1976.

[192] G.A. Seregin, Some remarks on variational problems for functionals with $LlnL$ growth, *Zapiski Nauchn. Sem. POMI*, 213 (1994), 164–174.

[193] A. H. Schatz, Pointwise error estimates and asymptotic error expansion inequalities for the finite element method on irregular grids: Part 1: global estimates, *Math. Comput.*, 67 (1998), 877–899.

[194] A. H. Schatz and L. B. Wahlbin, Interior maximum norm estimates for finite element methods, *Math. Comput.*, 31 (1977), 414–442.

[195] A. H. Schatz and L. B. Wahlbin, Maximum norm estimates for finite element method on plane polygonal domains. Part 1, *Math. Comput.*, 32 (1978), 73–109.

[196] A. H. Schatz and L. B. Wahlbin, Maximum norm estimates for finite element method on plane polygonal domains. Part 2, refinements, *Math. Comput.*, 33 (1979), 465–492.

[197] A. H. Schatz and J. Wang, Some new error estimates for Ritz–Galerkin methods with minimal regularity assumptions, *Math. Comput.*, 65 (1996), 213, 19–27.

[198] M. Schultz and O.Steinbach, A new a posteriori error estimator in adaptive direct boundary element methods: the Dirichlet problem, *Calcolo*, 37 (2000), 79–96.

[199] M. Schulz and W. L. Wendland, A general approach to a posteriori error estimates for strictly monotone and Lipschitz continuous nonlinear operators illustrated in elasto-plasticity, in *Proceedings of 2nd European Conference on Numerical Mathematics and Advanced Applications, Heidelberg, 1997, World Sci. Publishing, River Edge, New York, 1998*, 572–579.

[200] M. Schultz and W. Wendland, *On an adaptive finite element method for elasto-plastic deformation with hardening*, Preprint 1997/39, SFB 259, Stuttgardt, 1997.

[201] S. L. Sobolev, *Some Applications of Functional Analysis in Mathematical Physics*, Izdt. Leningrad. Gos. Univ., Leningrad, 1955 (in Russian translated in *Translation of Mathematical Monographs, Volume 90* American Mathematical Society, Providence, RI, 1991).

[202] E. Stein and S. Ohnimus, Coupled model- and solution-adaptivity in the finite element method, *Comput. Methods Appl. Mech. Engrg.*, 150 (1997), 327–350.

[203] E. Stein, F.J. Bartold, S. Ohnimus and M. Schmidt, Adaptive finite elements in elastoplasticity with mechanical error indicators and Neumann-type estimators, in *Advances in Adaptive Computational Methods in Mechanics, Eds., P. Ladevéze and J. T. Oden, Elsevier, New York, 1998*, 81–100.

[204] G. Strang and G. Fix, *An analysis of the finite element method*, Prentice Hall, Englewood Cliffs, 1973.

[205] T. Strouboulis and J. T. Oden, A posteriori error estimation of finite element approximations in fluid mechanics, *Comput. Meth. Appl. Mech. Engrg.*, 78 (1990), 201–242.

[206] R. Temam, Problèmes mathématique en plasticité, Bordas, Paris, 1983.

[207] K. Tsuruta and K. Ohmori, A posteriori error estimation for Volterra integro-differential equations, *Mem. Numer. Math. No. 3* 1976, 33–47.

[208] N. N. Uraltseva, On the regularity of solutions to variational inequalities *Uspekhi Mat. Nauk*, 42 (1987), 6, 151–174 (in Russian).

[209] R.S. Varga, *Matrix Iterative Analysis*, Prentice–Hall, Englewood Cliffs, New Jersey, 1962.

[210] A. Veeser, Efficient and reliable a posteriori error estimators for elliptic obstacle problems, *SIAM J. Numer. Anal.*, 39 (2001), 146–167.

[211] R. Verfürth, A posteriori error estimators for the Stokes equations, *Numer. Math.*, 55 (1989), 309–326.

[212] R. Verfürth, A posteriori error estimates for nonlinear problems. Finite element discretisations of elliptic equations, *Math. Comput.*, 62 (1994) 445–475.

[213] R. Verfürth, *A review of a posteriori error estimation and adaptive mesh-refinement techniques*, Wiley and Sons, Teubner, New-York, 1996.

[214] A. Vesser, Efficient and reliable a posteriori error estimators for elliptic obstacle problems, *SIAM J. Numer. Anal.*, 39 (2001), 1, 146–167.

[215] M. I. Vishik, The method of orthogonal projections for selfadjoint equations, *Sov. Math. Dokl*, 1947, 56.

[216] L. B. Wahlbin, *Superconvergence in Galerkin Finite Element Methods*, Lecture Notes in Mathematics 1605, Springer-Verlag, 1995.

[217] J. Wang, Superconvergence analysis of finite element solutions by the least-squares surface fitting on irregular meshes for smooth problems, *J. Math. Study*, 33 (2000), 3, 229–243.

[218] H. Weil, The method of orthogonal projections in potential theory, *Duke Math. J.*, 7 (1940), 411–444.

[219] J. R. Whiteman and G. Goodsell, A survey of gradient superconvergence for finite approximations to second order elliptic problems on triangular and tetrahedral meshes, in *The Mathematics of Finite Elements and Applications VII, Ed. J. R. Whiteman, Academic Press, 1991*, 55–74.

[220] W. H. Young. On classes of summable functions and their Fourier series, *London Roy. Soc. Proc. (A)*, 87 (1912), 225–229.

[221] S. Zaremba, Sur un probleme toujours possible comprenant, a titre de cas particuliers, le probleme de Dirichlet et celui de Neumann, *J. math. pures et appl.* 6 (1927), 2, 127–163.

[222] E. Zeidler, *Nonlinear functional analysis and its applications. I. Fixed-point theorems*, Springer-Verlag, New York, 1986,

[223] O. C. Zienkiewicz, Achievements and some unsolved problems of the finite element method, *Int. J. Numer. Meth. Engrg.*, 47 (2000), 9–28.

[224] O. C. Zienkiewicz, B. Boroomand and J.Z. Zhu, Recovery procedures in error estimation and adaptivity: Adaptivity in linear problems, in *Advances in Adaptive Computational Methods in Mechanics, Ed. P. Ladevéze and J. T. Oden, Elsevier, New York, 1998*, 3–23.

[225] O. C. Zienkiewicz and K. Morgan, *Finite elements and approximation*, Wiley and Sons, New York, 1983.

[226] O. C. Zienkiewicz and J. Z. Zhu, A simple error estimator and adaptive procedure for practical engineering analysis, *Internat. J. Numer. Meth. Engrg.*, 24 (1987) 337–357.

[227] O. C. Zienkiewicz and J. Z. Zhu, Adaptive techniques in the finite element method, *Commun. Appl. Numer. Methods*, 4 (1988), 197–204.

[228] O. C. Zienkiewicz and J. Z. Zhu, The superconvergent patch recovery and a posteriori error estimates. Part 1: The recovery technique, *Int. J. Numer. Meth. Engrg.*, 33 (1992), 1331–1364.

[229] M. Zlámal, Some superconvergence results in the finite element method. Mathematical aspects of finite element methods, in *Proc. Conf., Consiglio Naz. delle Ricerche (C.N.R.), Rome, 1975, Lecture Notes in Math. 606, Springer-Verlag, Berlin, 1977, 353–362.*

Notation

$:=$	equals by definition
$\forall$	for all
$\exists$	exists
$\exists!$	exists and is unique
$\equiv$	identity sign
$\to$	tends to, converges
$\rightharpoonup$	weakly converges
$\Longleftrightarrow$	equivalent
$\Rightarrow$	implies
$\mathbb{N}$	space of natural numbers
$\mathbb{R}$	space of real numbers
$\mathbb{R}_+$	set of nonnegative real numbers
$\mathbb{R}^n$	space of real vectors of the dimension n
$\mathbb{M}^{n \times n}$	space of real $n \times n$ matrixes
$\emptyset$	empty set
$\mathcal{B}(x, \rho)$	ball of the radius ρ centered at x
Ω, ω	sets in $\mathbb{R}^n$
$\partial\Omega$	boundary of Ω
$\operatorname{diam}\Omega$	diameter of the set Ω
$\lvert\Omega\rvert$	Lebesgue measure of the set Ω
ν	exterior normal to $\partial\Omega$
$\mathbb{I}$	unit matrix (tensor)
$a \cdot b$	scalar product of vectors: $\sum_i a_i b_i$
$a \otimes b$	tensor product of vectors: $\sum_{i,j} a_i b_j$
$\sigma : \varepsilon$	scalar product of tensors: $\sum_{i,j} \sigma_{ij} \varepsilon_{ij}$
$\operatorname{tr}\sigma$	trace: $\sum_i \sigma_{ii}$
σ^D	deviator: $\sigma - \frac{1}{n}\operatorname{tr}\sigma\mathbb{I}$

Δ	Laplace operator; $\Delta\phi = \frac{\partial^2\phi}{\partial^2 x_1} + \ldots \frac{\partial^2\phi}{\partial^2 x_d}$
∇	gradient operator; $\nabla\phi = \left(\frac{\partial\phi}{\partial x_1}, \ldots \frac{\partial\phi}{\partial x_d}\right)$
div	divergence operator; $\operatorname{div} v = \frac{\partial v_1}{\partial x_1}, \ldots \frac{\partial v_d}{\partial x_d}$
X, Y, V	functional spaces
X^*, Y^*, V^*	topologically dual spaces
0_X	zero element of X
P^k	space of the polynomial functions of the degree k
$L_p(\Omega)$	space of functions integrable with power p over Ω
$\|\cdot\|_{p,\Omega}$	norm of $L_p(\Omega)$
$\|\cdot\|_\Omega$	norm of $L_2(\Omega)$
$\|\cdot\|_{\infty,\Omega}$	supremum norm
$W_p^l(\Omega)$	Sobolev space of of functions, whose derivatives are integrable with power p up to order l
$\mathcal{L}(X,Y)$	space of linear bounded operators acting from X to Y
$\|\cdot\|_{l,p,\Omega}$	norm of $W_p^l(\Omega)$
$H(\operatorname{div},\Omega)$	space of square-integrable functions with square integrable divergence
$\|\cdot\|_{\operatorname{div},\Omega}$	norm of $H(\operatorname{div},\Omega)$
$H^{-1}(\Omega)$	space conjugate to $\overset{\circ}{H}^1(\Omega)$
$\operatorname{conv} K$	convex envelope of the set K
$\operatorname{epi} F$	epigraph of the functional F
$\operatorname{dom} F$	the set of v such that $F(v) < +\infty$
$\partial F(x)$	subdifferential of the functional F at x
F^*	Polar functional (Young–Fenhel conjugate) to F; $F^*(v^*) := \sup_v\{\langle v^*, v\rangle - F(v)\}$
F^{**}	Bipolar functional (double conjugate) to F $F^{**}(v) := \sup_{v^*}\{\langle v^*, v\rangle - F^*(v^*)\}$
$D(v,v^*)$	compound functional $D(v,v^*) := F(v) + F^*(v^*) - \langle v^*, v\rangle$
$\partial F(v)$	subdifferential of F at v

Index